Reloading Data Protection

Serge Gutwirth • Ronald Leenes • Paul De Hert
Editors

Reloading Data Protection

Multidisciplinary Insights and Contemporary Challenges

Springer

Editors
Serge Gutwirth
Law, Science, Technology
and Society (LSTS)
Faculty of Law and Criminology
Vrije Universiteit Brussel
Brussels, Belgium

Paul De Hert
Law, Science, Technology
and Society (LSTS)
Faculty of Law and Criminology
Vrije Universiteit Brussel
Brussels, Belgium

Ronald Leenes
Tilburg Institute for Law, Technology
and Society (TILT)
Tilburg University
Tilburg, Netherlands

ISBN 978-94-007-7539-8 ISBN 978-94-007-7540-4 (eBook)
DOI 10.1007/978-94-007-7540-4
Springer Dordrecht Heidelberg London New York

Library of Congress Control Number: 2013947601

Printed on acid-free paper

Springer is part of Springer Science+Business Media (www.springer.com)

Foreword

On 23 January 2012, the European Commission presented its proposal for a new "Data protection package". With the new package, the Commission seeks to strengthen data protection for the European citizens and set a standard for the rest of the world. The proposal has indicated clear directions for the future of data protection on the agenda. These include a 'right to be forgotten', stricter rules regarding online profiling, and stronger sanctions for non compliance. Although the proposal clearly follows the avenue already taken with Directive 95/46/EC, it also introduces controversial new elements as well as uncertainty. And thus, as expected, the new and re-considered, directions put forward by the Commission have spurred and extensive highly challenging process of discussion, negotiation and lobbying in 2012 and 2013. At the time of writing this foreword (June 2013), the European Parliament is discussing the almost 4,000 amendments that were tabled as a result of these discussions and everyone in the field is eagerly awaiting the outcome of the parliamentary decision making.

The sixth annual CPDP conference, held in Brussels on 23–25 January 2013, was sharply influenced by the fresh release of the European Commission's new plans and proposals, and quite some attention was given to the new or reformulated concepts of the package. This "reloading" of data protection, or even its "rebooting", has given a boost to reflection and research on the subject. This book volume bears witness to this reloading of data protection.

The present book is one of the products of the sixth edition of the annual Brussels based international International Conference on Computers, Privacy and Data Protection. CPDP 2013, was held under the same title: *Reloading data protection*. The conference welcomed 750 participants at 'our' venue—the magnificent *Les Halles*, while another 1,200 people were reached through free public events organized in the evenings, also in Brussels. The 3 day conference offered participants 45 panels and several workshops and special sessions, with 199 speakers from academia, the public and private sectors, and civil society.

This volume brings together 16 chapters offering conceptual analyses, highlighting issues, proposing solutions, and discussing practices regarding privacy and data protection. The first section of the book, provides an overview of developments in data protection in different parts of the world. The second section focuses on one of the most captivating innovations of the data protection package, namely how to forget and the right to be forgotten in a digital world. The third section proposes five chapters on a recurring, and thus, obviously still important and disputed theme of the CPDP-conferences : the surveillance, control and steering of individuals and groups of people and the still more performing tools (data mining, profiling, convergence) to realise those objectives, and this with illustrations from the domain of law enforcement and smart surveillance. The book concludes with five chapters that aim at increasing our understanding of the changing nature of privacy (concerns) and data protection.

The chapters in this volume stem from two tracks. Nine chapters (3, 4, 5, 10, 11, 12, 13, 14, and 15) originate from responses to the conference's call for papers and have thus already been presented during the conference. The remaining chapters (1, 2, 6, 7, 8, 9, and 16) were submitted by invited speakers in the months following the conference. All the chapters of this book have been peer reviewed and commented on by at least two referees with expertise and interest in the subject matter. Since their work is crucial for maintaining the scientific quality of the book we would explicitly take the opportunity to thank them, *ad nominatim*, for their commitment and efforts: Claudia Aradau, Petra Bard, Rocco Bellanova, Laurent Beslay, Diana Alonso Blas, Caspar Bowden, Ian Brown, Lee Bygrave, Johann Čas, Helena Carrapiço, Claudia Diaz, Rodrigo Firmino, Gus Hosein, Simone Fischer-Hübner, Catherine Flick, Marieke de Goede, Gloria González Fuster, Antonella Galetta, Seda Gürses, Dara Hallinan, Marit Hansen, Hans Hedbom, Hielke Hijmans, Gerrit Hornung, Julien Jeandesboz, Christopher Kuner, Eleni Kosta, Marc Langheinrich, Daniel Le Métayer, Tobias Mahler, Gary Marx, Lucas Melgaço, Charles Raab, Joseph Savirimuthu, Dimitra Stefanatou, Anton Vedder, John Vervaele and Tal Zarsky.

May this book meet the reader's expectations and contribute to the quality of the, today particularly actual and pertinent, debate about the next steps of the becoming of privacy and data protection.

Serge Gutwirth
Ronald Leenes
Paul De Hert

Contents

Contributors

Marion Albers is Professor of Public Law, Information and Communication Law, Health Law and Theory of Law at Hamburg University. She studied law, sociology and political science at the universities of Berlin and Bielefeld and received her Ph.D. in law with a thesis on crime prevention and provisions for prosecution. Her postdoctoral thesis (Habilitation) focused on questions of informational self-determination. She was assistant at the Federal Constitutional Court in Karlsruhe. From 2002–2005 she served as an expert in the Advisory Committee of the Bundestag (German Parliament) for Ethics and Law of Modern Health Care. Her main areas of research include Fundamental Rights, Information Law and Data Protection, Health Law and Biolaw, Police Law and Law of Intelligence Services, Theory and Sociology of Law. She is engaged in several, partly interdisciplinary projects dealing with legal aspects of privacy, data protection, biobanks and health data, surveillance problems or measures against the financing of terrorism.
e-mail: marion.albers@web.de

Colin Bennett received his Bachelor's and Master's degrees from the University of Wales, and his Ph.D. from the University of Illinois at Urbana-Champaign. Since 1986 he has taught in the Department of Political Science at the University of Victoria, where he is now Professor. From 1999–2000, he was a fellow at Harvard's Kennedy School of Government. In 2007 he was a Visiting Fellow at the Center for the Study of Law and Society at University of California, Berkeley. In 2010, he is Visiting Professor at the School of Law, University of New South Wales. His research has focused on the comparative analysis of surveillance technologies and privacy protection policies at the domestic and international levels. In addition to numerous scholarly and newspaper articles, he has published five books, including *The Privacy Advocates: Resisting the Spread of Surveillance* (The MIT Press, 2008), and policy reports on privacy protection for Canadian and international agencies. He is currently the co-investigator of a large Major Collaborative Research Initiative grant entitled "The New Transparency: Surveillance and Social Sorting."
e-mail: cjb@uvic.ca

Dr. Paul Bernal is a Lecturer in Information Technology, Intellectual Property and Media Law at the University of East Anglia Law School. His background is unusual for a legal academic: his original degree from Cambridge University was in

mathematics, he qualified as a Chartered Accountant and has worked in business and in the voluntary sector, working in mental health and criminal justice, before returning to academia. He has a Ph.D. from the LSE, based on research into data privacy and autonomy, and in particular the role of commercial data gathering in the internet as it develops. His current research is centred around privacy and human rights, particularly on the internet, and includes such elements as the role of communications surveillance, the interactions between businesses and governments in relation to data privacy, children's rights on the internet, and Data Protection reform. Paul also specialises in the burgeoning area of social media—both as an academic looking into the roles played by social networks and social networkers and the laws and practices that have an impact upon them, and as a personal, legal and political blogger and tweeter. blog: http://paulbernal.wordpress.com/, Twitter: @paulbernalUK
e-mail: P.A.Bernal@lse.ac.uk

Denis Butin is a postdoctoral researcher in the Privatics team at Inria (Lyon, France). His research currently focuses on security policy languages and accountability by design. He holds a Ph.D. in computer science from Dublin City University, where he worked on the application of formal methods to electronic voting protocol analysis. Earlier, he earned a Master‚Äôs degree in mathematics and computer science at the University of Tours. He is involved with the Security chapter of the European FI-WARE project.
e-mail: mchicote@dc.uba.ar

Marcos Chicote is a research intern at Inria (Lyon, France). His areas of interest include software engineering, automatic program analysis and program verification and has broad experience in industrial software development. He holds a Master's degree in Computer Science from the University of Buenos Aires.
e-mail: denis.butin@Inria.fr

Liane Colonna is a second-year doctoral candidate at the Swedish Law and Informatics Research Institute located at Stockholm University. The working title of her research project is the "Legal Implications of Data Mining in the European Union and the United States." She is a member of the New York bar and has an LLM in European law from Stockholm University.
e-mail: liane.colonna@juridicum.su.se

Vlad Coroama is a postdoctoral researcher in the Center for Industrial Ecology, University of Coimbra, Portugal. For more than a decade, his research revolved around the relation between ICT and sustainability. While in recent years he focused on the environmental dimension, Vlad is also interested in the societal effects induced by an expanding ICT monitorisation of everyday life. Vlad holds a Ph.D. in Computer Science from the ETH Zurich, Switzerland. Vlad can be reached at vlad.coroama@dem.uc.pt
e-mail: vlad.coroama@dem.uc.pt

Els De Busser studied Law at Antwerp University and obtained an additional degree in Criminology and an Advanced Master's degree in European Criminology and Criminal Justice Systems from Ghent University, Belgium. From March 2001 to October 2009, she worked as a researcher and professor's assistant in the field of European Criminal Law at Ghent University, Institute for International Research on Criminal Policy where she defended her Ph.D. entitled 'EU internal and transatlantic cooperation in criminal matters from a personal data perspective. A substantive law approach' in May 2009. In November 2009, she joined the European Criminal Law section of the Max Planck Institute in Freiburg, Germany. Her research and publications focus on international cooperation in criminal matters and data protection.
e-mail: e.busser@mpicc.de

Paul De Hert is professor of law at the Faculty of Law and Criminology of Vrije Universiteit Brussel. He is the Director of the research group on Fundamental Rights and Constitutionalism (FRC) and senior member of the research group on Law, Science, Technology & Society (LSTS). Paul De Hert is also associated-professor Law and Technology at the Tilburg Institute for Law and Technology (TILT)
e-mail: Paul.de.hert@vub.ac.be

Danilo Doneda is a professor of Civil Law at the Law School of FGV Direito Rio and a researcher at the Centre for Technology and Society (CTS) in the same institution. He holds a Ph.D. in Civil Law from the State University of Rio de Janeiro with a thesis about data protection later published as a book, a Master degree from the same institution and a Law degree from the Federal University of Paraná. He also works in the Brazilian Ministry of Justice as General Coordinator for Market Studies and Monitoring at the National Department for Consumer Protection.
e-mail: danilo@doneda.net

Rachel Finn is an Associate Partner at Trilateral Research & Consulting since 2010. Her research expertise includes the social effects of surveillance; new surveillance technologies; surveillance and the law; crime, deviance and social control; risk and security; and identity-based social exclusion. She also conducts research on stakeholder engagement mechanisms and privacy impact assessment methodologies, and advises on ethics, policy and implementation. Rachel has published articles in peer-reviewed journals and is co-authoring a book for Routledge on the social impacts of 'new technologies' of surveillance. She has a Ph.D. in sociology from the University of Manchester, UK.
e-mail: rachel.finn@trilateralresearch.com

Antonella Galetta is Ph.D. researcher at Law, Science, Technology and Society (LSTS) of the Vrije Universiteit Brussel, Faculty of Law and Criminology. Her research interests focus on law, privacy, data protection, technology and surveillance studies. She is currently doing research in the framework of the FP7 project IRISS, Increasing Resilience in Surveillance Societies. She holds a B.A. in Law (University of Macerata), a M.A. in European and International Studies (University of Trento)

and a M.A. in International Relations (University of Bologna). She has working experiences at the European Parliament and European NGOs.
e-mail: antonella.galetta@vub.ac.be

Dorota Głowacka Lawyer at the Helsinki Foundation for Human Rights, Poland dealing with freedom of expression and right to privacy issues. Coordinator of the HFHR's 'Observatory of Media Freedom in Poland' programme. Ph.D. candidate at the Public International Law Department, Law Faculty, University of Lodz, Poland.
e-mail: d.glowacka@hfhr.org.pl

Serge Gutwirth is a professor of human rights, legal theory, comparative law and legal research at the Faculty of Law and Criminology of the Vrije Universiteit Brussel.(VUB), where he studied law, criminology and also obtained a postgraduate degree in technology and science studies. Gutwirth founded and still chairs the VUB-research group Law Science Technology & Society (http://www.vub.ac.be/LSTS). He publishes widely in Dutch French and English. Amongst his recent co-edited publications are *Safeguards in a world of ambient intelligence (Springer, 2008), Profiling the European citizen* (Springer 2008), *Reinventing data protection?* (Springer 2009), *Data protection in a profiled world* (Springer, 2010) and *Computers, privacy and data protection: an element of choice* (Springer, 2011) and European Data Protection: in good health? (2012). Currently, Serge Gutwirth is particularly interested both in technical legal issues raised by technology (particularly in the field of data protection and privacy) and in more generic issues related to the articulation of law, sciences, technologies and societies.
e-mail: serge.gutwirth@vub.ac.be

Beata Konieczna Researcher and lecturer of the Faculty of Law and Administration at the University of Cardinal Stefan Wyszyński University in Warsaw, Poland. A member of the Board of the Legal—Information Scientific Center. Research interests are: personal data protection, computerization of public administration. The author of scientific articles devoted to protection of personal data.
e-mail: beata-konieczna@wp.pl

Paulan Korenhof is a Ph.D. student at the Tilburg Institute for Law, Technology and Society (TILT, Tilburg University) and Privacy & Idenity Lab (PI.Lab, a collaboration between Radboud University, SIDN, Tilburg University and TNO). As part of PI.Lab she works as a guest researcher at the digital security department of the Radboud University. She holds a masters degree in both Public Law and Philosophy. Paulan focuses her research on the problems that are associated with the so-called "Right to be Forgotten" by looking at them from a meta-perspective.
e-mail: p.e.i.korenhof@tilburguniversity.edu

Marc Langheinrich is an assistant professor in the Faculty of Informatics at the Università della Svizzera Italiana (USI) in Lugano, Switzerland, where he heads the Research Group for Ubiquitous Computing. Marc received his Ph.D. in Computer Science from the ETH Zurich, Switzerland, in 2005 with his work on "Privacy in Ubiquitous Computing". Marc is one of the authors of P3P, a W3C-standard for

privacy on the Web, and has published extensively on privacy aspects of ubiquitous and pervasive computing systems. Marc can be reached at langheinrich@acm.org
e-mail: langheinrich@acm.org

Ronald Leenes is professor in Regulation by Technology at TILT, the Tilburg Institute for Law, Technology, and Society (Tilburg University). His primary research interests are privacy and identity management, regulation of, and by, technology. He is also involved in research in ID fraud, biometrics and robotics. Ronald has worked on several EU projects in the privacy and identity domain, such as the EU FP6 PRIME project, EU FP7 PrimeLife and A4Cloud. He has co-edited numerous volumes on privacy and identity-management, including Digital Privacy: PRIME - Privacy and Identity Management for Europe (2011), Computers, privacy and data protection: an element of choice (Springer, 2011), European Data Protection: in good health? (Springer 2012), European Data Protection: Coming of Age (Springer 2013).

Daniel Le Métayer is Research Director for Inria (the French National Institute for Research in Computer Science and Control) and head of the Inria Project Lab CAPPRIS. CAPPRIS is an interdisciplinary initiative involving seven research teams working on various aspects of privacy. From 2000 to 2006, Daniel Le M√©tayer worked for Trusted Logic, a leading company in security and open middleware for embedded systems. Daniel Le Métayer has been involved in various international projects on IT security, software design and analysis, testing, etc. He has also served on programme committees of many IT international conferences and he has been the editor of special issues of computer science journals such as ACM Transactions on Software Engineering and Theoretical Computer Science.
e-mail: daniel.le-metayer@inria.fr

Laura Schertel Mendes is a Ph.D. candidate at the Humboldt University of Berlin, under the supervision of Prof. Stefan Grundmann, with a scholarship of the German Academic Exchange Service (DAAD). Her research addresses data protection in the private sector, focusing the problem of consent. She holds a Law Degree and a Master of Law from the University of Brasilia, Brazil. She worked from 2007 to 2010 in the Brazilian Ministry of Justice as General Coordinator for Market Studies and Monitoring at the National Department for Consumer Protection.
e-mail: lauraschertel@hotmail.com

Adam Molnar is a Ph.D. candidate in the Department of Political Science at the University of Victoria and a Researcher with the New Transparency project. His research interests focus on the legal, normative, and technical dimensions of digitally mediated surveillance and privacy, particularly in the areas of policing, national security, and public safety governance. His dissertation focuses on the legacies of security and policing operations associated with major sporting events, with an empirical focus on current trends in national security, including public order policing, civilian-military relations, disaster management, and a range of applications of surveillance technologies in contemporary policing.
e-mail: apm@uvic.ca

Anthony Morton is a Research Student in the Information Security Research Group at the Department of Computer Science, University College London (UCL), UK. He commenced his Ph.D. in 2010, having gained an M.Sc. in Information Security at UCL. He is a Chartered Engineer through the British Computer Society, and also has an MBA in Technology Management, an M.Sc. in Computing for Commerce and Industry and an M.A. in Classical Studies, all from The Open University. His Ph.D. research focuses on the influence of the privacy behaviour of technology services—consisting of a technology platform and providing organisation—on the construction of peoples' privacy concerns. Prior to commencing his studies at UCL, he was employed in the IT industry for 25 years in software development, technical management and consultancy roles.
e-mail: anthony.morton.09@ucl.ac.uk

Nóra Ní Loideain B.A., L.L.B., L.L.M. (Public Law) is a Ph.D. candidate and CHESS scholar at the Faculty of Law in the University of Cambridge. Her doctoral thesis concerns the Data Retention Directive (2006/24/EC), specifically the surveillance of communications data by law enforcement authorities and the right to respect to private life and correspondence in Europe. She has previously worked as a Legal Research Officer in the Office of the Director of Public Prosecutions and as a Judicial Researcher for the Supreme Court of Ireland. Her main research interests and publications are in the fields of EU law and policy-making; civil liberties and human rights, particularly under the EU and ECHR systems; and data protection.
e-mail: nl301@cam.ac.uk

Christopher Parsons is a doctoral candidate at the University of Victoria, Canada. Christopher's research, teaching, and consulting interests involve how privacy is affected by digitally mediated surveillance, and the normative implications that such surveillance has in (and on) contemporary Western political systems. His current research streams examine how and why Internet service providers use deep packet inspection technologies, the privacy and policy challenges raised by social media network communications, and difficulties concerning the use of electronic identity cards to access government services. In addition, he has published in the Canadian Journal of Law and Society, European Journal of Law and Technology, Canadian Privacy Law Review, CTheory, and has book chapters in a series of academic and popular books and reports. Christopher is a Ph.D. Candidate in the Department of Political Science at the University of Victoria, a Privacy by Design Ambassador and a Principal at BlockG Privacy and Security Consulting.
e-mail: Christopher@Christopher-Parsons.com

Wouter M.P. Steijn has graduated Developmental Psychology at Leiden University and is now a Ph.D. candidate at Tilburg Institute for Law, Technology, and Society at Tilburg University and partner of the Privacy & Identity Lab. His research is part of the multi-disciplinary project "Social dimensions of privacy" and investigates individuals' behaviour on social network sites and their privacy attitudes and conceptions. A special point of interest is age related differences.
e-mail: w.m.p.steijn@uvt.nl

John Vervaele is full time professor of economic and European criminal law at Utrecht Law School (the Netherlands) and professor of European criminal law at the College of Europe in Bruges (Belgium). He is vice-president of the AIDP, in charge of the scientific coordination of the world organization for criminal law. His scholarly work is dealing with collar crime and economic offences and European criminal law and procedure. The main topics in his research field are: enforcement of Union law; standards of due process of law, procedural safeguards and human rights; criminal law and procedure and regional integration; comparative economic and financial criminal law; terrorism and criminal procedure. He has realized a lot of research in these areas, both for Dutch Departments and European Institutions and worked as well as a consultant for them. He is regularly teaching as visiting professor in foreign universities, in Europe, the US, Latin America and China.
e-mail: J.A.E.Vervaele@uu.nl

David Wright is Managing Partner of Trilateral Research & Consulting, a London-based limited liability partnership, which he founded in 2004. Trilateral specialises in privacy, data protection, surveillance, security, risk and foresight issues. He has initiated and organised many successful consortia for European projects. Among recent projects is one on privacy and risk management for the UK Information Commissioner's Office. Another concerns privacy seals for the Institute for the Protection and Security of Citizens (IPSC) in Italy. He has published many articles in peer-reviewed journals. His most recent book is Privacy Impact Assessment, published by Springer in 2012.
e-mail: david.wright@trilateralresearch.com

Gabriela Zanfir holds an L.L.M. in Human Rights, obtained at the University of Craiova (Romania). She is currently a Ph.D. candidate at the same university, writing a thesis about the rights of the data subject in EU data protection law, with a focus on Romanian law. She is especially interested in the civil law mechanisms of protection of these rights and in conceptualizing the right to the protection of personal data as droit subjectif. Her research interests recently extended to cloud computing regulation. She was a visiting researcher for three months in 2012 at the Tilburg Institute for Law and Technology (The Netherlands). She is also a Teaching Assistant for the property law and torts courses.
e-mail: gabriela.zanfir@gmail.com

Part I
Data Protection in the World: Brazil and Poland

Chapter 1
Data Protection in Brazil: New Developments and Current Challenges

Danilo Doneda and Laura Schertel Mendes

1.1 Introduction

In the twentieth century, few legal concepts have transformed as much as that involving the right to privacy. The concept departed from discussions about the violation of privacy of celebrities photographed in embarrassing or intimate situations and reached discussions on massive data processing of millions of citizens by public and private entities through modern information technologies.[1] In this context, Stefano Rodotà affirms that privacy has been reinvented in the twentieth century, since this right has come to involve concepts such as transparency and control of personal data (beyond the right to be let alone and the notion of confidentiality), inducing the development of the right to data protection.[2]

This transformation has been observed since the 1970s in national data protection laws and in international treaties and agreements on the matter.[3] This legislative production started in Europe and North America as a response to the rise of electronic data collection and processing by governments and large companies.[4] Since then, technology and data protection laws have evolved, and the geographic boundaries of

---

[1] Simitis (1987, p. 709).
[2] See Rodotà (2008, p. 15).
[3] Regarding transnational policy instruments on data protection, including, e.g., Convention 108 of the Council of Europe, the Guidelines of the Organization of Economic Cooperation and Development (OECD), the Directive 95/46/EC of the European Union, see Bennett and Raab (2006, p. 83–115).
[4] Mayer-Schönberger (2001, p. 221).

D. Doneda (✉)
FGV Direito Rio-Praia de Botafogo,
190-13° andar-CEP 22250-900-Rio de Janeiro, RJ, Brazil
e-mail: danilo@doneda.net

L. S. Mendes
Humboldt-Universität zu Berlin,
Juristische Fakultät (Lehrstuhl für Bürgerliches Recht, Deutsches-, Europäisches- und Internationales Privat- und Wirtschaftsrecht),
Unter den Linden 6, 10099, Berlin
e-mail: lauraschertel@hotmail.com

S. Gutwirth et al. (eds.), *Reloading Data Protection*,
DOI 10.1007/978-94-007-7540-4_1, © Springer Science+Business Media Dordrecht 2014

data protection legislation have spread throughout the world.[5] It is clear today that the new frontiers are regions such as Asia and Latin America, where in the last decade several countries have updated their legislation to incorporate some degree of protection of personal data.

In Brazil, the concept of privacy and the instruments for its protection have undergone constant development in recent years by both courts and legislatures, to deal with the challenges of data processing. Although Brazil does not have a general data protection law as do several South American countries, a data protection framework is being developed from various elements such as the privacy rights provided in the Brazilian Constitution or several statutes that deal directly with personal data.

In fact, data protection is increasingly becoming a matter of autonomous regulation in Brazil, in relation to the constitutional right to privacy, what can be seen as a turning point in this matter. That is, more and more conflicts regarding data processing are being considered within a framework of transparency and control, rather than a privacy framework, which emphasizes opacity and confidentiality. This is due to the recent developments of the case law, the enforcement efforts of public authorities and the new acts issued in 2011 (i.e., the Credit Information Law and the Access to Information Law), which are the object of analysis of this paper.

As a consequence, one can observe the coexistence of two kinds of legal tools that deal with the flow of information in the Brazilian legal system, that is, privacy tools and data protection tools, in the words of Paul de Hert and Serge Gutwirth.[6] This paper concentrates on the data protection framework in Brazil, from its foundations to new developments, examining perspectives for further evolution and challenges to be faced. Rather than conduct a static analysis,[7] we aim to discuss the direction in which the Brazilian data protection framework is evolving.

The goal of this paper is, therefore, to analyze how data protection is guaranteed in Brazil, considering the recent development of new instruments and laws. The analysis is organized in three steps: (1) The first part addresses the foundations of data protection in Brazil, in particular, the constitutional provisions and the consumer protection code; (2) The second part addresses the new developments of data protection laws and instruments in the last years, particularly, the Credit Information Law and the Access to Information Law; (3) The third part analyses the challenges of guaranteeing data protection in Brazil and the tasks that must be carried out to improve data protection in the country.

---

[5] For an overview of the data protection legislation in the world, since the 1970s, see Table 5.1 "The diffusion of data protection legislation by region" in Bennett and Raab (2006, p. 127).

[6] According to them, privacy tools and data protection tools are complementary: while the former focuses more on opacity, the latter emphasizes control and transparency. See De Hert and Gutwirth (2006).

[7] Highlighting the non static feature of privacy and data protection, even within a more or less stable legal framework (e.g. the Data Protection Directive of 1995): Gutwirth et al. (2011), p. v.

1.2 Foundations of Data Protection in Brazil

A legal framework for data protection in Brazil has been developed from constitutional grounds up to specific legal measures in the last decades. Among several other sets of legislations that, in various ways, foresee some extent of generic privacy provisions, we will focus on the roots of the Brazilian data protection framework, based both on its constitutional grounds, and on the Consumer Protection Code.

1.2.1 Constitutional Protection and the Habeas Data Writ

The Brazilian Constitution directly addresses issues regarding information by providing for the fundamental rights of freedom of expression[8] and access to information and transparency.[9] In addition, it acknowledges the inviolability of private life and privacy[10] and also of telephonic, telegraphic and data communications,[11] and establishes that the home is the holy and inviolable refuge of the individual.[12] Furthermore, it provides for the writ of habeas data,[13] which gives citizens a way to access and correct data about themselves held by third parties.

The writ of habeas data was originally introduced in Brazil's 1988 Constitution and has since influenced several other Latin American countries to adopt similar provisions, to the extent that it was, at some point, taken as the root of a new Latin American data protection framework.[14] Habeas data also bears resemblance to the inscription of rights regarding privacy, data protection and computers in the new constitutional charts of two European countries that also were transitioning back to democracy in the 1970s after a period of dictatorship, i.e., Portugal and Spain.

The essence of Brazil's habeas data writ is to provide citizens with a tool to access and correct personal information stored by public bodies. It has been considered, as its legislative process indicates, to be an instrument much needed in the political situation in which it arose, when Brazil (like several countries in the region) was in transition to a democratic political regime.[15] At this time citizens needed a tool to access information that the military dictatorship had gathered about them,[16] and the habeas data was envisaged as this instrument. This means that the main inspiration for Brazil's writ of habeas data wasn't the legal framework about data protection

---

[8] Art. 5°, IX; art. 220, Federal Constitution.

[9] Art. 5°, XIV; Art. 220; Art. 5°, XXXIII; Art. 5°, XXXIV, Federal Constitution.

[10] Art. 5°, X, Federal Constitution.

[11] Art. 5°, XII, Federal Constitution.

[12] Art. 5°, XII, Federal Constitution.

[13] Art. 5°, LXXII, Federal Constitution.

[14] See Pulcinelli (1999); Guadamuz (2000).

[15] See Barroso (1998, p. 211). See also Dallari (1997, p. 72), Barbosa Moreira (1998, p. 127).

[16] Stella Calloni. "Los archivos del horror del operativo Condor", in: <www.derechos.org/nizkor/doc/condor/calloni.html>.

that several European nations had developed by that time nor the U.S. legal privacy tradition, but rather the mentioned requirements of the country's political moment.[17]

In fact, habeas data was not proposed as a modern data protection tool nor did it develop into one over time. It is a relatively costly and slow writ—it must be presented by a lawyer and only after the plaintiff has already requested the data directly from the defendant without success. Instead of adapting habeas data to a more dynamic environment, other instruments were developed in Brazilian law to address the increase of electronic data processing.

1.2.2 Ensuring Data Protection Through the Consumer Protection Code

Although the Brazilian Constitution recognizes a variety of privacy rights as well as the habeas data writ, as seen above, data protection, in a modern sense, initially emerged in Brazil as a consumer protection issue. In fact, the Consumer Protection Code (Law 8.078 of 1990) provided a multifaceted framework in which privacy and data protection demands could develop and be addressed. As the evolution of the issue in other countries reveals, the right to data protection tends to emerge in those legal fields that are more likely to welcome the new social demands. This task fell in Brazil to the Consumer Protection Code, since it entails a variety of principle-based norms, which are broad enough to offer solutions to new conflicts related to information technology.

Consumer protection plays a central role in Brazil's legal system. The Consumer Protection Code was enacted to balance the information and power asymmetries between consumers and traders.[18] It establishes norms regarding private, procedural and criminal law, as well as provides for an administrative structure for the enforcement of consumer rights. Moreover, it organizes a National Consumer Protection System, to coordinate the more than 600 public bodies responsible for consumer protection at the federal, state and local levels, which operate as an extrajudicial dispute resolution structure. Nonetheless consumers can also seek redress in the judicial system, particularly in small claims courts.

The recognition of consumer protection as a constitutional matter is central to the Brazilian legal system. Article 170, V, of the federal Constitution foresees consumer protection as a principle of the economic order and Art. 48 of its temporary provisions stipulates an obligation of enacting a Consumer Protection Code. The Constitution establishes, moreover, in its chapter of fundamental rights that "the State shall promote, as provided by law, consumer protection" (Art. 5°, XXXII). This norm implies not only a subjective right, but also a duty to protect,[19] which is directed to the state as a whole—the executive, legislative and judiciary branches. The duty to protect can involve, for instance, the duty to interpret law, taking into

[17] Doneda (2006, p. 328).

[18] Marques et al. (2006, p. 33).

[19] Concerning the concept of the fundamental right as being the duty of the state to provide protection, see Pieroth and Schlink (2005, p. 23).

account the vulnerability of consumers and their need for protection, or the duty of the state to develop a regulatory system to protect consumers.[20]

Four pillars of the Brazilian consumer protection system explain how it could promote and enforce data protection standards: (a) specific regulations for consumer databases that address the rectification and notice process; (b) a broad clause governing damage claims (overall liability); (c) a public consumer redress structure, which includes both an administrative and a judicial system of redress (small claims courts); and (d) a broad conceptualization of who are consumers.

The Consumer Protection Code establishes, in its Art. 43, specific rights and safeguards regarding personal information stored in databases, namely: (a) consumers shall have access to all the personal information stored on databases (right of access); (b) all stored data shall be objective, accurate and in a comprehensible language (principle of data quality); (c) consumers shall be notified, through written communication, before the storage of any negative personal information (principle of transparency); (d) the party responsible for the database shall immediately promote the rectification or cancellation of any inaccurate data that is being stored (right of rectification and implicitly justified cancellation[21]); and (e) the time limit for storage of negative personal data is 5 years (right to forget). This norm, which was inspired by the U.S. Fair Credit Reporting Act,[22] clearly has many similarities with the fair information principles of data protection.[23]

These data protection standards provided by Article 43 gain relevance when associated with the general clause of overall strict liability established by Article 6, VI, and Article 14, of the Consumer Protection Code. In fact, courts have recognized a broad right to compensation, for instance, when negative personal data about a consumer is stored without previous notification or when a consumer's application for credit is refused, based on incorrect data. Since the Brazilian judicial system has a variety of small claims courts, which facilitate consumer litigation and dispense the need for hiring a lawyer, this single norm had a huge impact on the legal system. Furthermore, consumers may register their complaints against credit information databases at the Public Consumer Protection Bodies, which will handle the individual complaint through an extra-judicial conciliation procedure. The National Register of Consumer Complaints in Brazil (SINDEC) recorded in the year 2012 more than 20,000 complaints about problems regarding the inappropriate storage or processing of credit information.[24]

Finally, the Brazilian Consumer Protection Code establishes a broad concept of consumer, which allows its application in a variety of cases, beyond the strict

---

[20] Pieroth and Schlink (2005, p. 23).

[21] Gambogi Carvalho (2003, p. 77–119).

[22] See Herman Benjamin et al. (2005, p. 400).

[23] In a comparative study of the data protection policies of four countries (Sweden, the United States, West Germany and the United Kingdom), Bennett systematizes the Fair Information Principles in six principles: openness, individual access and correction, collection limitation, use limitation, disclosure limitation and security. See Bennett (1992, p. 101).

[24] The total of registers in the year 2012 was 2.031.289. The National Register of Consumer Complaints in Brazil (SINDEC) is a public database and can be accessed through the website: http://portal.mj.gov.br/sindec/.

contractual relation between consumers and traders. The conceptualization of consumer in the Code comprises four definitions: (a) according to the standard definition, consumer is any physical person or corporate entity who acquires or uses a product or service as a final user (Art. 2°): (b) consumer is also a collectivity of persons who participate in consumer relations (Art. 2, § 2°); (c) consumer is, furthermore, anyone who has suffered damages caused by a commercial activity (Art. 17) and (d) any person who is exposed to a commercial practice, such as advertising or databases is also considered a consumer (Art. 29).[25] This means that if any of these definitions fits the case, the Consumer Protection Code is applied.

For this reason, a person doesn't need to prove any contractual relation to exercise his rights to correction and disclosure of his personal information against a database. Furthermore, this means that consumer damage claims can be directed not only against the firm with which he has a contract, but also against the party responsible for the database. That is why the data protection norms of the Consumer Code have had a much broader application than the strict relation between consumers and traders, promoting a modernization that extended beyond consumer relations.[26]

1.3 New Developments in the Brazilian Data Protection Framework

As seen above, the Brazilian legal system has a variety of privacy and data protection instruments, found in both the Constitution and ordinary laws. While the Constitution provides, in addition to the habeas data writ, many confidentiality guarantees (inviolability of home, private life and privacy as well as the confidentiality of correspondence, and telephonic, telegraphic and data communications), the Consumer Protection Code establishes a specific data protection norm, based on the concept of notification, rectification and compensation. Although they play an important role in protecting privacy, some of these instruments were found to have limitations and needed to be complemented to meet new challenges and problems. Against this background, one can understand the recent developments in the Brazilian data protection system, namely the Credit Information Law and the Transparency Act, both issued in 2011.

1.3.1 The Credit Information Law

The Credit Information Law (Law 12.414 of 2011) aims to regulate credit information systems, especially, borrowers' payment histories. Under the Consumer Protection Code, there was no doubt about the lawfulness of recording "negative" data about a

[25] Marques (2011, p. 385, 386).

[26] Tepedino (1999, p. 199–216).

consumer, that is, information about consumer default. There was, however, legal uncertainty about storing borrowers' payment histories ("positive information"). It was therefore important for the Credit Information Law to provide detailed regulations concerning credit information databases, thus establishing a secure legal framework that simultaneously encourages data flow and protects personal data. Given the size and complexity of the law, and its accompanying regulation (Decree 7.829 of October 2012), it is not possible to analyze all its rules in detail. Rather, we will examine the main principles and norms, concerning data protection rights.

In summary, it can be said that this law established a variety of rules ranging from the creation of a payment history to the establishment of responsibilities in case of damages, determining, for instance, when a payment history can be created (Art. 4), what information can be stored (Art. 3, § 2 and § 3), what are the rights of the data subject (Art. 5), what are the duties of the data processor (Art. 6), who supervises the databases (Art. 17) and who is liable in case of damages (Art. 16). Regarding the type of its norms, one could say that the Credit Information Law corresponds to a typical U.S. regulation issue, i.e., credit reporting, although it has a European form. As we will see, many of its norms correspond to the principles provided in Convention 108 of the Council of Europe and in the European Directive 95/46/EC.

The key principle of the Credit Information Law is that the consumer should have control over his personal information and, therefore, over the creation and use of his payment histories. In this sense, the law grants the consumer power over the creation, transference and cancellation of his credit history. Consumer consent is, hence, the touchstone of this framework, as provided by Article 4. Furthermore, according to Article 5, consumers shall obtain the cancellation of the record upon request and, as determined by Article 9, the sharing of information is permitted only if expressly authorized by the consumer. As seen, the main goal of the act is to grant consumers control over the flow of personal information in the market.

Like the Consumer Protection Code, the Credit Information Law establishes the principle of quality or accuracy of personal data (Art. 3, § 1), as well as the rights to the access, rectification and cancellation of data (Art. 5, II and III). Furthermore, it grants the consumer access to the main criteria used in the credit rating process, that is, the consumer has the right to know the criteria upon which a calculation of credit risk is based (Art. 5, IV). In relation to risk assessment, the act grants consumers the right to ask for a review of any decision made exclusively by automated means (Art. 5, VI). This rule is comparable to Article 15 of the European Directive 95/46/EC[27] and aims to ensure the possibility of human intervention in a process of making decisions that can significantly affect his or her life.

A very important improvement made by the Credit Information Law was to provide an explicit legal basis for the purpose limitation principle in the Brazilian system, which was already implicit under the Consumer Protection Code. As established by

---

[27] According to Art. 15 of the Directive 95/46/EC, "Member States shall grant the right to every person not to be subject to a decision that produces legal effects concerning him or significantly affects him and which is based solely on automated processing of data intended to evaluate certain personal aspects relating to him, such as his performance at work, creditworthiness, reliability, conduct, etc."

the act, the principle of finality permeates the entire credit information system. Firstly, the act defines the strict scope of its application, which are solely databases related to risk assessment in credit and commercial transactions (Art. 2, I). Secondly, it establishes the right of the data subject to have the processing of personal information limited to the original purposes of collection (Art. 5, VII). Thirdly, Article 7 describes the purposes for which the data collected under this act can be used: either to conduct risk analysis or to assist decisions regarding the granting of credit or other commercial transactions that involve financial risk. This implies that these databases cannot be used for direct marketing or any other activity not mentioned in the law. In this context, one notices another similarity to the European Directive, particularly Article 6, 1, b, which determines that personal data should be "collected for specified, explicit and legitimate purposes and not further processed in a way incompatible with those purposes."

A key rule of the Credit Information Law is the prohibition against storage of sensitive and excessive information, as provided by Article 3, § 3. According to the act, excessive information is that which is not related to the credit risk analysis. The act describes as sensitive information that related to social and ethnic origin, health, genetic information, sexual orientation and political, religious and philosophical beliefs. This prohibition is based on the fact that the processing of some types of information can lead to discrimination, violating the principle of equality. Therefore, it is possible to make another parallel to the European Directive, namely, Article 8, which concerns the processing of special data categories.

Furthermore, the Credit Information Law stipulates a system of strict liability, in which all material and moral damages must be repaired (Art. 16), without needing to prove negligence or fault. This rule is in accord with the liability clause of the Consumer Protection Code and relies on the concept that the liability arises from the risk of the activity.

Finally, an important concept endorsed by the act is the need for data processing to be controlled by an administrative authority. Rather than creating an authority to fulfill this function, the Credit Information Law designates the existing public consumer protection bodies, at the federal, state and local levels, as responsible for the supervision (Art. 17). Moreover, it establishes that the administrative penalties provided by the Consumer Protection Code shall be applied as well. Both provisions are only to be applied when the data subject qualifies as a consumer.

1.3.2 The Access to Information Law

Although the 1988 Brazilian Constitution established transparency as a principle for public administration and granted every citizen the right to access information from public bodies[28], the statute that created the procedures for this access only entered into force in 2012.

[28] Art. 5°, XIV; art. 5°, XXXIII; art. 5°, XXXIV, Federal Constitution.

The Access to Information Law (Law 12.527 of 2011) regulates the general constitutional rules regarding this issue and was drafted to respond to a particular need: to define rules for the treatment of personal information processed by public bodies. Its main goal was, of course, to grant free access to public information, which includes a considerable amount of personal information that can be classified as such. Since there was no general rule for the specific protection of personal information in other statutes (although this protection can be derived from constitutional principles), the Access to Information Law had to include a specific topic about protection of personal data held by public bodies.

Protection of personal data in this law basically comprehends that data concerning a particular individual shall not be disclosed to third parties who have sought access to the information. However, such information can be disclosed if the data subject has provided consent, if the data was produced more than 100 years before the access request, or if it qualifies for one of the exemptions (in case of health requirements, relevant public research, in compliance with a warrant, is needed for the protection of human rights or in a case of preponderant public interest).

The legal basis mentioned by the law for protecting personal data is also particularly relevant. For the first time in the Brazilian legal system, the treatment of personal information was directly related not only to the protection of privacy but was also seen as a means to ensure individual freedom in general. In this sense, Brazilian law contemplated for the first time a modern and general statement of the data protection principles tied to individual freedoms in a broader sense—something that the habeas data writ, even with its genealogy, has never attained.

Looking more closely at Article 31 of the Access to Information Law (the article that deals with personal data), it becomes evident that the law treats the protection of personal data as secondary to the disclosure of information. This can be inferred both from the broad nature of the exemptions for the free access to personal data and from the lack of other specific measures for its protection (for example, sensitive personal data has no special level of protection).

The architecture of data protection present in the access to information law is no more than that strictly necessary for the harmonization of the access to information—which is the purpose of the statute—and the protection of personal data, considering that without some form of mandatory protection of personal data the statute could be found to be seriously lacking compliance with constitutional provisions. Even so, the secondary nature of the data protection provisions in this statute and the importance the regulation of personal data plays in the complete legal framework for information demonstrate the need for specific measures regarding personal data protection to be found outside the Access to Information Law.

Nevertheless, the access to information law has made a concrete contribution to the Brazilian data protection framework, and not only because of the specific provisions of its Article 31. The statute, by creating a simple process with time limits to order a public body to produce information after a request is made, has also developed an instrument that makes it easier for citizens to request their own personal information from public bodies, without the burden or inconvenience that could be faced if the request were made by general administrative means (which would lack the specific enforcement of the access to information law) or through a writ of habeas data (which, among other drawbacks, would require a lawyer).

This is a potential intersection between access to information and data protection statutes, which, according to David Banisar, is often used in countries that have no specific law to deal with personal information but have some kind of access to information mechanism or, in countries that have both statutes but in some way filter or adapt the access to personal information requirements to the access to information framework.[29]

1.4 Current Challenges of Guaranteeing Data Protection in Brazil

As analyzed in the previous section, new developments in ordinary law have complemented the legal foundations of data protection in Brazil, improving the instruments for dealing with data processing problems in the country. Nonetheless, there are still many challenges to be faced to adequately respond to the risks arising from data processing in a network society. These challenges can be divided into two categories: on the one hand, there are challenges related to enforcement, since there is already a data protection framework that needs to be implemented; on the other hand, there are regulation issues, since there is a lack of legislation in some areas, which must be addressed by the Congress.

1.4.1 Enforcement: The Role of the Judiciary and of the Consumer Protection Bodies

A systematic interpretation of the Consumer Protection Code and the Credit Information Law builds a framework for data protection in Brazil's private sector. A key element of this framework is the broad concept of consumer established by the Consumer Protection Code, so that its application is not limited to the person who acquires or uses a product or service as a final user, but applies to anyone who is exposed to a commercial practice or who has suffered damages caused by a commercial activity.[30]

Against this background, it is possible to outline the principles and procedures that private data controllers must meet, to comply with the data protection system in Brazil: (1) Transparency: all processing of personal data shall occur in a transparent way. Data controllers must assure that the data subject knows about the purpose of the collection and the use of the data, the kind of data being processed, and the identity of the data controller; (2) Control of personal information: a central element of data

[29] Banisar (2011).

[30] See Sect. 2.2.

protection is that the data subject should have control of his personal information. Consent is, therefore, the legal instrument that materializes this control and may be limited only in exceptional circumstances; (3) Purpose limitation principle: any processing of personal data must comply with the context in which data are collected. Thus, information collected for one purpose cannot be further processed in a way incompatible with those purposes; (4) Guarantee of the rights to access, rectification and cancellation: the data subject shall have free access to his data, should be able to rectify inaccurate and outdated information and should be able to cancel data that was stored improperly; (5) Special protection for sensitive data: personal information that could generate consumer discrimination should have stronger protection, such as data concerning religious and political choices, sexual preference, race, health and genetic data.

As can be seen, a framework exists for data processing in the Brazilian private sector, which corresponds to the main concepts of the Fair Information Principles, Convention 108 of the Council of Europe and Directive 95/46/EC. A current challenge in this field is, therefore, to enforce the existing norms, in order to guarantee protection for the data subject. There are many actors that are responsible for the implementation of data protection norms.

Primarily, the courts play an important role in this enforcement, interpreting and applying data protection instruments and concepts. In fact, a qualitative analysis of the Brazilian case law on data protection indicates that the decisions of the courts are moving from a strict view of the credit information issue to a broader perspective, in which the processing of personal data is understood as a general risk to a citizen's personality.[31] Two cases can illustrate this shift.

Well-known in this context is a 1995 decision of the Superior Court of Justice, under the leading opinion of the rapporteur, Minister Ruy Rosado de Aguiar, concerning time limitation for the storage of personal data. The court is the highest jurisdiction for non-constitutional cases in Brazil. In this case, the court decided that credit records about consumer default could not be stored for more than 5 years, as provided by the Consumer Protection Code, and not for 20 years, the period in which the debts prescribe, according to the Civil Code.[32] In this decision, the Court extended its analysis to the risks of the processing of personal data in general and not only to the credit reporting activity. This case was an innovation in Brazilian case law, because it drew attention to the risks arising from the data processing activity, by both the public and private sectors.

In recent years, issues concerning data protection on the Internet have entered the courts and compel the courts to find adequate solutions within the existing framework. A recent decision of the Superior Court of Justice, concerning a disclosure of a picture on a website, indicates how the problem of data protection on the Internet is increasingly gaining relevance in the Brazilian legal system. In this case, the court decided that the company that controlled the website was liable for the misuse of the

---

[31] Mendes (2011, p. 54).

[32] STJ, REsp 22.337–9/RS, 4.ª T., j. 13.02.1995, v.u., rel. Min. Ruy Rosado de Aguiar, DJ 20.03.1995

image and had to pay compensation for material and moral damages.[33] Central to the decision was the opinion of the rapporteur, which discussed the new challenges posed by the Internet to the legal system and recognized that technological innovations gave rise to the development of a new concept of privacy, based on the control of personal information by the individual.

In addition to the judiciary, the executive branch also has a very important role in enforcing data protection rights in the private sector. As mentioned before, Brazil's Consumer Protection System comprises more than 600 public bodies, at the federal, state and local levels. The Consumer Protection Code grants all of them the same legal powers, which range from receiving consumer complaints to applying administrative penalties to the companies, in case of non-compliance with the law.[34] Although there is no hierarchy among these public bodies, the National Secretary of Consumer Protection, which is part of the Ministry of Justice, performs the political coordination of the system.[35]

In fact, data protection is becoming an issue of public policy in Brazil and the consumer protection bodies are taking actions to enforce data protection rights within the existing framework. One interesting step, for instance, was the creation of a "do-not-call registry" in many states.[36] In São Paulo, the registry was created by a state law, which made the consumer protection body (Procon São Paulo) responsible for its management and supervision.[37] Furthermore, the National Secretary of Consumer Protection is working at many levels to enforce data protection rights of consumers. It published, for instance, a study on consumer right to data protection in Brazil, as an effort to stimulate discussion on this issue[38] and added data protection as a subject of the training courses to the staff of the consumer protection bodies.[39] Concerning the supervision activities, it has the power to investigate practices, which indicate violation of data protection and privacy rights of consumers. An example of an ongoing investigation is the Phorm-case.[40] The company is being investigated for suspected privacy violation caused by its behavioral advertising system.

---

[33] STJ, REsp 1.168.547/RJ, 4.ª T., j. 11.05.2010, v.u., rel. Min. Luis Felipe Salomão, DJe 07.02.2011

[34] Art. 55 and 56 of the Consumer Protection Code.

[35] Art. 106 of the Consumer Protection Code.

[36] This measure is currently available in the states of Mato Grosso do Sul, Paraná, Rio Grande do Sul, Alagoas and São Paulo. See the following websites: <http://www.procon.pr.gov.br/modules/conteudo/conteudo.php?conteudo=485>; <http://www.proconbloqueio.rs.gov.br>; <http://www.procon.ms.gov.br/index.php?templat=vis&site=115&id_comp=2309&id_reg=96052&voltar=home&site_reg=115&id_comp_orig=2309>; <http://naoperturbe.itec.al.gov.br>.

[37] http://www.procon.sp.gov.br/BloqueioTelef/

[38] <http://portal.mj.gov.br/main.asp?Team = {B5920EBA-9DBE-46E9-985E-033900EB51EB}>

[39] <http://portal.mj.gov.br/main.asp?ViewID ={3DB528D3-F9F0-4B22-AA4B-6CF6BBA31173}&params = itemID = {FDD46AEE-F356-420E-A868-C18A4BC52E98};&UIPartUID = {2218FAF9-5230-431C-A9E3-E780D3E67DFE}>

[40] http://www.senado.gov.br/noticias/opiniaopublica/inc/senamidia/notSenamidia.asp?ud=20100630&datNoticia=20100630&codNoticia=409485&nomeOrgao=&nomeJornal=O+Globo&codOrgao=47&tipPagina=1; http://veja.abril.com.br/agencias/ae/economia/detail/2010-06-29-1132438.shtml

1.4.2 *Regulation: The Need of Comprehensive and Sectorial Data Protection Laws*

Although some problems regarding data protection in Brazil require enforcement measures, as seen above, there are some issues that can only be adequately addressed by a broad regulation such as a comprehensive data protection act. This would increase the legal certainty of business activities involving the processing of personal data and guarantee wider protection to individuals against the risks to privacy arising from data processing. This explains why there have been many attempts to create a general legal framework for data protection in Brazil.

In spite of some legislative activity around bills that addressed the issues of data protection in the last decade, to this day no comprehensive data protection bill has reached the final stages of deliberation in either of Brazil's federal legislative bodies. In fact, until recently, none of the few data protection bills proposed[41] even contemplated all of the usual components of a general data protection bill, such as its application to both the public and private sectors or the prevision of a public authority to enforce its rules.

Since 2005, however, the Brazilian government has pondered the prospect of a general data protection bill, after the Argentine government proposed, in a Mercosur working group, the establishment of rules governing data protection in the region to improve citizenship and commerce. As a result of the debate generated at the time, the Brazilian Ministry of Justice drafted a data protection bill and submitted it to public consultation over an online platform in late 2010.[42] During the process of public consultation, the draft bill received more than 800 proposals from public and private entities.[43] The federal government is now expected to present formally a bill to Congress.[44]

The publically available version of the draft bill[45] has its structure based on standard data protection principles that, in a broad way, are akin to those present in international documents such as Convention 108 of the Council of Europe or Directive 95/46/EC. It includes, for instance, provisions about transborder data flow and contemplates the creation of a public authority responsible for enforcing the law. The structure of the draft bill indicates the influence of established national data protection statutes, such as the Italian, German, Portuguese and Spanish ones. Moreover, Brazilian laws, such as the Consumer Protection Code and the Competition Act, influenced the draft.

---

[41] Bills such as PLS 321 of 2004 or PLC 4060 of 2012.

[42] The public discussion is still available in read-only mode at: <http://culturadigital.br/dadospessoais/>

[43] <http://portal.mj.gov.br/main.asp?View={08DEBD27-66DA-4035-BE88-27126C102E22}&Team=&params=itemID={53B2C85F-206D-4DCC-A3D0-85E8E38F6D41};&UIPartUID=2218FAF9-5230-431C-A9E3-E780D3E67DFE}.>

[44] <http://www.tiinside.com.br/15/02/2013/governo-prepara-projeto-de-lei-para-protecao-de-dados-na-web/ti/325360/news.aspx>.

[45] <http://culturadigital.br/dadospessoais/files/2011/03/PL-Protecao-de-Dados_.pdf>.

Overall, the architecture of the data protection framework contemplated in the draft is for a general law that applies to both the public sector and to companies. It is based on a unified and centralized scope rather than on sectorial provisions and relies on unified rules to be applied to the whole country rather than on empowering states and local authorities. Finally, its provisions are directly based on constitutional principles for protecting individuals and personal freedom.

Much of these specifications are not to be considered as options that the legislator could freely choose. Since the Brazilian civil law derives directly from continental European models[46], it tends to privilege a systematic and centralized approach to the regulation of fundamental rights, in contrast to options such as a sectorial approach or even solutions strongly based on self-regulation. In addition, the characteristics of the Brazilian federation require a law of federal scope rather than a regional one, due to the specific nature of Brazil's federal system.

Considering the recent comprehensive reform of the European data protection framework[47], proposed by the European Commission, it is interesting to analyze if and how it is influencing the current efforts of developing new legislation on data protection in Brazil. Examining the Brazilian draft bill, it is possible to notice that some of the proposed norms are comparable with the articles of the European Regulation proposal, such as the breach notification (Art. 27, draft bill) and the norm regarding the binding characteristic of self-regulated codes (Art. 45, draft bill)[48]. Therefore, although it is not possible to establish a direct relation between both processes, we can see that the formulation of the draft bill of data protection has clearly taken into account the new developments in Europe.

Meanwhile, the data protection scenario in the region has changed since 2005. Several Latin American countries have adopted a general data protection law: in addition to Argentina, which pioneered the issue, Mexico, Uruguay, Colombia, Peru and others have statutes governing the area.[49] In Brazil, the lack of a broad regulation in this field has increasingly been considered as a problem both by citizens and by companies: on the one hand, citizens are more and more aware of the risks of an uncontrolled data flow, as issues such as identity theft and commercial abuse of personal data have gained visibility[50]; on the other hand, compliance with international standards concerning the international transfer of personal data and a strong

[46] René (2002).

[47] <http://ec.europa.eu/justice/newsroom/data-protection/news/120125_en.htm>.

[48] See the draft at: <http://culturadigital.br/dadospessoais/files/2011/03/PL-Protecao-de-Dados_.pdf>

[49] Regarding the development of data protection legislation in Latin America, see http://www.redipd.org/ (Ibero-American Network of Data Protection).

[50] The Press is increasingly reporting on these matters. See, e.g.: <http://www1.folha.uol.com.br/mercado/1182808-crescem-as-fraudes-com-uso-do-cpf-alheio-um-terco-dos-casos-envolve-telefonia.shtml>; http://www1.folha.uol.com.br/folha/dinheiro/ult91u418838.shtml.

set of rules governing data protection in general are considered a necessity for the development of new businesses in the country, such as cloud computing.[51]

In this context, a comprehensive data protection act is seen at the same time as a way of ensuring more protection to the citizens, concerning the processing of their personal data, and increasing legal certainty to the companies, regarding how to process and use personal data within the legal framework. Furthermore, analyzing the current data protection framework in Brazil, we see that there are some challenges to be faced, which would need a sectorial approach. Three of these challenges, both in the public and in the private sector, will now be highlighted.

First of all, problems related to data protection in the Internet need a special attention of the regulators. Problems concerning data protection in social networks, cookies, behavioral advertising, cloud computing as well as problems related to privacy on smart phones demand a specific approach. It is clear that these are all transnational problems, and as such they need a supranational response. Nonetheless, it is important to address these questions, in order to solve the problems and demands in the national level. The proposed law 2126 of 2011, commonly referred to as the Civil Framework for the Internet, deals, among many issues, also with data protection on the Internet.[52] It aims to establish a set of rights to all Internet users in Brazil and announces as guiding principles both the "protection of privacy" and "protection of personal data, under the terms of the law." In this context, both the Civil Framework for the Internet and the Data Protection draft bill are certainly important steps in this direction, although they don't address all these specific questions regarding online privacy.

The second challenge involves guaranteeing privacy in specific sectors, such as, for example, the health sector. Health information is a very sensitive kind of personal data and can cause much harm to the data subject if used in an inadequate form. Considering that a huge part of the medical information system is already automatized in Brazil, it would be necessary to establish specific rules that define the legal use and flow of this information. Even more concerns raise the use of genetic data. Unlike many countries that regulate genetic tests, limiting direct–to–consumer genetic testing[53], e.g., Germany, Portugal, France and Switzerland, Brazil lacks a specific regulation on the matter, even though Law 12.654 of 2012 deals specifically with genetic profiling, as it creates the National Genetic Profiling Database for criminal enforcement purposes.

The third challenge concerns data protection in the public sector, particularly, in regard to the flow of personal data from public databases to private data processors as well as the investigation activities of public bodies, using new technologies. Concerning the first issue, there isn't a broad regulation, but only governmental decrees that limit the transfer of personal data from public databases. Specially the federal

[51] The Ministry of Science and Technology has stated the necessity of a data protection law to develop the cloud computing sector in the country: <http://convergenciadigital.uol.com.br/cgi/cgilua.exe/sys/start.htm?infoid=31438&sid=97;> <http://www.sucesuba.org.br/brasil-se-prepara-para-dar-um-salto-em-cloud-computing>.

[52] This bill is a result of years of public discussions held by the federal government in cooperation with the legal and internet scholars from the Fundação Getúlio Vargas, in the Rio Law School.

[53] See Borry et al. (2012).

governmental database of the "Bolsa Família", an important governmental assistance program, entails very sensitive information about a huge share of the Brazilian population and demands strong safeguards to protect the citizens against discriminatory acts and negative decisions that could affect significantly their life. The data protection law could be an adequate instrument to address this issue.

Another problem related to the public sector involves the investigative activities of public bodies through new technologies, which bear large risks to the individuals. The Law 10.217/2001, for example, which aims to repress organized crime, allows technical surveillance provided that there is a warrant issued by a judicial authority. However, this is the only condition established for the legitimacy of the surveillance acts, and the law lacks other kinds of protection, like a wiretapping report, such as required in other countries' legislation.[54] A specific norm that covers data protection and privacy issues in investigations of public bodies would be appropriate to address these problems.

In all of these cases, a comprehensive data protection law as well as sectorial laws are essential instruments to meet the challenges and risks posed by data processing through new technologies. Above all, the creation of a central data protection authority seems to be crucial in order to centralize the discussions on these matters, propose new regulations and enforce the existing norms.

1.5 Conclusion

It is clear that Brazil's data protection framework has evolved continuously in recent years, so as to protect the individual's personality in different contexts and to establish legal certainty concerning particular kinds of data processing. The basis for this protection, the privacy rights established by the Brazilian Constitution, the habeas data writ and the Consumer Protection Code, have been complemented by new legislative developments, particularly the recent acts regarding credit information and access to information.

The Credit Information Law provided a legal foundation for storing and processing borrowers' payment histories, establishing the consent of the data subject as the touchstone of this system and imposing strict requirements for the processing of personal data, which are comparable with the principles and rights of Convention 108 from the Council of Europe and the Data Protection Directive. The Access to Information Law has improved the data protection framework both by complementing the habeas data writ with a material right to access information (reinforcing the right of the data subject to access personal information stored in public databases), and by providing minimum data protection rules, which establish consent or a specific legal provision as requirements for the disclosure of personal data.

[54] See, e.g., the U.S. Wiretap Act, 18 USC § 2519—Reports concerning intercepted wire, oral, or electronic communications.

Based on this analysis, one can notice that there are two kinds of challenges to be faced, in order to guarantee data protection rights in Brazil: on the one hand, the judiciary and the executive branches have to enforce the existing framework by applying the norms through a systematic interpretation[55], as well as by implementing public policies on data protection[56]; on the other hand, there are many challenges, which could only be faced through regulation initiatives, especially the enactment of a comprehensive data protection law and of sectorial acts.[57]

Thus, in any case, it can be observed that data protection is becoming an autonomous field in Brazil and gaining relevance within the legal system. It is therefore to be expected that the latest developments in information technology will increasingly demand the creation of new data protection instruments in Brazil to deal with the risks to individual privacy presented, for example, by the ubiquitous data processing,[58] the Internet of things,[59] online searching[60] and the web 2.0. The current revision of the European Directive[61] and the white paper on digital privacy,[62] released by the U.S. government, both in the year 2012, indicate the need to create new legal instruments concerning privacy and data protection to deal with technological progress and globalization. There is no doubt that the Brazilian data protection system must continue to develop, in order to guarantee fundamental rights and legal certainty in a networked society.

References

Banisar, David. 2011. The right to information and privacy: balancing rights and managing conflicts. http://ssrn.com/abstract=1786473. Accessed 10 Sept 2013.

Barbosa Moreira, José Carlos. 1998. O Habeas Data brasileiro e sua lei regulamentadora. In *Habeas data*, ed. Teresa Arruda Alvim Wambier, São Paulo: RT.

Barroso, Luís Roberto. 1998. A viagem redonda: habeas data, direitos constitucionais e provas ilícitas. In *Habeas data*, ed. Teresa Arruda Alvim Wambier, São Paulo: RT.

Bennett, Colin. 1992. *Regulating privacy: Data protection and public policy in Europe and the United States*. Ithaca: Cornell University Press.

Bennett, Colin, and Charles Raab. 2006. *The governance of privacy: Policy instruments in global perspective*. London: MIT Press.

Borry, Pascal, et al. 2012. Legislation on direct-to-consumer genetic testing in seven European countries. *European Journal of Human Genetics* 20, 715–721.

---

[55] See Sect. 4.1.

[56] See Sect. 4.1.

[57] See Sect. 4.2.

[58] Concerning the concept of ubiquitous data processing, see Mattern (2008).

[59] Regarding the problem of online searching by public authorities and the solution of the German Constitutional Court, See Hoffmann-Riem (2009, p. 519).

[60] See Hoffmann-Riem (2008, p. 1011).

[61] <http://ec.europa.eu/justice/newsroom/data-protection/news/120125_en.htm>

[62] <http://www.whitehouse.gov/the-press-office/2012/02/23/fact-sheet-plan-protect-privacy-internet-age-adopting-consumer-privacy-b>

Dallari, Dalmo de Abreu. 1997. El Hábeas Data en Brasil. Ius et Praxis 3 (1): 71–80. http://www.redalyc.org/articulo.oa?id=19730108.
De Hert, Paul, and Serge Gutwirth. 2006. Privacy, data protection and law enforcement. Opacity of the individual and transparency of power. In *Privacy and the criminal law*, eds. Eric Claes, Antony Duff and Serge Gutwirth. Antwerp: Intersentia.
Doneda, Danilo. 2006. *Da Privacidade à Proteção de Dados Pessoais*. Rio de Janeiro: Renovar.
Gambogi Carvalho, Ana Paula. 2003. O consumidor e o direito à autodeterminação informacional. *Revista de Direito do Consumidor* 46: 77.
Guadamuz, Andres. 2000. Habeas data: The Latin-American response to data protection. 2 *The Journal of Information, Law and Technology*. http://www2.warwick.ac.uk/fac/cos/law/elj/jilt/2000_2/guadamuz/. Accessed 10 Sept 2013.
Gutwirth, Serge, Yves Poullet, Paul De Hert, and Ronald Leenes. 2011. Preface. In Computers, privacy and data protection: An element of choice, eds. Serge Gutwirth, Yves Poullet, Paul De Hert and Ronald Leenes, Dordrecht: Springer.
Herman Benjamin, Antonio et al. 2005. *Código brasileiro de Defesa do Consumidor comentado pelos autores do anteprojeto*. Rio de Janeiro: Forense Universitária.
Hoffmann-Riem, Wolfgang. 2008. Der grundrechtliche Schutz der Vertraulichkeit und Integrität eigengenutzter informationstechnischer Systeme. *Juristen Zeitung* 21: 1009–1022.
Hoffmann-Riem, Wolfgang. 2009. Grundrechts- und Funktionsschutz für elektronisch vernetzte Kommunikation. *Archiv des öffentlichen Rechts* 134.
Marques, Claudia Lima. 2011. *Contratos no Código de Defesa do Consumidor. O Novo Regime das Relações Contratuais*. São Paulo: Revista dos Tribunais.
Marques, Claudia Lima, Antônio Herman Benjamin, and Bruno Miragem. 2006. *Comentários ao Código de Defesa do Consumidor*. São Paulo: Revista dos Tribunais.
Mattern, Friedemann. 2008. Allgegenwärtige Datenverarbeitung—Trends, Visionen, Auswirkungen.... In *Digitale Visionen: Zur Gestaltung allgegenwärtiger Informationstechnologien*, eds. Alexander Roßnagel, Tom Sommerlatte, and Udo Winand, Berlin: Springer.
Mayer-Schönberger, Viktor. 2001. Generational development of data protection in Europe In *Technology and privacy: The new landscape*, eds. Philip Agre and Marc Rotenberg, Cambridge: MIT Press.
Mendes, Laura Schertel. 2011. O direito fundamental à proteção de dados pessoais. *Revista de Direito do Consumidor* 79: 45–81.
Pieroth, Bodo, and Bernhard Schlink. 2005. *Grundrechte—Staatsrecht II*. Heidelberg: C. F. Müller.
Pulcinelli, Oscar. 1999. *El habeas data en Indoiberoamérica*. Bogotá: Temis.
Simitis, Spiros. 1987. Reviewing privacy in an information society. *University of Pennsylvania Law Review* 135 (3): 707.
René, David. 2002. *Os grandes sistemas do direito contemporâneo*. São Paulo: Martins Fontes.
Rodotà, Stefano. 2008. *A vida na sociedade da vigilância: a privacidade hoje*. ed. Maria Celina Bodin de Moraes, trans. Danilo Doneda and Luciana Cabral Doneda. Rio de Janeiro: Renovar.
Tepedino, Gustavo. 1999. As relações de consumo e a nova teoria contratual. In *Temas de direito civil*. Rio de Janeiro: Renovar.

Chapter 2
The Effectiveness of Redress Mechanisms. Case study—Poland

Dorota Głowacka and Beata Konieczna

2.1 Redress Mechanisms Research Study—General Remarks

This article is a result of the national study conducted within the project "Data protection: redress mechanisms and their use"[1] run by the European Union Agency for Fundamental Rights ("FRA") within its FRANET multidisciplinary research network.[2] The aim of the project is a comparative overview and analysis of existing procedures and the legal consequences of data protection violations in the EU Member States. The FRA's previous engagement in data protection related research—such as the report on "Data Protection in the European Union: the role of National Data Protection Authorities"[3] as well the special Eurobarometer survey on attitudes towards data protection in the EU[4]—highlighted that redress mechanisms in the area of data protection, even though usually available, are not widely used. The Eurobarometer survey revealed, for example, that only 33 % of Europeans were aware of the existence and competences of the national data protection authority.

The amount of information available at the EU level is still insufficient to fully understand the poor use of redress mechanisms revealed by these researches. The FRA project attempts, therefore, to provide insight into why the existing redress measures in different EU Member States are not applied to their full extent. The answer to this question is particularly important with regard to the planned EU data protection reform, which may bring some innovation in this area compared to the

[1] European Union Agency for Fundamental Rights.

[2] FRANET is composed of National Focal Points in each EU Member State and Croatia, the aim of this network is to provide the Agency with comparable socio-legal data on fundamental rights issues to facilitate the FRA's comparative analyses at EU level.

[3] European Union Agency for Fundamental Rights (2010).

[4] Special Eurobarometer 359 (2011).

D. Głowacka (✉)
Zgoda 11, 00-018 Warsaw, Poland
e-mail: d.glowacka@hfhr.org.pl

B. Konieczna
e-mail: beata-konieczna@wp.pl

S. Gutwirth et al. (eds.), *Reloading Data Protection*,
DOI 10.1007/978-94-007-7540-4_2, © Springer Science+Business Media Dordrecht 2014

existing Directive 95/46/EC of the European Parliament and of the Council on the protection of individuals with regard to the processing of personal data and on the free movement of such data ("Data Protection Directive"). The results of the FRA research may therefore help to evaluate the improvements currently outlined in the European Commission's proposal on data protection reform presented in January 2012.[5]

The FRA project will compare the relevant legislation and redress mechanisms concerning data protection in 16 Member States. Using both desk and social field-work research, the research aims to capture the experiences and views of the main actors concerned, such as—*inter alia*—individuals who suffered from data protection violations, legal practitioners dealing with the subject within their professional duties, and the data protection authority's staff. The project's final report will be published by the FRA in 2013.

This article will briefly summarize the most interesting preliminary findings of the study conducted in Poland.[6] It will present both the general characteristics of the Polish study as well as certain phenomena specifically related to this particular national research. It will also discuss the most interesting views and experiences voiced by respondents participating in the project, which were used to identify common problems with regard to data protection provisions.

The general conclusion from the research is that the effectiveness of redress mechanisms in Poland is quite low. The article will attempt to identify the most serious barriers to efficient redress mechanisms as well as to recommend remedies. Hopefully, the conclusions drawn from the research may have a universal interest and provide some insight into what measures may be needed to improve the use of and access to redress for other countries facing similar encumbrances and in the context of the planned EU data protection reform.

2.2 Redress Mechanisms in Poland—Overview

The right to privacy and the principle of personal data protection were introduced to the Polish legal system in article 47[7] and article 51[8] of the country's Constitution of 2 April 1997.[9] These constitutional standards were further elaborated in the Personal

[5] European Commission (2012).

[6] The fieldwork was conducted between January and August 2012 by the Helsinki Foundation for Human Rights based in Warsaw which is FRA's "National Focal Point" in Poland.

[7] Article 47 of the Constitution: "Everyone shall have the right to legal protection of his private life and family life, of his honour and good reputation and to make decisions about his personal life".

[8] Article 51 of the Constitution: "1. No one may be obliged, except on the basis of statute, to disclose information concerning his person. 2. Public authorities shall not acquire, collect or make accessible information on citizens other than that which is necessary in a democratic state ruled by law. 3. Everyone shall have a right of access to official documents and data collections concerning him. Limitations upon such rights may be established by statute. 4. Everyone shall have the right to demand the correction or deletion of untrue or incomplete information, or information acquired by means contrary to statute. 5. Principles and procedures for collection of and access to information shall be specified by statute".

[9] The Constitution of the Republic of Poland (1997).

Data Protection Act of 29 August 1997.[10] The provisions of the Personal Data Protection Act also implement the Data Protection Directive into the Polish legal system. The concept of personal data protection entitles any person whose rights have been infringed to request a review of the legality of the processing of their data. Under Polish law, there are three procedures for seeking redress in the field of personal data protection: administrative, civil and criminal. The choice of the procedure depends on the particular circumstances of the case. At some instances all these remedies can be applied simultaneously.

2.2.1 *Administrative Procedure*

The Code of Administrative Procedure[11] in principle is the legal basis for the proceedings conducted by the Inspector General for the Protection of Personal Data ("IGPPD") pursuant to the Personal Data Protection Act. The IGPPD is an independent, central body of state administration appointed and dismissed by the Sejm of the Republic of Poland (lower chamber of the Parliament) upon the approval of the Senate (upper chamber of the Parliament). The general tasks of the IGPPD encompass the supervision of the compliance of data processing with the provisions of the Personal Data Protection Act. The proceedings initiated by the IGPPD may be launched upon receiving a complaint, or *ex-officio*. The proceedings should be concluded with an administrative decision determining whether a data protection violation has occurred.

In case of any breach of the provisions on personal data protection, the IGPPD shall, by its administrative decision, restore the proper legal state. In particular, the IGPPD may order to remedy the negligence, to complete, update, correct, disclose, or not to disclose personal data, to apply additional measures protecting the collected personal data or to erase personal data.[12] Parties that are not satisfied with the administrative decision have 14 days, from the date of delivery, to apply to the IGPPD for the re-examination of their case.[13] Further on, if the IGPPD determines to uphold the original challenged decision, the party has 30 days to file a complaint with the Provincial Administrative Court in Warsaw. A judgment delivered by a first-instance court may be challenged by lodging a cassation complaint with the Supreme Administrative Court. The complaint must be made within 30 days from the day a copy of the reasoned decision is served on the parties. The proceedings before the administrative courts are governed by the Administrative Courts Procedure.[14]

---

[10] Personal Data Protection Act (1997).

[11] The Code of Administrative Procedure (1960).

[12] Article 18 of the Personal Data Protection Act.

[13] Article 21 (1) of the Personal Data Protection Act: "A party may submit a motion to have their case reconsidered with the Inspector General".

[14] The Administrative Courts Procedure Act (2002).

The IGPPD cannot impose financial penalties for data violations. In order to increase its effectiveness, the IGPPD was afforded a power to enforce duties imposed by virtue of its decisions. This includes the authority to impose a coercive fine on entities that fail to perform administrative decisions issued by the IGPPD.[15] The maximum fine amounts to PLN 10,000 zlotych (around 2,500 €) for natural persons and PLN 50,000 zlotych (around 12,500 €) for legal persons. In addition, the IGPPD may intervene with various authorities and entities in order to ensure efficient protection of personal data. It may also request competent authorities to issue or to amend legal acts in cases relating to personal data protection. Importantly, the entity shall give an answer in writing to any address or request made by the IGPPD within 30 days following its receipt.[16]

2.2.2 *Criminal Procedure*

Chapter 8 of the Personal Data Protection Act contains provisions on the criminal liability for offences defined in articles 49–54a of the Act. They include, *inter alia*, offences such as: disclosure or providing access to data to unauthorized persons, processing personal data in a data filing system where such processing is forbidden, violation of the obligation to protect data against unauthorized access, damage or destruction, failure to provide data subjects with obligatory information about their rights with regard to data processing, or hindering the performance of inspection activities performed by the IGPPD. The offences are prosecuted by the law enforcement bodies in accordance with the general procedural rules laid down in the Code of Criminal Procedure.[17] Under the applicable criminal provisions laid down in the Personal Data Protection Act, the possible criminal sanctions, depending on particular offences are a fine and restriction or deprivation of liberty. The term of imprisonment ranges from less than a year, if the infringing party acted inadvertently, up to 3 years, if the offence concerns sensitive data.

2.2.3 *Civil Procedure*

Unauthorised processing of personal data may also be a cause of action in a case involving an infringement of personal rights, which include protection of privacy and personal data. Pursuant to article 23 of the Civil Code[18], personal rights (such as health, freedom, dignity, freedom of conscience, surname, pseudonym, or image) are protected in civil law, notwithstanding the protection granted by virtue of other

[15] Article 12 (3) of the Personal Data Protection Act.

[16] Article 19a of the Personal Data Protection Act.

[17] The Code of Criminal Procedure (1997).

[18] The Civil Code (1964).

legal enactments. Under article 24 of the Civil Code, the legislator allowed for several kinds of remedies in the event of personal rights violation. The claimant may seek the remedy of the results of the infringement through—for example—a publication of apologies. They may also seek compensation in the case when a moral and pecuniary damage occurred (they may also ask the defendant to make a donation for a charitable or community purpose). Civil proceedings are commenced with a plea filed with the court by the claimant (the data subject) against the defendant (person responsible for the data protection violation). Against the judgment of the first instance court, the parties can lodge an appeal with the second instance court and then also a cassation appeal (last resort appeal) with the Supreme Court. In the civil proceedings concerning the infringement of personal rights, the claimant's position is stronger as they may benefit from the "presumption of unlawfulness" of the defendant's action. This means that the claimant needs to only prove that his personal rights have been infringed, putting forward evidence confirming the infringement. It is the defendant, on the other hand, who must satisfy the court that his or her actions have been taken legally (e.g. there existed a legal basis for the processing of a claimant's personal data).

2.3 Characteristics and Methodology of the Research Study in Poland

The fieldwork research on the redress mechanisms in the field of data protection took place between January and August 2012 and comprised 36 one-to-one interviews with three groups of respondents: 15 individuals who have sought redress ("complainants"), 15 individuals who have experienced data protection violations but have not sought redress on that account ("non-complainants"), and judges and prosecutors conducting cases which involve data protection violations. Three focus group discussions were also conducted as part of the research. The targeted groups included the IGPPD's staff, members of non-governmental organisations working in the field of data protection, and advocates and legal counsellors specialising in data protection law.

In the group of complainants, out of 15 respondents—10 initiated criminal proceedings, 7 commenced administrative proceedings and 4 brought civil actions (some of the respondents initiated more than one procedure at the same time). In the non-complainants group, the respondents most frequently stated that they had considered taking administrative actions (9 out of 15 respondents). The second most popular measure was alternative dispute resolution procedures, such as settlements (indicated by 8 respondents). Only 3 respondents considered bringing their case before a criminal or civil court. Respondents were allowed to list more than one potential redress procedure.

Respondents pointed out a series of events in which they had faced personal data violations. The most common situations concerned unlawful personal data processing by governmental bodies as well as private financial institutions and marketing companies. Personal data violations very often included disclosing data for commercial purposes, resulting *inter alia* in unwanted telemarketing calls to the respondents' private phone numbers. Many respondents also complained about breaches involving the use of the Internet, such as their personal data being posted on-line or unsolicited information being sent to their e-mail addresses. Moreover, there were some violations mentioned which were committed by employers from both the private and public sector with regard to the processing of a broader range of employees' personal data (including for example biometric data) than prescribed in Polish employment law. There was also one reported case in which personal data were unlawfully collected and processed by one of the intelligence agencies and one case in which there was an excessive use of CCTV cameras.

The general starting point for a reflection on redress mechanisms in most respondents' cases was a subjective sense of harm and a feeling that the boundaries of an acceptable level of interference with their privacy had been crossed. The greater the sense of infringement, the more willing they were to seek redress. One of the respondents pointed to the "aggregation effect" of data violation. This means that a decision to take specific legal steps is taken on the basis of a subjectively understood aggregate of data violations. One of the main motivating factors for the respondents to seek redress was a desire to counteract the abuse of public power, publicize the problem and draw attention to the violating of personal data protection law by the state and local authorities; another important motivation was a desire to obtain compensation for a data protection violation. The compensations did not have to be financial, but could be limited to an obligation to publish apologies, for example. Finally, the respondents claimed that what drove them to take legal measures was very often a personal aversion towards the breaching party (for example an ex-employer) and the desire to "punish" them by imposing some kind of penalty on them for their misconduct.

2.4 Barrier No. 1: A Lack of Education on and Low Awareness of Data Protection

The first barrier to effective redress mechanisms identified by the results of the study is the low awareness and lack of education on data protection issues among data subjects. The main reason for not seeking redress in the case of non-complainants was that very often they were unable to define if they actually had faced a data protection violation. They could not always link the negative emotions they experienced with regard to a particular situation with the infringement of their right to privacy. They were unable to assess if the harm they suffered was "serious enough" in order to legitimately claim their rights or whether it fell within the scope of the data protection law. Some respondents were also not aware of the existence and the characteristics of a personal data protection body (this applies not only to the individuals suffering from data protection violations, but sometimes even the members of the judicial community).

Regarding the complainants who could recognize data protection violations, the vast majority did not have sufficient information about the redress mechanisms that could be applied in their cases. They were usually not aware of different redress mechanisms at their disposal (civil, penal or administrative) and what outcome they could expect from particular procedures. The lack of knowledge in this respect very often led to the very low level of satisfaction with regards to the results of the proceedings they had decided to initiate. Even if the outcome turned out to be positive for the respondents, it did not meet their expectations and eventually caused disappointment.

For example, for those complainants who sought redress intending to obtain compensation or "punish" the data controller responsible for the data protection violation, the administrative procedure before the data protection authority was unsatisfactory. For many complainants application to the IGPPD seemed the most "natural choice" as they perceived it as the most well-qualified and specialized body in the area of data protection issues. At the same time, there is a low public awareness with regard to the IGPPD's actual competences, which are more limited than many respondents expected. Most of the respondents initiating administrative procedure before the IGGPD were not aware that it is not competent to award compensation for moral damage (which is a competence reserved by the civil court). Furthermore the IGGPD is not intended and authorized to impose criminal penalties on individuals liable for unlawful data processing (which is a competence reserved by the criminal court). With regard to criminal proceedings, if the IGGPD decides that an action or omission of a person responsible for violation contains all requisite elements of the offence laid down in the Personal Data Protection Act, it can only report the offence to the law enforcement body, which may then take further actions. What is more, the IGPPD's competences are even more limited if in the meantime the data controller remedies the violation. The IGPPD may intervene only with regard to existing violations.

That is why in many respondents' opinions the administrative proceedings conducted by the IGPPD followed by the judicial administrative proceedings carried out by the administrative courts fail to provide complete protection, as they do not have the effective compensatory or punitive function. The vast majority of respondents were not aware these results could be achieved through civil or criminal proceedings before civil or criminal courts and not by the administrative procedure.

The low level of public awareness regarding data protection issues and redress mechanisms results from a lack of education and lack of easy access to information. The main and primary source of knowledge for the respondents was the Internet. However, many of them stated that the online sources are fragmented, unclear and not sufficiently accessible. According to the respondents, there is no website which would offer a comprehensive guide to the available redress mechanisms, a list of the stages of the proceedings, and the rights and obligations of individuals pursuing legal redress procedures. For example the respondents noted that the website of the IGPPD does not entirely serve this purpose, even though it contains an on-line educational platform. They noted that the way it is written is too formal and in language which is not user-friendly; moreover, it is incorrectly positioned on the web in search engine results and therefore difficult to access. On the other hand it was not very common for the respondents to obtain knowledge on the area of data protection through professional legal assistance, which was believed to be excessively costly.

To sum up, despite the existence of three different kinds of procedures, which in some instances can be applied simultaneously, their existence does not improve the effectiveness of the redress mechanisms. This is because people seeking redress for data protection violations are not aware of the existence of these measures; they do not distinguish them, they do not always understand their characteristics or they do not have sufficient knowledge about how they should be applied.

2.5 Barrier No. 2: The Insufficient Enforceability of Data Protection Law

The second barrier weakening the effectiveness of the redress mechanisms in Poland mentioned by most of the respondents was the insufficient enforceability of data protection law. This applies especially to the criminal procedure that was evaluated as the least effective by both the complainants and focus group interview participants such as the IGPPD's staff and judges.

The complainants who had tried to initiate the criminal proceedings reported that they had had the impression that criminal judges and prosecutors were not familiar with personal data law, tended to be unwilling to investigate such cases, and underestimated its serious character. According to the complainants and the IGPPD's staff, it is mainly the prosecutors who are not prepared to apply data protection provisions. The study revealed that a low level of expertise in the field of personal data protection among law enforcement authorities is the result of the absence of any extended coverage of this area during legal vocational courses or subsequent training. These educational shortcomings may generate a negative attitude towards prosecuting data protection cases. That is why most respondents' cases concerning data protection violations were discontinued by prosecutors as "negligible social harm"; only in a few cases did the prosecution eventually bring the indictment to the court. This view was also confirmed by the annual report on the IGPPD's activities for 2011.[19] The report states that the analysis of the pre-trial proceedings conducted by the prosecution leads to a conclusion that a very small number of cases end up in court. According to the report, there were only two sentencing judgments by the criminal courts in 2011 resulting from the IGPPD's notification of crimes in previous years (in 2010 the IGPPD made 23 notifications, in 2009—27, in 2008—31). The prosecution does not provide data on the general number of crime notifications concerning data protection violations.[20]

In the prosecutors' opinion there is a whole range of factors contributing to the negative perception of the activity of the prosecution service and its preparation for handling data protection cases. The prosecutors claimed that criminal provisions

[19] IGPPD (2012, p. 243–244).

[20] The response of the Prosecutor General to the Helsinki Foundation for Human Rights (2012).

contained in the Data Protection Act are poorly drafted in the first place. In the majority of cases they cannot be applied because the reported crimes do not contain all the requisite elements of an offence laid down in the law. Moreover, prosecutors consider data protection provisions to be "detached from real life". The prosecutors also complained about the lack of professional training on data protection issues and specifically noted the need to organize training courses on this topic to enhance the competence of the prosecution service. They also stated that they consider the criminal liability for data protection violations extensively severe, disproportional and therefore expressed the view that these kinds of cases should be handled under administrative and not criminal law. Especially due to the limited resources at the prosecution office forcing them to focus on "more serious" crimes. Some of the prosecutors (and other respondents) even suggested that the criminal provisions should be abolished from the Personal Data Protection Act.

According to the members of the focus group interviews, the most effective legal remedy to data protection violations is the civil procedure. That is because it may bring a desirable and satisfactory outcome to the claimant (financial compensation, apologies, etc). Moreover in cases involving the protection of personal data (falling under the heading of personal rights), the claimants have a much better position at the trial as they benefit from the "presumption of the unlawfulness" of the infringement. But the most effective civil measure is at the same time the least popular one. The respondents' top two choices in selecting a redress mechanism were the criminal and administrative procedures, whereas the civil procedure was considered an alternative that might be used if the first two mechanisms fail. Only 4 out 15 complainants took advantage of civil law instruments. The respondents perceived the civil procedure as the most expensive, complex, and long-lasting process (which is not always true) and were not aware of its positive aspects. They also perceived the civil procedure as the most engaging for the party and they were anxious that they would be "left on their own" during the trial (especially if they could not afford a professional legal aid). For the criminal or administrative procedure, the respondents believed they would be offered "support" from the IGPPD or the prosecution during the proceedings.

2.6 Barrier No. 3: The Data Protection Law—its Complexity and Limited Scope

The third barrier which lowers the effectiveness of the redress mechanisms that was identified by the respondents is the complexity of the data protection law, namely: the formalism and lack of transparency of the procedures as well as its extensive duration (and the fact that it is very hard to assess the length of the proceedings). The language of the Personal Data Protection Act itself was also described by most of the individual respondents as difficult and discouraging.

The problem of the complexity of the procedures was very well reflected by the information provided by the representative of the IGPPD's staff who claimed that around 80 % of the complaints submitted to the data protection authority for the first time contain formal defects and they have to be returned to the complainants in order to be remedied. This applies not only to the complaints filed by individuals, but also by professional attorneys. Only after the formal defects had been remedied, could the complaint be re-submitted and further proceeded with, which significantly prolongs the procedure. Most of the respondents also noted that they were unable to use the Internet document filing system on the IGPPD's website because it seemed too complicated. Furthermore the respondents questioned the mechanism of the IGPPD's reconsideration of its own decisions, which is obligatory before filing the complaint to the administrative court. Many respondents claimed it was not effective and a time-consuming legal remedy which unnecessarily extends the procedure.[21]

Another problematic issue indicated by the respondents were the numerous limitations to the inspection powers of the IGPPD under the Polish Data Protection Act. Examples of these jurisdictional gaps that were specifically reported during the study were the IGPPD's inability to verify whether an apostate's records have been deleted from parish records and the lack of a mechanism for establishing the scope of data processed by the Central Anti-corruption Bureau ("CBA", one of the Polish intelligence agencies).

Over 30 individuals who made an act of apostasy[22] from the Catholic Church answered the call to participate in the research. A member of this group was interviewed as part of the complainants' group. He described situations where parish priests refused to delete the apostates' personal data from parish records. At the same time he claimed that the apostates were unable to complain to the IGPPD whose jurisdiction is excluded with regard to processing data for the purposes of a church or religious association.[23] Therefore under the current Polish Data Protection Law the IGPPD is not competent to issue an administrative decision obliging the church to remove or correct the personal data processed within its books. Some of the apostates formed an active group aimed at effecting change in the applicable provisions governing the issue of leaving the Catholic Church. One of the group's initiatives was the creation of a website which contains a comprehensive, step-by-step guide to the

[21] Pursuant to the Code of the Administrative Procedure and the Administrative Courts Procedure the IGPPD must issue an administrative decision within 30 days after receiving a complaint. In the course of 14 days after the date when the original decision is served on the party, the party may ask for reconsideration of the case by the IGPPD. Only after the second decision is issued and served on the party, the party has 30 days for challenging the decision before a Provincial Administrative Court.

[22] Apostasy is currently understood as a conscious, voluntary and public defection from the Church. On 29 September 2008 the Polish Episcopal Conference laid down a procedure for persons wanting to make an act of apostasy in one of the Polish dioceses. Under this procedure an aspiring apostate shall deliver a statement drafted personally to the parish priest in his or her place of residence in the presence of two adult witnesses. The priest has an obligation to confirm the receipt of such a statement with his handwritten signature and a parish seal and to send the same to the diocesan curia which further instructs the priest to make an appropriate entry in the parish register of baptism. The note that a person has left the Church should be permanently entered on the margin of this person's certificate of baptism kept in the register of baptism. Under the ecclesiastical law an entry of baptism must not be deleted from the register of baptism.

[23] Article 43 (2) of the Personal Data Protection Act.

process of apostasy, accessible for everyone.[24] They also launched a social campaign advocating legislative changes to the Personal Data Protection Act that would allow the IGPPD to control the processing of personal data by religious institutions.

The "apostates' case" was clearly a Polish phenomenon. In this context it is interesting to note that pursuant to the Data Protection Directive, the inspection powers of data protection authorities may be subject to limitations in the areas of public security, economic or financial interests and crime prevention. The Directive provides for no such exemption in the case of the processing of data for the purposes of a church or a religious association, whereas the Polish Data Protection Act expressly waived the IGPPD's jurisdiction in this respect. The problem was reported to the European Commission[25] and the Commission noticed it and obliged itself to verify incompatibility of the Polish act with the Data Protection Directive. The Commission also noted that religious associations processing personal data should not receive any special privileges as they do not fall within any exceptions established under the directive. Moreover, the Court of Justice of the European Union clarified in the *Bodil Lindqvist* case (C-101/01)[26] that activities of a religious nature were not covered by the scope of such exceptions.

Another example of the limited powers of the IGPPD is the case of a journalist whose various telecommunications data, including phone records and location data for 6 months between 2005–2007 (some constituting personal data), were unlawfully acquired by one of the intelligence agencies—the CBA. The journalist was known for writing about high-profile and scandalous operations of the CBA. In this case the journalist could not complain to the IGPPD which has no jurisdiction over intelligence agencies under Polish law.[27] Eventually he filed a civil suit with the court claiming there was a violation of his personal rights such as the right to privacy, freedom of communication and, above all, the right to the freedom of expression because the CBA's conduct posed a threat to the journalist-source confidentiality. The journalist won the case in the court of first instance.[28] The court confirmed that the practice of misusing data retention, by collecting and reviewing the claimant's telecommunications data, is unlawful despite broad competences of the agency in this respect and exclusion from the IGPPD's control with regard to the processing of personal data. The court stated that by accessing the journalist's phone records the public authorities had clearly interfered with his constitutional freedoms. Such interference should be possible solely when it is clearly permissible under the law, appropriately justified and proportionate in comparison to the benefits expected to be obtained (e.g., in the case of a serious crime). The court also confirmed that the journalist's phone bills should be protected under the regulations concerning the right to privacy and the journalistic sources of information.

---

[24] Apostazja.info.

[25] Complaint ref. no. CHAP 2011(776) (2011).

[26] The judgement of the Court of Justice of the European Union (2003).

[27] Article 43 (2) of the Personal Data Protection Act.

[28] The judgement of the Warsaw District Court (2012). The judgement is not final. The appeal will be examined in April 2013.

2.7 Barrier No. 4: Insufficient Access to Professional and Effective Legal Aid

Respondents were in general reluctant to use paid professional legal aid in cases related to redress sought in the field of data protection. Parties to all kinds of proceedings usually sought a remedy acting in person unless they were forced to comply with the legal representation requirements in the appellate procedure (for example in the case of complaining to the court of final resort such as the Supreme Court or the Supreme Administrative Court). Among the respondents who sought remedies for personal data protection violations, only one hired and paid a lawyer. Other respondents who used professional legal aid received *ad-hoc pro-bono* legal support. The main reason for this situation indicated by the respondents was the relatively high costs of legal representation. However, it must be noted that the majority of respondents seeking redress, even though they had not received legal advice, felt that such assistance was needed. The substantial expenses related to using the services of an advocate or a legal counsellor therefore remain a barrier to accessing redress mechanisms in the field of data protection for data subjects. On the other hand it should be noted that according to lawyers participating in the focus group interviews, the data controllers use paid legal assistance at all procedural stages.

At the same time, the quality of legal representation in personal data protection cases is relatively low, with the notable exception of a number of law firms with a record of top-class service in this area. The study has shown that few lawyers specialize in data protection in Poland and that they are concentrated around big cities, especially the capital Warsaw. The specialized law firms, according to both the IGPPD's staff and the judges, offer very high quality services but mostly for companies and institutions that process data and not for data subjects. Therefore, access to experts in the field for the latter is problematic.

The availability of free legal aid is also limited. This is due to the more general problem of a lack of effective free legal assistance mechanisms in Poland. During court proceedings it is possible to apply for a court-appointed lawyer after meeting the criteria prescribed by law (for example financial criteria; these criteria apply to all kinds of cases). Nevertheless, the attorneys appointed as part of the state legal aid scheme do not always have the required expertise in the area of personal data protection. There is no free legal advice granted by the state in the pre-trial stage. In many areas of law, this gap is filled by non-governmental organizations providing *pro bono* legal aid. Regarding data protection issues though, there is no organization in Poland that would offer comprehensive, free-of-charge legal assistance and would be specialized solely in this field. The existing organizations that deal with personal data focus mainly on advocacy activities. There are also a number of associations offering general *pro bono* legal advice in most typical criminal or civil cases, but these are not sufficiently prepared for handling data protection issues. On the other hand there are a number of organizations providing counselling in more specific areas such as asylum seekers' law or consumer rights, often operating in partnership with central authorities such as the Office of Competition and Consumer Protection. However, with regard to data protection, no similar initiative has been so far observed.

2.8 Conclusions and Recommendations

The most alarming conclusion that transpires from the interviews with the respondents who participated in the Polish part of the "Data protection: redress mechanisms and their use" study is the low awareness of the available redress mechanisms in the data protection field. The research revealed that the theme of personal data protection has not become common knowledge despite 15 years of the operation of the Personal Data Protection Act in Poland. Most of the respondents were unable to define the data protection violation or seemed unaware of the possible redress measures. Those with a more developed legal awareness were not decided as to which procedure (administrative, criminal, civil) is the "right one" for a given case. The problem of low awareness was reported not only by the individuals who experienced unlawful data processing, but also all the other groups of respondents such as judges, prosecutors, advocates, legal counsellors, the members of civil society organisations and law enforcement bodies.

Therefore in the light of the results of the study and presented observations on the availability and effectiveness of personal data protection redress mechanisms, the following actions may be taken in order to improve the current situation: (1) awareness-raising and educational activities, (2) certain legislative reforms, (3) easier access to effective legal aid.

As a part of the awareness-raising and educational activities performed in the area of data protection, members of the public should learn more about the existence and functioning of the IGPPD. In this respect, particular emphasis should be put on providing information on the areas of the most frequent personal data protection violations (Internet and banking sector, healthcare, etc.), on the entities that most often commit violations (both in private and public sector) and on the essence of the violation and its consequences.

Moreover, it is very important to promote sources where individuals can find comprehensive information on data protection and redress mechanisms. According to the results of the study, a knowledge access point for citizens and legal advice providers could be the website of the IGPPD. However, the website should be better promoted and positioned through the browsers and adjusted to the needs of data subjects (for example, it should be written in more transparent and simpler language). It should also include more practical guidance for potential complainants, who could learn how to identify a data protection violation and what steps to take in order to seek redress. The guide should assist complainants in selecting the type of proceedings (administrative, judicial-administrative, criminal, civil) and should present a detailed description of the course of each procedure and the obligations of the parties.

Another proposal mentioned by the respondents was to introduce the basics of data protection as a subject of instruction to at least a secondary education curriculum. Educational activities could be supported by far-reaching social campaigns. Furthermore, training opportunities should be offered to advocates and legal counsellors, judges and staff of law enforcement authorities, non-governmental organisations and data controllers.

As regards the legislative reforms, many respondents proposed abolishing the penal procedure which, according to the study, is at the moment the least effective remedy for data protection violations. However, this should not be done without at the same time implementing more effective measures aimed at improving the enforceability of data protection law within civil and administrative procedures.

For instance, in the case of administrative proceedings, it would be important to strengthen the position and competences of the IGPPD. The majority of respondents see the IGPPD as an independent, although not always effective body, which is a result of numerous limitations stemming from the regulation. It follows from the respondents' opinions that administrative proceedings conducted by the IGPPD followed by the judicial administrative proceedings carried out by administrative courts fail to provide complete protection. One of the proposals was therefore to introduce financial administrative sanctions imposed by the IGPPD, which could replace the existing criminal sanctions. Other proposals concerned appointing the court of general jurisdiction as the court competent to hear appeals against IGPPD's decisions (instead of the administrative court) or creating a specialised court in this respect. The specialized court could operate in a similar manner as the existing Court of Competition and Consumer Protection in Poland, reviewing appeals lodged—for example—against the decisions of the President of the Office of Competition and Consumer Protection. What is more, dropping certain limitations to the IGPPD's inspection powers with regard to, for example, religious institutions should be also taken into consideration.

Another significant issue is an organisational reform of the IGPPD office which—according to the project—was said to be understaffed, overloaded with work and underequipped. The statistics published on the IGPPD's website reveal that the number of cases processed by the Polish data protection authority is increasing year-by-year. One of the reasons for this is an increasing volume of complaints and requests for interpretation. In 2007 the number of complaints addressed to IGPPD was 796 while in 2011 it reached 1272. As regards the requests for interpretations, in 2007 there were 1298 requests lodged, while in 2011 the number of requests reached 3935.[29] Therefore, additional human and financial resources within the IGPPD's office are needed. In this context it is important to note that the recent amendment to the Personal Data Protection Act[30] enabled the IGPPD to create local branches. This development was considered a very good idea which was supposed to "decentralize" the IGPPD service and make it more available to individuals from outside Warsaw. Unfortunately there were no budgetary means provided to implement this organisational reform and so far the local branches have not been established.

Easier access to effective legal aid should be also provided, especially as regards legal advice at the pre-trial stage. This proposal involves not only—as mentioned above—trainings for professional lawyers or empowering the IGPPD office in providing legal assistance on a broader scale, but also creating a non-governmental

[29] Own work developed on the basis of data available at the IGPPD's official website (2013).

[30] The Act amending the Personal Data Protection Act and Certain Other Acts (2010).

organisation specialized in *pro bono* data protection legal services, which at present does not exist in Poland.

Finally, it should be noted that one of the main current barriers for data protection redress mechanisms are jurisdictional issues related to processing of personal data in the cyberspace. This aspect was not a subject to a detailed analysis in the study. Nevertheless it was briefly mentioned by some of the respondents, especially lawyers practicing in the data protection field, as an emerging matter for example with regard to the the use of social network services. Polish citizens rank fourth in Europe in the use of social networks (85 % of internet users use these services and this number keeps growing).[31] Still the most popular social network services in Poland are operated by foreign companies.[32]

There are no official statistical data available on the number of data protection cases including a cross-border element.[33] At the same time media more and more often inform about cases of Polish citizens who come across jurisdictional issues while attempting to bring a legal action against a social network or a search engine run by an overseas companies.[34] In the study, the respondents noted that jurisdictional conflicts and limited jurisdiction of the national authorities with regard to the cross-border data processing will become one of the most significant challenges for the efficient legal protection. Therefore the legal solutions aiming to improve the effectiveness of redress mechanisms should include also the question of international cooperation between the data protection authorities, law enforcement agencies and the internet industry.

References

Legal Acts

Personal Data Protection Act, Poland. August 29, 1997. Journal of Laws 2002 No. 101, item 926 as amended (Ustawa o ochronie danych osobowych).

The Administrative Courts Procedure Act, Poland. August 30, 2002. Journal of Laws 2002, No. 153, item 1270 as amended (Ustawa—Prawo o postępowaniu przed sądami administracyjnymi).

The Act amending the Personal Data Protection Act and Certain Other Acts, Poland. October 29, 2010. Journal of Laws 2010 No. 229, item 1497.

The Code of Administrative Procedure, Poland. June 14, 1960. Journal of Laws 2000 No. 98, item 1071, as amended (Kodeks postępowania administracyjnego).

The Constitution of the Republic of Poland. April 2, 1997. Journal of Laws 1997 No. 78, item 483.

The Civil Code, Poland. April 23, 1964. Journal of Laws 1964, No. 16, item 93, as amended (Kodeks cywilny).

The Code of Criminal Procedure, Poland. June 6, 1997. Journal of Laws 1997, No. 88, item 555, as amended (Kodeks postępowania karnego).

[31] Ministry of Administration and Digitalisation (2012, p. 165).

[32] Wirtualne media.

[33] For example such data are not provided in the IGPPD's annual reports

[34] For example: Metro; Dziennik.pl; Dziennik Łódzki.

Judgments

Bodil Lindqvist case, Court of Justice of the European Union. November 6, 2003. Case no. C-101/01.
X V. Centralne Biuro Antykorupcyjne, Warsaw District Court. April 26, 2012. Case no. IIC 626/11.

Reports, Surveys, Communications

European Commission. 2012. Communication from the commission to the European parliament, the council, the European economic and social committee and the committee of regions, safeguarding privacy in a connected world, a European data protection framework for the 21st Century, COM (2012). 9. Brussels.
European Union Agency for Fundamental Rights. 2010. *Data protection in the European union: The role of national data protection authorities*. Luxembourg: Publications Office of the European Union.
Inspector General for the Protection of Personal Data. 2012. The annual report of the inspector general for the protection of personal data for 2011. Warsaw: 2012. http://www.giodo.gov.pl/data/filemanager_pl/sprawozdaniaroczne/2011.pdf.
Ministry of Administration and Digitalisation, Społeczeństo informacyjne w liczbach. 2012 (Information society in numbers), Warsaw: 2012. http://mac.gov.pl/dzialania/jest-nowy-raport-spoleczenstwo-informacyjne-w-liczbach.
Special Eurobarometer 359. 2011. Attitudes on Data Protection and Electronic Identity in the European Union. Brussels.

Press Releases

Metro. Polak pozywa Facebooka. Bobowski z Sosnowca kontra Zuckerberg z Santa Clara (The Polish sues Facebook. Bobowski from Sosnowiec versus Zuckerberg from Santa Clara). http://metromsn.gazeta.pl/Wydarzenia/1,127307,13033164,Polak_pozywa_Faceboka__Bobowski_z_Sosnowca_kontra.html. Accessed 11 Dec 2012.
Dziennik Łódzki. 25-letnia łodzianka walczy z Google o usunięcie wulgarnego filmu (25-year old women from Lodzi battles against Google about an obscene video). http://www.dzienniklodzki.pl/stronaglowna/355791,25-letnia-lodzianka-walczy-z-google-o-usuniecie-wulgarnego,id,t.html?cookie=1. Accessed 12 Jan 2011.
Dziennik.pl. Tymochowicz pozywa Googla. Chce 20 mln dolarów (Tymochowicz sues Google. He demands 20 mln dollars compensation). http://wiadomosci.dziennik.pl/media/artykuly/423765,piotr-tymochowicz-pozywa-google.html. Accessed 2 April 2013.
Wirtualne media. Czy czeka nas zmierzch portali społecznościowych (Is this fading of the social networks?). http://www.wirtualnemedia.pl/artykul/czy-czeka-nas-zmierzch-portali-spolecznosciowych#. Accessed 12 Aug 2012.

Websites

Apostazja.info. Apostazja.info—Jak wypisać się z kościoła katolickiego. www.apostazja.info. Accessed 3 March 2013.
European Union Agency for Fundamental Rights. Data Protection: redress mechanisms and their use. http://fra.europa.eu/en/project/2011/data-protection-redress-mechanisms-and-their-use. Accessed 3 March 2013.

Inspector General for the Protection of Personal Data (IGPPD), Official website. www.giodo.gov.pl. Accessed 3 March 2013.

Other

Complaint to the European Commission, ref. no. CHAP 2011(776). 2011. Wrong implementation of Directive 95/46/EC—Article 43(2) of the Polish Personal Data Protection Act—lack of powers of the Polish Data Protection Supervisory Authority. (March)

The response of the prosecutor general to the Helsinki foundation for Human Rights' motion for public information. 2012. (8 May).

Part II
Forgetting and the Right to be Forgotten

Chapter 3
Forgetting, Non-Forgetting and Quasi-Forgetting in Social Networking: Canadian Policy and Corporate Practice

Colin J. Bennett, Christopher Parsons and Adam Molnar

> *"You may not realize it, but whenever you go online, you're building an identity through the words and images you post and the activities you do. This can become part of your reputation, and it can be a lasting one. Once personal information goes online, it may be difficult to delete. While you may be able to delete it in one place,there may be cached versions or copies stored elsewhere that you cannot control. Digital storage is cheap and computer memory is plentiful–and unlike people, the Net never forgets" (Jennifer Stoddart, Canadian Privacy Commissioner, January 28th, 2011)*

The Canadian Privacy Commissioner's remarks above encapsulate a common perception about the online capture and retention of personal data. At some point in the last 10 years, it is argued, the economics and technical practicalities of retaining personal information have come to outweigh the arguments and potentials for deletion or erasure.[1] Whatever the institutional motivations, it is just easier to retain data than to get rid of it. Hence the dominant discourse about the "Net never forgetting" translates into strong warnings from privacy regulators and advocates about being extra careful about posting any information online, if you do not want it to come back and haunt you later in life.

Research for this paper was funded through the Office of the Privacy Commissioner of Canada's contributions program. None of the results are necessarily reflective of the Office's positions, and all research was conducted independently of the Commissioner. We thank Brittany Shamess and Michael Smith for research assistance.

[1] Victor Mayer-Schonberger, *Delete: The Virtue of Forgetting in the Digital Age (Princeton, NJ: Princeton University Press, 2011).*

C. J. Bennett (✉)
Department of Political Science, University of Victoria, Victoria,
British Columbia V8W 3R4, Canada
e-mail: cjb@uvic.ca

C. Parsons
The Citizen Lab, Munk School of Global Affairs, University of Toronto,
M5S 1A3 Toronto, Ontario, Canada

A. Molnar
Surveillance Studies Centre, ℅ Department of Sociology,
Queens University k7L 3N6 Kingston, Ontario, Canada

S. Gutwirth et al. (eds.), *Reloading Data Protection*,
DOI 10.1007/978-94-007-7540-4_3, © Springer Science+Business Media Dordrecht 2014

Almost as an antidote to these technical realities, the European Union has proposed in its new Draft Regulation[2] a "right to be forgotten." Using this provision, individuals could force an organization to delete personal data stored about them. Social networks that make such data public will be liable if it is subsequently republished by third-parties, and will be required to "take all reasonable steps, including technical measures" to inform third-parties to delete the information. The Article is included in the inventory of "rights of the data subject" and has intellectual roots in French law, which recognizes *le droit a l'oubli.* The right is not, as was originally proposed, limited to user-generated and -published data. It is broader, relating to any data concerning an individual, even if it has been generated or transmitted by someone else. This has significant implications for data controllers because they are expected to take all reasonable steps to meet individuals' requests, for themselves and for third-parties. Requests must be fulfilled "without delay", though exceptions exist for journalistic and artistic purposes, for complying with legal obligations, and when retaining the data is needed for proof of accuracy.

This provision has spawned an extraordinary amount of legal analysis and social criticism in a relatively short time, even though the right is far from new and is rooted in many legal provisions at national and European levels.[3] For some analysts, however, it has become a fundamental threat to freedom of expression, a tool of censorship, and an attack on search and archiving services. It has been seen as both "reactionary and fashionable" (van Hoboken 2011).[4] In many respects, this provision has become a lightening rod, symbolizing what many corporate interests regard as an overly intrusive, heavy-handed and unworkable European regulation (Fleisher 2011). It has also inspired commentary about the clash between European "protectionist" and American "free speech" values. Jeffrey Rosen, for example, has asserted that this right is "[t]he biggest threat to free speech on the Internet in the coming decade".[5]

The intensity of American opposition to this proposal is explained by fears of the extensive costs and practical complications that would arise if European citizens suddenly could erase their personal data, regardless of whether it had been posted by them, or by third parties. The intensity of this debate has sometimes overlooked, however, the perspectives of non-European countries with comprehensive data protection laws, and their experiences with enforcing consumers' rights of deletion and erasure against US corporations. The Canadian experience is especially illuminating in this regard. As Canadian life is more generally and immediately influenced by the

---

[2] European Commission, COM (2012) 11 final, "Proposal for a regulation of the European Parliament and of the Council on the protection of individuals with regard to the processing of personal data and onthe free movement of such data (General Data Protection Regulation)", (Brussels, 25 January 2012). The right to be forgotten is introduced in art. 17.

[3] Ambrose, M. and Ausloos, J, "The Right to be Forgotten Across the Pond" (paper presented at the Telecommunications Policy Research Conference, September 21, 2012) accessed October 20, 2012, http://papers.ssrn.com/sol3/papers.cfm?abstract_id=2032325.

[4] Joris van Hoboken, "9 Reasons Why a 'Right to be Forgotten' is Really Wrong," *Joris van Hoboken: about search engines, digital civil rights and more,* December 11, 2011, http://www.jorisvanhoboken.nl/?m=201112.

[5] Rosen, Jeffrey, "The Right to be Forgotten," *Stanford Law Review* 64 Online (2012): 88–92.

actions and policies of public and private organizations south of our border, there is a perennial cultural sense of "being on the front line." Long before the advent of "cloud-computing," Canadians have been accustomed to having their personal data processed in the United States. Canada has a long history of having to grapple with the legal, regulatory and technological challenges of enforcing Canadian privacy rules against US corporations and government agencies.

Moreover, there have been some recent success in enforcing Canadian privacy rules in a number of high-profile cases. Most notably, the Office of the Privacy Commissioner of Canada (OPC) investigated Facebook over the company's handling, disclosure, and retention of Canadian subscribers' personal information, and found Facebook in contravention of several provisions of Canada's private sector privacy law, the Personal Information Protection and Electronic Documents Act (PIPEDA). In this and other cases, the practical realities of enforcing rights of deletion and erasure have been tested and exposed. We argue that such cases carry important lessons for the contemporary debate about the "right to be forgotten."

This chapter analyzes some of the practical realities around deleting the personal data of Canadian users on predominantly US social networks. We first discuss the extent to which Canadians, under Canadian privacy law, can demand access to, and deletion of, their personal information retained by corporate actors. This entails an examination of Canadian privacy law, regulation and the related decisions by the OPC. We also consider the extent to which such a right could be exercised against companies that are based outside Canada. We then turn to corporate organizational practices. Our analysis of the privacy policies of over 20 social networking sites (SNSes) reveals that a range of qualified commitments and non-commitments are provided concerning the deletion and erasure of personal data. The right to delete personal data is also challenged by the practices and policies of Law Enforcement Authorities (LEAs). Law enforcement access to social networking data has become a significant policy issue in the face of recent debates over 'lawful access' legislation, which would impose data retention and disclosure requirements on telecommunications service providers, as well as 'open source' data collection, which is used in the course of routine policing investigations.

The ability of Canadians to be 'forgotten' by SNSes, therefore, confronts some complicated technical realities and organizational incentives. This chapter demonstrates that there is forgetting, non-forgetting and quasi-forgetting within the social networking environment, and in the context of the longstanding struggle to enforce Canadian privacy rights against US corporations. These practical realities hold important lessons for European efforts to shape its own data protection rules and enforce them against social networking companies.

3.1 Is There a Canadian "Right to be Forgotten"?

While the European debate concerning the Right to be Forgotten rages on, it can be instructive to turn to the Canadian setting to understand how core principles of 'forgetting' are already instantiated in other jurisdictions. So, while Canadian laws, policies, and decisions are not necessarily directly equivalent to the proposed European right, the Canadian example provides a good proxy to understand how 'forgetting' can play out in a Western democratic nation with relatively strong privacy laws.

Several Canadian privacy laws govern federal/provincial jurisdictions and public/private sectors. Though there are some gaps in coverage, the system has been judged "adequate" under the provisions of the Data Protection Directive, and therefore a safe harbor for the export of data relating to European citizens. Canada's public sector laws include both the 1982 Privacy Act, that regulates federal agencies, as well as provincial Information and Privacy Acts. Each of these stipulates that personal information should only be held as long as necessary to fulfill a legitimate statutory purpose. These laws require the establishment of "retention schedules" and demand the creation of secure and reliable record destruction measures.

With respect to social networking, the Personal Information Protection and Electronic Documents Act (PIPEDA) is the more relevant statute, as the principal legal instrument governing the private sector. Several of its provisions might add up to the equivalent of a "right to be forgotten." Schedule One (4.5) of the legislation states that "[p]ersonal information shall not be used or disclosed for purposes other than those for which it was collected, except with the consent of the individual or as required by law. Personal information shall be retained only as long as necessary for the fulfillment of those purposes." It also requires that organizations "develop guidelines and implement procedures with respect to the retention of personal information" (4.5.2). Furthermore, "personal information that is no longer required to fulfill the identified purposes should be destroyed, erased, or made anonymous. Organizations shall develop guidelines and implement procedures to govern the destruction of personal information" (4.5.3).

Hence, data erasure is not articulated as a right of the data subject but as an obligation of the data controller. Deleting or erasing data that is no longer needed to fulfill identified purposes is seen as a feature of "good" data protection practices and governance, and inextricably linked to questions of whether the data is still needed to meet stated, and identified, purposes. Such an analysis invariably leads to questions about individual consent, where another provision (Principle 4.3.8) may apply: "an individual may withdraw consent at any time, subject to legal or contractual restrictions and reasonable notice, and that the organization shall inform the individual of the implications of such withdrawal."

Thus the request to delete personal data can be interpreted as a "withdrawal of consent" and may appear as a legal equivalent to the "right to be forgotten." If such a right can be perceived in Canadian law, however, it only really applies when organizations collect data *about* an individual and retain it longer than required to fulfill identified purposes. Such a right is also interpreted within the larger framework of the "reasonable person" test under what is (essentially) a consent-based statute.

3.1.1 *The Enforcement of Canadian Privacy Law Against Social Networks*

Though Canadian law is written to require the deletion of some user data, and permit Canadian citizens to retract their consent concerning their data, there is still the issue of enforcing the law and responding to Canadians' wishes. With respect to many corporations operating in Canada, this creates practical instead of jurisdictional problems because of the extra-territorial reach of Canadian law.

Most of the SNSes used by Canadians have, at best, limited physical presences in Canada. This minimality of presence, however, does not mean that Canadian law does not apply to foreign companies less than domestic companies providing similar services to Canadians. This conclusion was reached in a case involving the US profiling company Accusearch, wherein the Federal Court of Canada (2012) insisted that the OPC had jurisdiction over the relevant privacy complaint insofar as a real and substantial connection could be found between the entity or the actions complained of, and Canada. As a result of this decision, the OPC's website emphasizes that:

> Where the Privacy Commissioner has jurisdiction over the subject matter of the complaint but the complaint deals with cloud computing infrastructure and thus is not obviously located in Canada, current jurisprudence is clear that the Privacy Commissioner may exert jurisdiction when assessment indicates that a real and substantial connection to Canada exists.[6]

Accordingly, the OPC has investigated Facebook, Google, Netflix, WhatsApp and other US-based companies regardless of their having a physical presence in Canada. In a famous and wide-reaching decision, the OPC asserted that Facebook violated provisions of PIPEDA, including section 4.5.3. The violation related to the confusing distinction between the deactivation of an account and the permanent deletion of data related to an account. The OPC wrote,

> [u]nder Facebook's current account deactivation policy, the personal information of users who have deactivated their accounts is retained indefinitely. Indefinite retention is a contravention of Principle 4.5 and 4.5.3 [. . .] a reasonable person would not consider it appropriate for Facebook to continue to retain indefinitely the personal information of a user who has deactivated his or her account and not reactivated it for a long time.[7]

Facebook was asked to implement a retention policy and inform users about it, and to delete personal information linked to deactivated accounts from Facebook's servers after a reasonable length of time. While Facebook did add information about account deletion to its privacy policy, the company did not develop a retention policy for deactivated accounts.

Jurisdictional issues could not be raised when the OPC investigated a complaint about the practices of Nexopia, a Canadian SNS directed towards young people.

---

[6] "Reaching for the Cloud(s): Privacy Issues related to Cloud Computing," Office of the Privacy Commissioner of Canada, accessed March 29, 2010, http://www.priv.gc.ca/information/pub/cc_201003_e.asp.

[7] "Report of the Findings into the Complaint filed by CIPPIC against Facebook Inc." Office of the Privacy Commissioner of Canada, (*paragraph 245*), July 16, 2009, http://www.priv.gc.ca/cf-dc/2009/2009_008_0716_e.pdf.

The complaint covered virtually every aspect of Nexopia's practices, including its retention of users' and non-users' personal data. Nexopia admitted to lacking internal policies and procedures for the retention, backup and destruction of its records. The company also confirmed that it retained users' and non-users' personal information in its database and archives since the website's inception in 2003. The OPC wrote that "it is clearly misleading to provide a "Delete Account" option—which states that specific personal information will be deleted—when in fact the information will be retained indefinitely in the website's archive."[8] Despite most of the complaints being considered "well-founded," Nexopia rejected some of the recommendations on technical grounds.

These interpretations of PIPEDA suggest that Canadians can legally tell these services to permanently and thoroughly delete their account information, notwithstanding technical difficulties and occasional need to retain the data for reasons of law enforcement (see below). Thus, the right of a user to request the permanent deletion of all user-generated data seems settled, at least in the eyes of the OPC. In this sense, there is a "right to be forgotten" in Canadian law. However, there have yet to be tests as to whether an individual can request the deletion of data that has been reposted by another user, or the deletion of data posted by a third-party. Such "take-down" requests, strenuously resisted by companies like Google may also pose real challenges under Canadian privacy law given that typical requests for the deletion of personal data assume a dichotomy between the "individual" and the "organization."[9]

Hence, PIPEDA only goes so far and Canadian citizens are then dependent on the range of ambiguous commitments to deletion, partial-deletion and non-deletion within the corporate privacy policies of mainly American companies. As we demonstrate below, these networks' own corporate practices often try to set the terms of how these matters *will* operate, regardless of the guidance provided by federal regulators or national laws.

3.2 Organizational Practices and Data Deletion

Canadians are prolific users of social networking services, with 60 % of online Canadians—and thus 50 % of all Canadians—being members of a social networking service.[10] Our analysis of these services' privacy policies reveals that companies seek to limit jurisdictional review of their practices while establishing company-specific data retention and disclosure policies. The companies also try to limit non-Americans' capacity to restrict the retention and revelation of their personal information. Together, these practices challenge Canadian privacy law, including rights of deletion and erasure.

[8] "Report of the Findings into the Complaint filed by CIPPIC against Nexopia,"Privacy Commissioner of Canada, (*paragraph 58*), accessed March 1, 2012, http://www.priv.gc.ca/cf-dc/2012/2012_001_0229_e.asp#summary.

[9] Peter Fleischer, "Foggy Thinking about the Right to Oblivion," *Peter Fleischer: Privacy...?* March 9, 2011, http://peterfleischer.blogspot.com/2011/03/foggy-thinking-about-the-right-to-oblivion.html.

[10] "Canada's Love Affair with Online Social Networking Continues," *Ipsos Reid*, 2011.

3.2.1 Jurisdiction and Complaints

Canada's privacy regime has successfully influenced the privacy behaviors of major global social networking companies.[11] Despite this track record, however, only one company in our sample, Club Penguin, specifically states its compliance with Canadian privacy law.[12] Of note, Club Penguin was a Canadian company that was subsequently acquired by Disney. Most other social networks (Blizzard,[13] Facebook,[14] Google,[15] LinkedIn,[16] LiveJournal,[17] MySpace,[18] Twitter,[19] Zynga[20]) emphasize that they comply with selected American statutes, such as the Child Online Protection Act, and some with the EU-US Safe Harbour Framework. Several companies stress their compliance with California law (Blizzard,[21] Facebook,[22] Tumblr,[23] Zynga[24]). Nexopia,[25] Yahoo!'s Flickr,[26] and Instagram[27] all fail to note which privacy laws and international guidelines they will comply with.

These companies often declare the jurisdictions and courts through which all legal proceedings must be conducted. Save for Yahoo!,[28] Nexopia,[29] and Plenty of Fish (a

[11] "Facebook breaches Canadian privacy law: commissioner." Canadian Broadcasting Corporation (CBC), *CBC News: Technology and Science*, July 16, 2009, Accessed October 17, 2012. http://www.cbc.ca/news/technology/story/2009/07/16/facebook-privacy-commissioner.html.

[12] "Club Penguin Privacy Policy," Last modified January 11, 2012, http://www.clubpenguin.com/privacy.htm.

[13] "Blizzard Entertainment® Online Privacy Policy." Last modified March 25, 2011, http://us.blizzard.com/en-us/company/about/privacy.html.

[14] "Facebook Data Use Policy", last modified June 8, 2012, http://www.facebook.com/full_data_use_policy.

[15] "Google Privacy Policy," last modified July 27, 2012, http://www.google.ca/intl/en/policies/privacy/.

[16] "LinkedIn Privacy Policy," last updated June 16, 2011, http://www.linkedin.com/static?key=privacy_policy&trk=hb_ft_priv.

[17] "LiveJournal Privacy Policy," last modified December 12, 2010, http://www.livejournal.com/legal/privacy.bml.

[18] "MySpace Privacy Policy." Last updated October 1, 2012, http://www.myspace.com/Help/Privacy.

[19] "Twitter Privacy Policy," last modified May 17, 2012, http://twitter.com/privacy.

[20] "Zynga Privacy Policy," last modified September 30, 2011, http://company.zynga.com/privacy/policy.

[21] "Blizzard Entertainment® Online Privacy Policy."

[22] "Facebook Data Use Policy."

[23] "Tumblr Privacy Policy," last modified March 22, 2012, http://www.tumblr.com/policy/en/privacy.

[24] "Zynga Privacy Policy."

[25] "Nexopia Privacy Policy," last modified November 2, 2009, http://www.nexopia.com/privacy.

[26] "Yahoo! Privacy Policy," last modified April 23, 2010, http://info.yahoo.com/privacy/ca/yahoo/.

[27] "Instagram Privacy Policy," last accessed October 28, 2012, http://instagram.com/legal/privacy/.

[28] "Yahoo! Privacy Policy."

[29] "Nexopia Privacy Policy."

Canadian dating social network),[30] which recognize Canadian courts, all claims must go through either American federal or the state courts of California or New York. Only Zynga, a social gaming company, explicitly recognized European jurisdictions, stating that non-US citizens would "agree to submit to the personal jurisdiction of the courts in Luxembourg."[31]

As noted in the previous section, American social networking companies must meet the requirements spelled out in PIPEDA. These requirements, however, have not led all companies to *actually respect or comply with* Canadian law. One corollary of being able to delete one's data is, of course, to discover what that data is, in the first place, and to access it if necessary. Schedule One (Section 4.9) of PIPEDA is clear:

> Upon request, an individual shall be informed of the existence, use, and disclosure of his or her personal information and shall be given access to that information. An individual shall be able to challenge the accuracy and completeness of the information and have it amended as appropriate.

We have, therefore, asked various SNSes to provide comprehensive records of the information they held on researchers in the course of our work.

Few companies have responded to these requests, and those that did either refused to provide any information or failed to comprehensively provide it. Furthermore, while basic data the subscriber generated may have been disclosed, most of associated metadata was not. Facebook, as an example, excludes *at least* 20 metadata items from their self-download feature, including data that the company collects on users (e.g. phone numbers) when other people in the user's network synchronize a device (e.g. iPhone) with Facebook, information logs on user "likes," and browser information that is logged when a user accesses a Facebook service. Twitter, similarly, excludes a significant amount of metadata; it provides five fields of data for each tweet, whereas circa 2010 there were 59–60 lines of data associated with each Tweet. Google's checkout service, similarly, lacks detailed metadata associated with communications.[32] Tumblr was the only company that both responded and refused to provide data. Instagram did respond however, providing a sample of user data that, like Facebook, was very limited in its scope.

The nondisclosure of *any* information is problematic, but the failure to disclose metadata linked with communications is also problematic because, when aggregated, metadata *is* content. Such data can include geographic coordinates, communications patterns, relative periods of activity, where such activity takes place, on what devices activities are linked to, and together this data can be used to impute relative affluence and technical sophistication based on communications tools. Further, when these

[30] "Plenty of fish Terms of Use Agreement," Last updated November 2, 2011, http://www.pof.com/terms.aspx.

[31] "Zynga Privacy Policy."

[32] "Social Networking and Canadian Privacy Law: Jurisdiction, Retention, and Disclosure," Christopher Parsons, Brief to Parliamentary Access to Information, Privacy and Ethics Committee, December 23, 2012, http://www.parl.gc.ca/Content/HOC/Committee/411/ETHI/WebDoc/WD5706433/411_ETHI_PSM_Briefs/ParsonsChristopherE.pdf.

kinds of data are cross-referenced between users they can provide deeper insight of the *community*. Moreover, by aggregating information across users in a community it is possible to impute further information about specific individuals. Consequently, failures to reply and/or comprehensively provide data are significant insofar as they speak to the relative unwillingness of social networking companies to fully comply with non-American privacy-related laws.

The most egregious example, Tumblr, stated that it "will not be providing the information you requested. Tumblr is a U.S.-based company with its headquarters in New York. It does not have a corporate presence in Canada and, therefore, it does not fall under the jurisdiction of PIPEDA or Canada's Office of the Privacy Commissioner." In a subsequent follow-up, after we had further explained the company's obligations under PIPEDA, the company reiterated: "We appreciate your interest in engaging in a legal discussion about the scope and reach of PIPEDA, but our prior correspondence stands."[33] The stated requirement to work through New York courts is interesting, given that Tumblr's privacy policy only recognizes the California Civil Code (S. 1798.83–1798.84) and acknowledges that California residents are entitled to ask for information about the categories of subscriber data the company is sharing with affiliates and third-parties.[34]

A further condition for the successful exercise of deletion rights is that corporations are supposed to provide recourse to individuals when they have concerns or complaints concerning a business's data handling processes.[35] In our analysis of major social networks we found that individuals may have challenges alerting a social networking company to their concerns about how the company may be retaining, processing, or disclosing their personal information. Of our sample, only three companies—Plenty of Fish, Reddit, and World of Warcraft—published their privacy officers' contact information. Most other companies had somewhat ambiguous contact forms or physical address information. Few companies had clear complaints or resolution processes. This said, two services, LiveJournal and MySpace, recognize the uniqueness of EU subscribers, with the former providing an EU mailing address for complaints and the latter encouraging Europeans to submit questions using the company's online form, or by mail. Tumblr also stands out, insofar as the published mailing address is exclusively for California residents.

Only Instagram entirely lacked a complaints mechanism though, in subsequent research, we found that its staff did return a truncated version of basic account information based on the user request. Specifically, user information was sent in an email attachment that included the following: a user ID number; username; first name; last name; email address; gender; birthday; phone number; biography; a user entered website address; an indication of whether the account is private/public; whether the account is 'active'; date user joined the service; signup IP address;

---

[33] Corporate counsel for Tumblr, Personal e-mail with Christopher Parsons.

[34] "Tumblr Privacy Policy."

[35] "OPC Guidance Documents: A Guide for Businesses and Organizations," last modified March 31, 2010, http://www.priv.gc.ca/information/guide_e.asp#015.

all user relationships, incoming requests, and followers, and whether these accounts are themselves public or private; and all user published media during the date range that the account has been active. Interestingly, the top of the file is listed as "[username]_subpoena_01/15/13", which indicates that this could be an identical format that is shared with authorities under lawful access, and specifically subpoena, circumstances.

3.2.2 How Social-Networking Services Understand Retention and Disclosure

Jurisdictional and complaint issues aside, a simple examination of how social networking companies state they retain data is revealing. As an example, Google recognizes that, after a user deletes account information, the company may not immediately delete data and that it may not remove data from their backup systems.[36] Such claims are worrying given the long-term retention problems surrounding Street View data and the revelation that actual retention periods remain ambiguous.[37] While Facebook states that it typically takes a month to delete data—with some information remaining in backup logs up to 90 days—the company's success in actually deleting data, such as photos uploaded to the site, has long been questionable.[38] Companies such as Yahoo! and Foursquare offer commitments similar to those of Facebook. Foursquare also notes that, even after subscribers delete information, "copies of that information may remain viewable elsewhere, to the extent it has been shared with others, distributed pursuant to privacy settings, or copied or stories by other users".[39] Tumblr parallels this statement, informing subscribers that even when deleting their accounts' content, public activity, such as posts that were 'liked' or shared, will remain stored on servers and accessible to the public.[40]

For other services the 'deletion' of subscriber data may largely amount to hiding the information from public viewers. LiveJournal, for example, recognizes that, while individuals can delete their account and accompanying information, data may take an unspecified amount of time to delete and the company may choose to retain the information to the extent necessary to protect the company's legal interests, comply with court orders, et cetera.[41] The inclusion of 'et cetera' leaves open the

---

[36] "Google Privacy Policy."

[37] "Google: Didn't delete Street View data after all," Yahoo! News, July 27, Accessed October 17, 2012. http://news.yahoo.com/google-didnt-delete-street-view-data-175540701–finance.html.

[38] "Three years later, deleting your photos on Facebook now actually works," Cheng, Jacqui, Ars Technica, August 16, 2012, accessed October 17, 2012, http://arstechnica.com/business/2012/08/facebook-finally-changes-photo-deletion-policy-after-3-years-of-reporting/.

[39] "Foursquare Labs, Inc. Privacy Policy," last modified July 13, 2012, https://foursquare.com/legal/privacy.

[40] "Tumblr Privacy Policy."

[41] "LiveJournal Privacy Policy."

full range of possible motivations to retain data in contravention of a subscriber's request. In the case of Meetup, the company reserves the right to retain information that the user requests removed if retention is needed to resolve disputes, troubleshoot problems, or enforce the terms of service. Regardless, the company promises, "your information is never completely removed from our databases due to technical and legal constraints (for example, we will not remove your information from our backup stores)."[42] Nexopia offers similar 'guarantees' as Meetup, insofar as Nexopia states that individuals ought not expect that their personal information will be completely removed from their systems following a deletion request.[43]

Given that many of these services function as platforms, and thus allow other developers to capture, process, and retain users' generated data, there is the potential for 'deleted' data on the platform (e.g. Facebook, Twitter, LinkedIn, Foursquare) to be retained indefinitely by third-party developers without technically-rigorous ways for the platform to enforce a users' deletion request on the third-party. Companies such as Club Penguin, Yahoo!, Google, and Apple[44] reserve the right to share collected or contributed information within and across their corporate organizations, and most social networks include provisos that they 'may' (read: will and do) share information with analytics companies and associated advertisers. Significantly, when we examined the social networking services using Ghostery, a tool that identifies web trackers, we found that all services with the exception of Facebook and Google revealed the presence of third-party analytics and and/or advertising services. Facebook and Google, of course, use their own backend analytics and advertising systems and thus do not need to rely on third-parties for such services.

3.2.3 Organizational Implications for 'Forgetting'

Current organizational practices may limit the practical instantiation of attempts to request that personal data be deleted. Few social networking services guarantee that data will, certifiably, be deleted and tend to offer either broad exceptions under which data will be retained or state outright that it *will not* be deleted. Given that most networks let individuals over the age of 13 join and use the services, this means that youths' personally identifiable information may also be retained indefinitely. Retained data could be retained indefinitely for 'legitimate' business purposes, purposes that the user may have consented to upon accepting the Terms of Service associated with the SNS. Moreover, even if a controller could successfully delete the data from their systems (and, it should be noted, few subscribers will be able to ascertain 'success' given both the lack of access to social networking services' data centers and their common lack of sufficient technical, temporal, and fiscal resources

[42] "Meetup Privacy Policy Statement," last modified May 23, 2010, http://www.meetup.com/privacy/.

[43] "Nexopia Privacy Policy."

[44] "Apple Privacy Policy," last modified updated May 21, 2012, http://www.apple.com/privacy/.

to mount independent forensic investigations) the data may remain in the databases of third-parties associated with the services' development platform. Comprehensive deletion of data held by these third-parties must rely on more than the 'good will' that companies such as Facebook have historically espoused towards their developer community;[45] subscribers must trust in Facebook, and in Facebook's trust in others, rather than certifiably knowing that their data is actually, meaningfully, going to be deleted.

Ultimately, while there is some degree to which subscribers can be 'forgotten' by these services today, successfully being forgotten is muddied by difficulties in ascertaining the data that organizations hold on individuals, in networks (not) adhering to relevant and applicable laws, in varying and unclear corporate retention periods, and in the limited capacities for subscribers to scrub data from third-parties that capture, process, or retain their personal information. The challenges facing individuals who seek to enforce their right to be forgotten are compounded when we turn to the capture, processing, and retention of social networking data for law enforcement purposes.

3.3 Lawful Enforcement Access to Social Networking Services

Social media provides Law Enforcement Authorities (LEAs) a burgeoning stream of information for detecting, preventing, and investigating potentially suspicious activities. Our research reveals how and why Canadian LEAs are using SNSes as proxy organizations to monitor, collect, and retain subscriber data. The circulation of data between SNSes and LEAs further challenges the implementation of rights of deletion, insofar as 'forgotten' corporate data may be "remembered" indefinitely by public bodies.

3.3.1 Information Sharing Protocols Between LEAs and SNSes

Access to private companies' digital records is a common expectation in contemporary law enforcement activities. Every SNS included in our analysis made mention that they will, under certain legal conditions, share information with LEAs or other public authorities. Many, if not all, have some form of 'law enforcement compliance' information that details the types of data available to LEAs, as well as detailed protocols for LEAs to follow to access user data. A small sample of these guides have been made public through leaks or FOIA requests, and they offer insights into the privacy and data management relationships between SNSes and LEAs.

SNSes make a range of information available to LEAs. For example, Facebook will provide authorities with "user contact info" (name, birth date, email address(s),

[45] Katherine Losse, *The Boy Kings: A Journey into the Heart of the Social Network* (New York: Free Press, 2012), 148.

physical address, city, state, zip, phone, registered mobile phone number, work phone, screen name (usually for AOL Messenger/iChat), and website), "group contact info" (a list of users currently registered in a specific group), "user neoprint" (a term for an expanded view of a user profile), "user photoprint" (a compilation of the photos a user has uploaded but not deleted), and "IP logs" (time/date stamps that note when user has logged in, the source IP address, and Internet Service Provider identified with the user Id)[46, 47]. Facebook's security team can also retrieve information for law enforcement that is not explicitly noted in their handbook's description of available data.[48] Similarly, Yahoo!'s compliance guide notes the availability of similar information, such as subscriber information, IP logs, photos, email and other private communication, group content (including email addresses of members), and metadata such as geo-locational information.[49]

For law enforcement, there is often a lag between *requesting* stored communications and SNSes *providing* the requested data. One consequence of this lag has been sharing protocols, typically referred to as 'preservation requests'. Several SNSes, including MySpace, Facebook, Yahoo! (Flickr), and LinkedIn, honour requests from law enforcement to preserve data, typically for up to 90 days. These requests provide sufficient time for LEAs to assemble necessary legal documents (e.g. subpoenas, court orders, search warrants) to access the preserved data.

3.3.2 *Investigative Instruments LEAs Use to Access Social Networking Data*

Canadian LEAs' investigative strategies differ according to whether information is publicly available or is stored on (typically American) servers. In the context of SNSes, publically available information is user generated content that law enforcement can access without court order because it is set to 'public' or 'friend of friend' viewing. Canadian LEAs are increasingly collecting such publicly available data when private information is not required for their investigations.[50] As an example, information is being collected using Facebook search, which provides authorities with public information from open profiles and public groups. Data collected from such public sources facilitates network-analysis and provides more complete pictures of individuals and their social circles.[51] Our interviews have revealed how Facebook's "self-download" feature, ostensibly meant to enhance

---

[46] "Facebook Subpoena/ Search Warrant Guidelines," Facebook, 2008.

[47] Toronto Police Services, Personal Interview with Adam Molnar, October 5, 2012.

[48] "Facebook Subpoena/ Search Warrant Guidelines," p. 7.

[49] "Yahoo! Privacy Centre," last modified April 23, 2010, http://info.yahoo.com/privacy/ca/yahoo/.

[50] "Social Media Sites: New Fora for Criminal, Communication, and Investigation Opportunities," Public Safety Canada, August 2011, last accessed on October 28, 2012, http://www.sfu.ca/iccrc/content/PS-SP-socialmedia.pdf.

[51] "Social Media Sites: New Fora for Criminal, Communication, and Investigation Opportunities."

subscribers' access to their private data, is being used to provide evidence to law enforcement, with one officer referring to this practice as a "best-practice"[52].

Private data is predominantly user generated but is stored privately on a user profile or includes non-publicly viewable metadata that the SNS collects when the user interacts with the service (e.g. geo-locational, facial recognition 'prints'). Where LEAs want access to private data they often first send a (legally) non-binding email requesting the data. When the SNS asks for, or requires, LEAs to submit requests using formal legal documents then either domestic or international legal instruments are used. Many American SNSes (e.g. Facebook, Google, and Twitter) explicitly honour Canadian court orders if they present an "equivalent authority"[53] to US court orders or administrative subpoenas. In Canadian law, production orders are used to request and compel communication records from SNSes. In the case of Facebook, their Ontario office functions as their Canadian hub for lawful access requests. Per Canadian legal requirements, such requests to this office must come from Ontario-based LEAs. Consequently, non-Ontario LEAs must be "backed" by Ontario officials.[54] These cross-provincial jurisdictional difficulties may be 'remedied' by Canada's proposed 'lawful access' legislation. Proponents of the legislations claim that the legislation will bolster the use and effectiveness of production orders by removing provincial jurisdictional barriers and creating new production orders to capture "traffic data" and "subscriber and/or service provider information",[55] though critics argue the legislation will instead facilitate SNS-linked 'fishing expeditions' and be used to monitor Canadians.[56]

While Canadian production orders are accompanied by judicial authorization, the orders are not always respected by SNSes;[57] in such cases LEAs can use Mutual Legal Assistance Treaties (MLATs) to retrieve information stored on US servers. MLATs facilitate cooperation between LEAs of different countries, and outline jurisdictional territories, associated investigative protocols, and conditions of sharing information and physical evidence linked to the particular investigation. Canadian LEAs initiate MLATs so that American authorities can compel American-based SNSes to preserve and provide data sought by the Canadians. While the MLAT process may result in the disclosure of US-based data, they are a cumbersome legal instrument and take from 6–8 months to "as long as never"[58] to complete. The lengthy processing times and jurisdictional challenges involved with lawful access to "private" user information through MLAT processes has placed a premium on acquiring as much SNS subscriber information using domestic—open source and legal instrument—methods.

---

[52] Toronto Police Force, Personal Interview with Adam Molnar, October 2, 2012.

[53] "Facebook Subpoena/ Search Warrant Guidelines," Facebook, 2010.

[54] Vancouver Police Department, Personal Interview with author, October 10, 2012.

[55] "Lawful Access—Consultation Document," Department of Justice, last modified August 3, 2012. http://justice.gc.ca/eng/cons/la-al/d.html.

[56] "Canadian Social Media Surveillance: Today and Tomorrow," Parsons, Christopher, Technology, Thoughts, and Trinkets, May 28, 2012, accessed January 27, 2013, http://www.christopher-parsons.com/blog/technology/canadian-social-media-surveillance-today-and-tomorrow/.

[57] Vancouver Police Department, Personal Interview with author, October 20, 2012.

[58] Fenton, Mark. Personal Interview with author, October 2012.

3.3.3 Data Management and Policing Operational Databases in Canada

Contemporary criminal justice practices largely depend on the efficacy of digital information management systems. LEAs want to build pictures of suspicious activity over time, from "pre-crime" to "post-crime." Consequently, information and data retention are integral to the stated intent to "detect, prevent, and investigate" such activity. Canada's national police rely on two primary operational databases to provide digital storage and access of information related to their investigations, the Canadian Police Information Centre (CPIC) and the Police Reporting and Occurrence System (PROS). CPIC holds more than "10 million records and processed more than 200 million queries through 40,000 access points in 2009". PROS is a "records management system containing information on individuals who have come into contact with police, either as a suspect, victim, or offender" and is meant to "record all aspects of an investigation".[59] PROS integrates the RCMP with 23 police partner agencies and processes about 1.6 million occurrence files per year. Significantly, the PROS database mandate would permit the collection, retention and sharing of public and non-public information gleaned from SNSes. CPICs rigid data structures, on the other hand, limit the integration of such information.

Both of these databases are administered by the federal RCMP and are subject to Canada's federal public sector laws that include the 1982 Privacy Act. As previously mentioned, this legislation stipulates that these databases are bound by "retention and destruction schedules" to ensure that any personal information not be held any longer than needed to fulfill a legitimate statutory purpose. When records are no longer associated with an active investigatory file, data must be permanently deleted.

An OPC audit of these databases in 2011 revealed significant variances in the practical implementation of these obligations. The OPC found that "the RCMP had yet to formally establish MOUs with approximately 25 % of the police agencies that access CPIC" and consequently could not prevent several agencies from disseminating details on "convictions, discharges, or pardons to employers without the informed consent of the prospective employee".[60] An audit of the PROS database reflected that, though a comprehensive privacy policy and set of operating procedures existed, serious problems concerning management of, and access to, the data persisted. Specifically, the OPC found that personal information was being held in the PROS for longer than allowable under the Canadian Privacy Act. Further, the RCMP could not prove that they performed the necessary reviews to guarantee that policies governing personal information in the database were being met. As a result, if misuse of the database to occur, it would be difficult to investigate transgressions.[61]

[59] "Audit of Selected RCMP Operational Databases," Privacy Commissioner of Canada, 2011 http://www.priv.gc.ca/information/pub/ar-vr/ar-vr_rcmp_2011_e.asp, p. 7.

[60] "Audit of Selected RCMP Operational Databases," p. 4.

[61] "Audit of Selected RCMP Operational Databases."

Consequently, while consumers may try to delete personal information from their social networks, LEAs may retain, circulate, and process this information without the citizen's knowledge. The practical implications of the collection and retention of data by Canadian and non-Canadian LEAs undermines the exercise of deletion and erasure rights, and any hope that a right to be forgotten will be a comprehensive right; instead, it might better be understood as a right to be quasi-forgotten, with 'forgetting' being dependent on the circumstances and particularities associated with each subscriber's account in relation to particular public or private organizational incentives governing the management of that data.

3.4 Conclusion

While Jennifer Stoddart was noted as stating the "Net never forgets" in the epigraph to this chapter, our analysis of the major SNSes operating in Canada demonstrates that forgetting occurs along a multidimensional continuum. At a policy level, there are commitments to deletion, partial deletion, and non-deletion. None of these practices constitutes 'forgetting.' Rarely has a SNS committed to the total and thorough erasure of all data relating to users. Even more rarely has that erasure occurred. Those commitments and non-commitments may, or may not, be reflected in actual organizational practices and technical capabilities.[62]

A distinction must be made between what a social network service forgets, and forgetting social networking information. Thus, when Facebook, for example, deletes your information, the RCMP does not necessarily do the same. Deletion is not the same as "forgetting." Deletion takes place in the context of powerful institutional expectations, motivations, and legacies. The privacy policies we surveyed reveal that companies each engage in a process of "quasi-forgetting," where promises of erasure or deletion are hedged by a number of conditions relating to the timing of the deletion, the inability to guarantee the behavior of third-parties (including law enforcement), the need to retain for unspecified legal purposes, the technical complexities, and the realities of data analytics.

"Quasi-forgetting", therefore, is reflected in the following rhetorical devices:

- Forgetting: but not yet
- Forgetting: but only for what we deem to be personally identifiable information
- Forgetting: but not information that your friends have said or shared about you
- Forgetting: but only for us, not for others
- Forgetting: but we need to cover our legal backs
- Forgetting: but we cannot guarantee complete erasure
- Forgetting: but not for third-party analytics

[62] For another detailed review, see "The Right to be Forgotten Across the Pond," Ambrose, M. and Ausloos, J. Paper presented at the Telecommunications Policy Research Conference, September 21, 2012. Accessed online, October 20, 2012, http://papers.ssrn.com/sol3/papers.cfm?abstract_id=2032325.

These exceptions and qualifications are readily apparent, and often readily admitted to by SNSes. They constitute the "known unknowns." Beyond these, there may be a range of unintended effects of personal data retention within a social-networking environment that are even less understood and controlled for—the "unknown unknowns" of networked communications.

Of course, these conclusions stem from an assessment of corporate willingness and ability to implement the specific provisions contained in Canada's PIPEDA. Regardless of whether the scattered provisions in Canadian law add up to the equivalent of a "right to be forgotten," our analysis suggests that the legal and policy dilemmas that have shaped the international debates about the 'right to be forgotten' require a more nuanced appreciation of current erasure and deletion practices, and of the technical conditions and organizational incentives that underpin them. Recent research by Bigo et al. has further clarified the relative immunity of extra-jurisdictional organizations, and the undermining of individual citizen's rights, as controllers of data that has originated from users in opposing jurisdictions.[63] Our work correlates with Bigo et al.'s findings insofar as current reliance on cloud computing infrastructures threaten to, or already are threatening, data protection legislation. Such threats, ultimately, risk undermining the legally instantiated rights of individual citizens.

The "right to be forgotten" is not just a debate for the lawyers, in other words. And it is not just a debate for European regulators, privacy advocates, and American companies. To the extent that European policy has expressed, and continues to express, the de facto standard for the global communication of personal data, the political, social, economic and legal consequences of these transatlantic tussles can have profound consequences for other countries and for their systems of personal data protection. The current controversy has opened up an interesting debate about corporate responsibilities for the deletion and erasure of personal data on the Internet. It should force all corporations to consider their compliance with existing regimes, and to put clear procedures in place to take action when an individual asks: "Please delete my data."

References

Ambrose, M., and J. Ausloos. 2012. The right to be forgotten across the pond. Paper presented at the Telecommunications Policy Research Conference, September 21, 2012. http://papers.ssrn.com/sol3/papers.cfm?abstract_id=2032325. Accessed 20 Oct 2012.

Apple. 2012. Apple privacy policy. http://www.apple.com/privacy/. Last modified 21 May 2012.

Bigo, Dider, et al. 2012. Fighting cyber crime and protecting privacy in the cloud. Study prepared for European Parliament Directorate-General for Internal Policies, 2012. http://www.europarl.europa.eu/committees/en/studiesdownload.html?languageDocument=EN & file=79050.

---

[63] "Fighting cyber crime and protecting privacy in the cloud," Didier Bigo et al, 2012, study prepared for European Parliament Directorate-General for Internal Policies, acessed on February 13th, 2013, http://www.europarl.europa.eu/committees/en/studiesdownload.html?languageDocument = EN&file = 79050.

Blizzard Entertainment. 2011. Blizzard entertainment® online privacy policy. Last modified 25 Mar 2011. http://us.blizzard.com/en-us/company/about/privacy.html.
Canadian Broadcasting Corporation (CBC). 2009. Facebook breaches Canadian privacy law: commissioner. *CBC News: Technology and Science*, July 16, 2009. http://www.cbc.ca/news/technology/story/2009/07/16/facebook-privacy-commissioner.html. Accessed 17 Oct 2012.
Cheng, Jacqui. 2012. Three years later, deleting your photos on Facebook now actually works. *Ars Technica*, August 16, 2012. http://arstechnica.com/business/2012/08/facebook-finally-changes-photo-deletion-policy-after-3-years-of-reporting/. Accessed 17 Oct 2012.
Club Penguin. 2012. Privacy policy. Last modified 11 Jan 2012. http://www.clubpenguin.com/privacy.htm.
Department of Justice. 2012. Lawful access—consultation document. Last modified 3 Aug 2012. http://justice.gc.ca/eng/cons/la-al/d.html.
European Union (EU). 2012. Proposal for a regulation of the European Union and the Council on the Protection of Individuals with respect to the processing of personal data and on the free movement of such data (General Data Protection Regulation). Published January 25, 2012. http://ec.europa.eu/justice/data-protection/document/review2012/com_2012_11_en.pdf.
Facebook. 2008. Facebook subpoena/search warrant guidelines.
Facebook. 2010. Facebook subpoena/search warrant guidelines.
Facebook. 2012. Data use policy. Last modified 8 June 2012. http://www.facebook.com/full_data_use_policy.
Federal Court of Canada. 2012. Philippa Lawson v. Accusearch Inc. and Federal Privacy Commissioner. Last modified 26 Oct 2012. http://reports.fja.gc.ca/eng/2007/2007fc125/2007fc125.html
Fleischer, Peter. 2011. Foggy thinking about the right to oblivion. Peter Fleischer: Privacy...? March 9, 2011. http://peterfleischer.blogspot.com/2011/03/foggy-thinking-about-the-right-to-oblivion.html. Accessed 17 Oct 2012.
Foursquare. 2012. Foursquare Labs, Inc. Privacy Policy. Last modified 13 July 2012. https://foursquare.com/legal/privacy.
Google. 2012. Privacy policy. Last updated 27 July 2012. http://www.google.ca/intl/en/policies/privacy/.
Instagram. Privacy policy. Last updated 30 Aug 2012. http://instagram.com/about/legal/privacy/.
Ipsos Reid. 2011. Canada's love affair with online social networking continues. Ipsos Reid.
LinkedIn. 2011. Privacy policy. Last modified 16 June 2011. http://www.linkedin.com/static?key=privacy_policy&trk=hb_ft_priv.
LiveJournal. 2010. LiveJournal privacy policy. Last modified 12 Dec 2010. http://www.livejournal.com/legal/privacy.bml.
Losse, Katherine. 2012. *The boy kings: A journey into the heart of the social network*. New York: Free Press.
Mayer-Schonberger, Victor. 2011. *Delete: The virtue of forgetting in the digital age*. Princeton, NJ: Princeton University Press.
Meetup. 2010. Meetup privacy policy statement. Last updated 23 May 2010. http://www.meetup.com/privacy/.
MySpace. 2012. Privacy policy. Last updated 1 Oct 2012. http://www.myspace.com/Help/Privacy.
Nexopia. 2009. Privacy policy. Last updated 29 Nov 2009. http://www.nexopia.com/privacy.
Parsons, Christopher. 2012a. Canadian social media surveillance: Today and tomorrow. Technology, thoughts, and trinkets. May 28, 2012. http://www.christopher-parsons.com/blog/technology/canadian-social-media-surveillance-today-and-tomorrow/. Accessed 27 Jan 2013.
Parsons, Christopher. 2012b. Social networking and Canadian privacy law: Jurisdiction, retention, and disclosure. Brief to Parliamentary Access to Information, Privacy and Ethics Committee, December 23, 2012. http://www.parl.gc.ca/Content/HOC/Committee/411/ETHI/WebDoc/WD5706433/411_ETHI_PSM_Briefs/ParsonsChristopherE.pdf. Accessed 19 Feb 2013.
Plenty of Fish. 2011. Plenty of fish terms of use agreement. Last updated 2 Nov 2011. http://www.pof.com/terms.aspx.

Privacy Commissioner of Canada. 2009. *Report of the findings into the complaint filed by CIPPIC against Facebook Inc.* July 16, 2009. http://www.priv.gc.ca/cf-dc/2009/2009_008_0716_e.asp#sect7a.

Privacy Commissioner of Canada. 2010. *Reaching for the cloud(s): Privacy issues related to cloud computing.* Last modified 29 Mar 2010. http://www.priv.gc.ca/information/pub/cc_201003_e.asp#toc5.

Privacy Commissioner of Canada. 2011. *Audit of selected operational databases.* Last modified 17 Nov 2011. http://www.priv.gc.ca/information/pub/ar-vr/ar-vr_rcmp_2011_e.asp.

Privacy Commissioner of Canada. 2012. *Report of the findings into the complaint filed by CIPPIC against Nexopia.* Last modified 1 Mar 2012. http://www.priv.gc.ca/cf-dc/2012/2012_001_0229_e.asp#summary.

Public Safety Canada. 2011. *Social media sites: New fora for criminal, communication, and investigation opportunities*, August 2011. http://www.sfu.ca/iccrc/content/PS-SP-socialmedia.pdf. Accessed 28 Oct 2012.

Rosen, Jeffrey. 2012. The right to be forgotten. *Stanford Law Review* 64:88–92.

Stoddart, Jennifer. 2011. The net never forgets: Remember to protect personal data. Website of the *Office of the Privacy Commissioner of Canada*, January 28, 2011. http://www.priv.gc.ca/resource/dpd/2011/index_e.asp. Accessed 17 Oct 2012.

Tumblr. 2012. Privacy policy. Last updated 22 Mar 2012. http://www.tumblr.com/policy/en/privacy.

Twitter. Twitter privacy policy. Last updated 17 May 2012. http://twitter.com/privacy.

van Hoboken, Joris. 2011. 9 reasons why a 'Right to be Forgotten' is really wrong. *Joris van Hoboken: about search engines, digital civil rights and more.* December 11, 2011. http://www.jorisvanhoboken.nl/?m = 201112.

Vinograd, Cassandra, and Raphael Satter. 2012. Google: Didn't delete Street View data after all. *Yahoo! News*, July 27. http://news.yahoo.com/google-didnt-delete-street-view-data-175540701–finance.html. Accessed 17 Oct 2012.

Yahoo!. 2010. Yahoo! privacy centre. Last updated 23 Apr 2010. http://info.yahoo.com/privacy/ca/yahoo/.

Zynga. 2011. Privacy policy. Last updated 30 Sept 2011. http://company.zynga.com/privacy/policy.

Chapter 4
The EU, the US and Right to be Forgotten

Paul Bernal

The so-called 'right to be forgotten' has been a subject of much debate on both sides of the Atlantic since Commissioner Viviane Reding announced her intention to introduce it in 2010.[1] What is seen by those proposing it on the European side to be a simple and logical extension of existing data protection principles is presented in the US as 'the biggest threat to free speech on the internet in the current decade'.[2] Both sides see themselves as protecting the rights of the ordinary people—the EU in the face of the potentially overwhelming power of the corporate internet behemoths, the US in the face of the excessive and controlling zeal of the European regulators.

This chapter looks at whether they might both be right in some ways—and both wrong in others. It will attempt to untangle the issues that really underlie both the 'right to be forgotten' and the arguments that are being made both in favour and against it. Is the dispute over the right to be forgotten really about freedom of expression—or are there other issues of as much or even greater importance? In essence, is it free *enterprise* rather than *free speech* that really lies behind the US resistance to the right to be forgotten?

The chapter will conclude with a look at how a way forward might be found for the right to be forgotten. This must begin with a better understanding of the underlying issues on both sides of the Atlantic and both sides of the debate—and the arguments being made on their basis. If a solution is to be found that supports the rights and needs of individuals without undermining either freedom of expression or the freedom and flexibility that has been crucial to the development of the internet, it will need to take these arguments fully into account.

That, however, may not be the end of the story. The debate—and the dispute—may become an academic one if the major providers on the Internet, Google and

[1] The Proposed Data Protection Regulation, dated 25/1/2012 is available online at http://ec.europa.eu/justice/data-protection/document/review2012/com_2012_11_en.pdf.
[2] Proposed Data Protection Regulation, Article 17.

P. Bernal (✉)
UEA Law School, University of East Anglia, Highsett 34, CB2 1NY Cambridge, UK
e-mail: P.A.Bernal@lse.ac.uk

S. Gutwirth et al. (eds.), *Reloading Data Protection*,
DOI 10.1007/978-94-007-7540-4_4, © Springer Science+Business Media Dordrecht 2014

others, decide to comply voluntarily with the wishes of the European regulators. Given the deep differences in views between European and American lawmakers and legal scholars, this may be the only way that the broad gap can bridged.

4.1 The European Perspective

The starting point for understanding the European perspective is to look at the proposed Data Protection Regulation,[3] where the right to be forgotten is set out. Article 17 (1) of the proposed regulation as it existed at the time of writing defined the 'right to be forgotten and to erasure' as follows:

> The data subject shall have the right to obtain from the controller the erasure of personal data relating to them and the abstention from further dissemination of such data[4]

The regulation goes on to say that this right can be applied where:

a. the data are no longer necessary in relation to the purposes for which they were collected or otherwise processed;
b. the data subject withdraws consent on which the processing is based according to point (a) of Article 6(1), or when the storage period consented to has expired, and where there is no other legal ground for the processing of the data;
c. the data subject objects to the processing of personal data pursuant to Article 19;
d. the processing of the data does not comply with this Regulation for other reasons.

Much of this is already built into the existing data protection regime—indeed, it can be argued that the right to be forgotten already exists under the current regime. Data is already only permitted to be held for a specific purpose[5] and for no longer than necessary,[6] the data subject is already entitled to object,[7] and data must be processed in accordance with the rules set down in the data protection regime.[8] The only part that appears new—at least in terms of the application of the right—is the idea that consent, once given, may be withdrawn. Though this may be technically 'new', it is something that can be argued to be implicit in a broader understanding of the nature of consent—and certainly something that fits logically into the idea of giving individuals more rights over 'their' data.

---

[3] Data Protection Directive ("DPD") (Directive 95/46/EC), downloadable from http://eur-lex.europa.eu/LexUriServ/LexUriServ.do?uri=CELEX:31995L0046:en:HTML, Article 6.1(b).

[4] DPD Article 6.1 (e).

[5] DPD Article 14.

[6] DPD Article 6.1 (a).

[7] As suggested, for example, in Bernal (2011).

[8] For more on the origins of the right in the French *le droit à l'oubli* and the Italian *diritto al' oblio* see ibid., Sect. 1.1.

4.1.1 A Right to be Forgotten—or a Right to Delete?

At first examination, therefore, the right does not look very much as though it's really about being 'forgotten'—or as any kind of threat to free speech. Indeed, it looks more like a right to delete[9] than any real kind of 'right to be forgotten'—the name comes to an extent from one part of the right's origins, in French and Italian law.[10] As Peter Hustinx, the European Data Protection Supervisor, put it in a speech to the Oxford Privacy Information Law and Society Conference in June 2012:

> There is also something of a mistranslation—*le droit à l'oubli* in French is not really the right to be forgotten, so there is an overstatement in the process. We got carried away.[11]

That is part of the problem—a 'mistranslation' from the French and Italian—but it masks what is a deeper issue: that there are two qualitatively different aspects to the right, based on its history and name. The old French *le droit à l'oubli* and the Italian *diritto al' oblio* were much more about 'forgetting'—about the rights of criminals with spent convictions to have those convictions 'forgotten' and not taken into account in how they are dealt with by the media and by potential employers and so forth. The new right, as set out in the regulation, is much more about the deletion of data, and applies to everyone, not just those who have some specific item or story that is being used inappropriately against them.

When understood as a right to delete/right to erasure rather than a real 'right to be forgotten', the right fits well with the rest of the regulation. It follows Article 15, which gives the data subject the right to access to personal data, and Article 16, which grants a right to rectify inaccurate data. A right to delete is just a small step further than these rights of access and rectification, and has very different origins from *le droit à l'oubli* and the *diritto al' oblio*—the problems encountered in dealing with personal data held on social networks.

4.1.2 The Role of Social Networks

European Commissioner Viviane Reding, who has been responsible for the reform of the data protection regime, has regularly hinted at the reason that the Commission is pushing the right: the role of social networks. On 30th November 2010, in the early stages of the debate, she made it quite clear:

---

[9] The speech can be accessed online here: http://www.edps.europa.eu/EDPSWEB/webdav/site/mySite/shared/Documents/EDPS/Publications/Speeches/2012/12-06-12_Speech_Oxford_EN.pdf.

[10] Speech: http://europa.eu/rapid/press-release_SPEECH-10-700_en.htm

[11] See http://www.nytimes.com/2008/02/11/business/worldbusiness/11iht-11facebook.9919316.html.

> I want to introduce the "right to be forgotten". Social network sites are a great way to stay in touch with friends and share information. But if people no longer want to use a service, they should have no problem wiping out their profiles.[12]

That has been one of the key drivers: how hard it has been to properly delete a Facebook account. Headlines like 'On Facebook, leaving is hard to do' in the New York Times in 2008 set the scene, with quotes like this:

> It's like the 'Hotel California,' "said Nipon Das, 34, a director at a biotechnology consulting firm in New York who tried, unsuccessfully, to delete his account this fall. "You can check out any time you like, but you can never leave.[13]

This, from the European perspective, represents a real problem—and one that needed addressing. As often seems to be the case with European regulators, when the problem isn't dealt with by the organisations concerned, the regulators decide to act, and when they act, they act with zeal. The current concerns with the so-called 'Cookie Directive',[14] for example, have followed a similar pattern: noises made by the EC about the problems with behavioural advertising, largely ignored or sidestepped by the behavioural advertising industry, resulting in a directive which has been seen by many as heavy-handed and counter-productive.[15] With Facebook seeming to do little to deal with the issue of account deletion, and not seeming to take European concerns as seriously as the Commission would like, the drive for the 'right to be forgotten' became more serious.

The proposed Regulation does make a specific attempt to address the impact of the right to be forgotten (and indeed all other aspects of the data protection regime) on freedom of expression, through Article 80(1), which requires that member states provide exemptions and derogations for data processing carried out 'solely for journalistic purposes or the purpose of artistic or literary expression'.[16] As shall be discussed in Sect. 3 below, however, this article has drawn significant criticism, particularly in terms of the limited understanding that it appears to display of what constitutes freedom of expression. The difference in understanding is, at least at a surface level, what causes the most division between the EU and the US.

---

[12] The 'e-Privacy Directive' (Directive 2002/58/EC), as modified by Directive 2009/136/EC, available online: http://eur-lex.europa.eu/LexUriServ/LexUriServ.do?uri=CONSLEG:2002L0058:20091219:EN:PDF.

[13] The UK's Information Commissioner, Christopher Graham, joined the criticism, suggesting the directive was 'dreamed up by politicians in Brussels' without the appropriate market research to back it up. See http://www.pcadvisor.co.uk/news/security/3381464/information-commissioner-criticises-dreamed-up-eu-cookie-directive/.

[14] Proposed Data Protection Regulation Article 80 (1). See also Sect. 3 below.

[15] http://blogs.law.harvard.edu/infolaw/2012/01/25/more-crap-from-the-e-u/

[16] http://www.techdirt.com/articles/20120129/23085517583/why-cant-europe-just-forget-ridiculous-idea-right-to-be-forgotten.shtml.

4.2 The US Perspective

The reaction to the proposals for the right to be forgotten in the US was quite dramatic. Headlines such as '*More crap from the EU*' by Jane Bambauer writing in Harvard's Info/Law blog[17] and '*Why Can't Europe Just Forget The Ridiculous Idea Of A 'Right To Be Forgotten'*'; by Mike Masnick in TechDirt[18] give a flavour of the overall reaction. Perhaps the most important, and most often quoted, response, came from Professor Jeffrey Rosen, writing in the Stanford Law Review online.[19] Rosen sums up his position like this:

> Although Reding depicted the new right as a modest expansion of existing data privacy rights, in fact it represents the biggest threat to free speech on the Internet in the coming decade.

Rosen's article notes the intellectual roots of the right to be forgotten—the French *le droit à l'oubli* and the Italian *diritto al' oblio*, as discussed above—rather than the practical roots in the difficulty people face in the deletion of social networking sites. He quotes the notorious case of the murderers of German actor Walter Sedlmayr attempting to remove the mention of their criminal history on Sedlmayr's Wikipedia page—which does indeed reveal a real attempt to 'rewrite history.'[20]

As Rosen puts it: "In theory, the right to be forgotten addresses an urgent problem in the digital age: it is very hard to escape your past on the Internet now that every photo, status update, and tweet lives forever in the cloud."[21]

That statement itself is revealing—particularly in terms of how the right has been presented. Is the right to be forgotten really supposed to be about escaping your past? The historical version—the *droit à l'oubli* and *diritto al' oblio*, and the right invoked by Sedlmayr's murderers—may well be, but is that what deleting a Facebook account is about? That is a more complex question, and one that will be discussed in more depth below.

One of the key objections raised by Rosen and others is that not only would the right to be forgotten allow people to have control over data that they themselves have placed on the internet, it would allow them to control data about them that other people have created or posted. Personal data, as Rosen correctly notes, is broadly defined as 'any information relating to a data subject'.[22] That, then, brings the 'free speech' issue to a focus. In these terms, if someone posts a 'story' about you, that story becomes 'personal data', and hence, using the right to be forgotten, you appear to have a 'right' to demand the deletion of that story. Looked at from this perspective, the right to be forgotten does indeed look like a tool of censorship, an attempt to

[17] Rosen (2012).

[18] See for example http://www.nytimes.com/2009/11/13/us/13wiki.html.

[19] Rosen (2012).

[20] Proposed Data Protection Regulation, Article 4(2).

[21] One of the ways that Rosen's article has been publicised: see https://www.privacyassociation.org/publications/2012_02_14_rosen_the_right_to_be_forgotten_could_close_the_internet.

[22] CTB v News Group Newspapers [2011] EWHC 1232 (QB).

allow the rewriting of history—and in direct contradiction of the all-important First Amendment of the US Constitution.

Some in Europe might wish to dismiss these stories as scaremongering, and headlines like 'the right to be forgotten could close the internet'[23] are not likely to help them engage in debate. However, it is important to understand that the objections are real, the examples provided by Rosen and others are real, and that the history of the misuse of 'privacy' to push forward censorship in other fields is real. In the UK courts in recent years, for example, privacy-related law has been used by footballers such as Ryan Giggs,[24] Rio Ferdinand[25] and John Terry[26] to attempt to keep their affairs from the public eye. The concern from the US is a real concern, and needs to be taken seriously.

4.3 Free Speech

Article 10 of the European Convention of Human Rights requires a right to freedom of expression, held in balance with Article 8, the right to a private life, and that balance is taken into account in the drafting of the proposed right to be forgotten. Specifically, Article 80 (1) of the proposed Data Protection Regulation states:

> Member States shall provide for exemptions or derogations from the provisions … …for the processing of personal data carried out solely for journalistic purposes or the purpose of artistic or literary expression in order to reconcile the right to the protection of personal data with the rules governing freedom of expression.

This article, however, has drawn criticism in two particular directions. Firstly, the wording does not seem to be sufficiently broad or inclusive to take into account either the interpretation of 'free speech' in the US or the development of new media and the internet. In the US tradition, free speech covers a great deal more than journalism and artistic and literary expression—and in the current state of the internet, many more people than journalists 'express' themselves. How would such a term deal with citizen journalists, or with bloggers, or with people writing product reviews on shopping websites or comments on message boards? In US terms, all of those people would expect to be covered by the First Amendment—and in Europe, on the surface at least, Article 10 would seem to apply.

Recital 121 in the proposed Regulation, discussing the exemptions and derogations goes some way to meet these criticisms. It suggests that member states should interpret the idea of journalism broadly:

> Member States should classify activities as "journalistic" for the purpose of the exemptions and derogations to be laid down under this Regulation if the object of these activities is the

---

[23] Ferdinand v Mirror Group Newspapers [2011] EWHC 2454 (QB).

[24] John Terry ("LNS") v Persons Unknown [2010] EWHC 119 (QB).

[25] Proposed Data Protection Regulation, recital 121.

[26] Proposed Data Protection Regulation Article 17(2).

> disclosure to the public of information, opinions or ideas, irrespective of the medium which is used to transmit them. They should not be limited to media undertakings and may be undertaken for profit-making or for non-profit making purposes.[27]

If this recital is itself interpreted broadly, it might meet many of the free speech objections, but to rely on the broad interpretation of a recital asking for a broad interpretation of a term in an article is unlikely to be considered robust enough protection for those who take free expression seriously. It also brings into focus the second equally fundamental issue that arises: who would determine what 'journalistic purposes or the purpose of artistic or literary expression' would be—and what would happen in case of doubt? Would the default be that data would be available for deletion or that free expression should take priority? As will be seen in Sect. 4 below, that question of defaults is a key difference between the approaches in Europe and the US.

There are other questions that arise from the 'free speech' debate. To what extent and in what situations can data be considered 'speech'? Is there any kind of qualitative aspect to it—or does it matter whether the data has been 'published' or made available in any direct way? These questions are much than just theoretical—they strike at some of the key issues surrounding control over data both on and offline. Looking at social networking, for example, the obvious examples that people generally talk about—indeed, that Rosen mentions in his piece on the right to be forgotten—are such things as embarrassing photographs or comments. These could be relatively easily described as 'speech', particularly in US terms, but they are not necessarily representative of the most important data held by Facebook, either for the individuals or for Facebook themselves. What is perhaps more valuable is less obvious data—social data, such as who you are 'friends' with, what kinds of people you interact with and in what way, what your taste in music might be, how long you spend online and so forth.

This is the data used by Facebook for profiling purposes—to target advertising amongst other things—but is never really 'published'. That is data that may in some senses be about 'the past', but has much more relevance to control and manipulation in the present and the future. Should this kind of data be protected by the First Amendment or Article 10 of the ECHR? It is hard to argue that either are really relevant at least to the principles of free expression: very little of this is ever really 'expressed'. This is the kind of data that individuals need to have control over—and need to be able to delete—if they are to have more autonomy both online and offline. That, ultimately, is the aim of data protection.

Another issue to consider is that of links—an issue clearly of great importance to organisations like Google. Should links to stories or to data be considered in the same terms as the stories or data themselves? This issue is being played out in related areas such as links to copyrighted material—to what extent are or should Google and their equivalents be responsible for what their links link to? With 'private' data the mechanics and issues have similarities to the copyright equivalents. Under the broad definition of personal data, it can be argued that a link to personal data is itself

[27] Bambauer (forthcoming 2014).

personal data, and hence subject to all the terms of the Data Protection Regulation, including the right to be forgotten.

Further to this, the draft regulation says that:

> Where the controller referred to in paragraph 1 has made the personal data public, it shall take all reasonable steps, including technical measures, in relation to data for the publication of which the controller is responsible, to inform third parties which are processing such data, that a data subject requests them to erase any links to, or copy or replication of that personal data. Where the controller has authorised a third party publication of personal data, the controller shall be considered responsible for that publication.[28]

This suggests that not only is the erasure of links part of the right to be forgotten, but it also places a burden on search engines and others to locate all those who are linked to in order to ensure that data that has already been disseminated can be located and erased. When this is considered in relation to data that is at least prima facie 'speech'—e.g. stories and pictures—this again looks very much like a potential tool of censorship and control. When considered in relation to other data—as noted above, data that is not really speech in any practical sense—it is again just a logical means of taking some kind of control back. Links, however, are an important issue in another key way, one that will be returned to in Sect. 5 of this Chapter: free enterprise.

In the US, however, neither the argument over stored data nor that over links will be easy to sustain. Indeed, it may well be that the US will consider the First Amendment even more broadly than before, particularly in respect of data. Jane Bambauer, in a forthcoming piece for the Stanford Law Review, effectively argues that almost all data should be given First Amendment protection. As she puts it:

> When the collection or distribution of data troubles lawmakers, it does so because data has the potential to inform, and to inspire new opinions. Data privacy laws regulate minds, not technology. Thus, for all practical purposes, and in every context relevant to the privacy debates, data is speech.[29]

The argument Bambauer makes is strong and sustained, supported by extensive US case law, and demonstrates the significance that the First Amendment plays in the US approach to data. Her conclusion is very direct: though providing First Amendment protection to all data could cause significant problems, free speech is more important, and in the end we will all realize this. Ultimately, she is calling for Europeans in particular to change their minds and realize that they are on the wrong track in pushing for data privacy, just as the noted American Jurist Oliver Wendell Holmes changed his mind over censorship as a result of his reflections about the Great War. Bambauer concludes that:

> The sanctity of a freely made mind requires protection not only for speech, but also for the digestion of raw facts.[30]

The parallels with Rosen's arguments are clear—but Bambauer's full, data based argument is a stronger and deeper one than the 'classical' free speech arguments

---

[28] Ibid. p. 62.

[29] Brin (1998).

[30] Bell and Gemmell (2009).

about censorship and the rewriting of history. There are also parallels between this argument and those of writers from Brin[31] in 1998 to Bell and Gemmell[32] in 2009 and onwards on the benefits of a 'transparent' society—that the advantages gained by the openness of data outweigh the problems caused by loss of privacy, and that as a result people should embrace that transparency. Bambauer's is however a specifically legally based argument, albeit from a somewhat extreme position.[33] Whether it could or should hold sway over the privacy arguments anywhere other than in the US is another matter—but it demonstrates some of the strength and depth of feeling in the US over the issues.

4.4 A Conflict of Approaches

On the surface at least, what underlies the dispute over the right to be forgotten is a conflict of cultures. Some of this conflict is obvious and often discussed: most directly the way that the European Union, in general, promotes privacy while the US, in general, prioritises free speech. In some ways this is more than just a question of priorities, but a question of defaults: in the US, freedom of speech is the default, and a very strong argument would need to be made for that default to be overridden. In the EU, though there is an official and explicit balancing operation between articles 8 and 10, privacy can sometimes appear to be the default. That could be seen to be the case here: Rosen, for example, believes that the right would mean that data could be deleted, and that data holders would have to 'prove' that they are entitled to the free expression exemption in Article 80 of the proposed regulation discussed in Sect. 3 above.

The argument over the free speech/privacy balance may appear simplistic on the surface, but it is important to understand how deeply ingrained these attitudes are in the way that issues are framed on both sides of the Atlantic. The debate in the US has very largely focussed on the free speech implications of the right—only rare articles such as Michael Hoven's for the Harvard Journal of Law & Technology, 'Balancing Privacy and Speech in the Right to Be Forgotten',[34] have put any focus on the privacy aspects—or the 'data' aspects of the proposed right. In the EU, meanwhile, the focus has been very much on privacy, and the free speech aspects have been largely dismissed as either being overblown or covered by the flawed free expression exemption.

If there is to be a way forward, something that will be returned to in Sect 6 of this Chapter, assumptions on both sides need to be more carefully examined and

---

[31] Bambauer's stance on the right to be forgotten can to an extent be gauged by the headline to the piece she wrote for Harvard's Info/Law blog noted above: '*More crap from the EU*'.

[32] Hoven (2012).

[33] It is already a legal right in a number of countries including Costa Rica, Estonia, Finland, France, Greece and Spain and has gained significant support.

[34] See http://news.cnet.com/8301-13506_3-57352967-17/vint-cerf-internet-access-isnt-a-human-right/.

challenged—but it is important to understand that the cultural differences between the US and the EU are greater than just a difference over the relative importance of privacy and free expression.

4.4.1 *A Difference in Approaches to 'Rights'*

The first difference is the approach to rights. The European approach is, in general, to have many rights, but held in balance—the idea of balancing the rights of privacy and of freedom of expression is one that fits well with the overall European approach to rights. The US approach is qualitatively different: there is a tendency to have fewer rights, but for those rights to be considered more powerful, more absolute. Freedom of expression again is an example: it isn't held in balance with privacy, it has primacy over privacy. It is considered the default rather than being held in balance.

This US approach to rights shows itself in a number of other ways. Firstly, the reluctance of the US to take the idea of economic, social and cultural rights seriously: from a traditional US perspective, those aren't 'rights', and certainly not in the same way that 'civil rights' or political rights are. The US ratified the International Covenant on Civil and Political Rights in 1992, but has still not ratified the International Covenant on Economic, Social and Cultural Rights. Even where the internet is concerned, while the idea of a 'right to internet access' has gained considerable currency,[35] Vint Cerf, known as the 'father of the internet' and an active campaigner for internet freedom has declared the reverse: that there is no human right to internet access.[36] His argument, that rather than being a right in itself, it is an enabler of rights, echoes the general US approach to rights. For something to be declared a 'right', it must be of fundamental importance, something that is practically possible, and something that isn't held in balance in its basic fform. Following this logic, the idea of a 'right to be forgotten' is at least prima facie flawed: of course we don't have a right to be forgotten.

4.4.2 *A Difference in Approaches to Regulation*

Secondly, there is a difference in the general approach to regulation. As noted in 1.2 above, European regulators are often zealous and direct: intervention and direct regulation are considered both normal and appropriate. In the US, there is a much more laissez faire approach, an encouragement of self-regulation and a reluctance to intervene in the market unless absolutely necessary. Haynes Stuart, for example suggests this kind of self-regulation in relation to a key aspect of the right to be forgotten: a framework for voluntary compliance with user requests for search result

[35] Stuart (2013).

[36] http://www.whitehouse.gov/sites/default/files/privacy-final.pdf.

deletion.[37] As Haynes Stuart points out, this is not just a philosophical argument: this kind of self-regulation may be the only way to meet the First amendment restrictions over demanding the deletion of speech—making such deletion legally enforceable would be likely to breech the First Amendment. The US laissez faire philosophy combines with its legal structure and historical practice to produce a very different regulatory approach from that in Europe.

That difference in regulatory approach is being played out at the moment in a some key fields related to the internet: most directly the potentially privacy-invasive tactics of behavioural advertisers. The 'Cookies Directive' referred to in 1.2 above was a response to these privacy-invasive actions: the response in the US was much less interventionist. So far it has involved two prongs: a 'consumer bill of rights' issued by the White House[38] and the 'Do Not Track' initiative, whereby technology providers and the advertising industry are intended to agree a common set of standards through which advertisers will agree to abide by consumers' decisions not to be tracked. Both of these initiatives rely to a great extent on goodwill and cooperation rather than heavy-handed legislation: the bill of rights appears largely aspirational, while Do Not Track is currently bogged down in disagreement as to both meaning and functionality, so much so that European Regulators are threatening even harsher regulation.[39] Neither the European approach—a heavy-handed directive—nor the US approach—soft touch aspiration and self-regulation—have really produced results yet in relation to online tracking but it is important to understand where these approaches come from. The bottom line is an attitude to business. The freedom of businesses to operate as they wish—the importance of free enterprise—is considered of far more importance in the US than it is in the EU.

4.5 Free Enterprise

Ultimately, it may well be the attitude to free enterprise that is of more importance to US resistance to the right to be forgotten than their attitude to free speech, however the debate has been framed both in public and in the academic literature. The biggest issue with the right to be forgotten may well be the practical one of how it might be implemented—indeed, whether its implementation is even possible—and what kinds of burdens it would place on businesses. Those burdens could be significant, particularly for search engines and social networks. For social networks it could require them to restructure their files in such a way as to make deletion even possible—something that would be challenging at best and probably very expensive.

[37] See for example http://www.theregister.co.uk/2012/10/11/regulators_threaten_do_not_track_standard/.

[38] Proposed Data Protection Regulation Article 3.

[39] Sanctions are detailed in Proposed Data Protection Regulation Article 79. Overall fines are envisaged as up to 2 % of annual worldwide turnover for some data protection breaches, and up to 1 % of annual worldwide turnover in relation to the right to be forgotten and erasure (Article 79 (5)).

For search engines the same—and possibly more so, depending on how the links issue discussed in Sect. 3 is resolved.

These challenges could mean significant extra costs—but they would also be significant restraints on the businesses' freedom to do business in their own way. This is challenging not only financially and technologically but also ideologically for a nation where free enterprise is of such paramount importance. Moreover, the businesses involved—from Facebook and Google downwards—have great lobbying power, particularly in the US, and are both able and willing to bring that power to bear. How much influence this lobbying has should not be underestimated. The fact that the proposed reform to the Data Protection Regime also includes an explicit extension of the scope of the regime to businesses worldwide who target their activities at consumers based in the EU,[40] as well as powerful new sanctions on those businesses including significant fines based on their global turnover[41] makes these burdens particularly worrying for the US businesses concerned—and has made their lobbying activities even more intense.

4.5.1 Enforcing 'Privacy by Design'?

The first question that many businesses may be asking about the right to be forgotten is how could and should they comply with the law, if it should come in. One possible answer, and the one most likely to be favoured by the European regulators, would be to implement some kind of real 'privacy by design'. Privacy by design has been pushed as a concept by regulators since its conception. Commissioner Reding has been a particular advocate—for example making it a centrepiece of her keynote address for Data Protection Day in 2010.

> Businesses must use their power of innovation to improve the protection of privacy and personal data from the very beginning of the development cycle. Privacy by Design is a principle that is in the interest of both citizens and businesses.[42]

The essence of privacy by design is that privacy, and privacy rights, must be built in to the design of any system from the outset—and as part of the fundamental design. In relation to the right to be forgotten, businesses would be expected to design their systems so that users' data is put together and linked together in a suitably compact and integrated form that it can easily be deleted. This in itself is nothing new: in order to properly meet the existing right of access to data, businesses should be able to present a user with their data in a useable form. In practice, that aim does not seem to have been realised to a great extent. In 2011, when Austrian student Max Schrems made a data access request, he was eventually presented with 1,200 pages

---

[40] See speech: http://europa.eu/rapid/press-release_SPEECH-10-16_en.htm.

[41] See for example http://www.guardian.co.uk/technology/2011/oct/20/facebook-fine-holding-data-deleted.

[42] The 'Europe vs. Facebook' campaign, whose website is http://europe-v-facebook.org/EN/en.html.

of data—revealing not only the quantity and nature of data held, but in how complex a form.[43] The Schrems case made the headlines, spawned an active campaign,[44] and led to an investigation by the Irish DPA—and added fuel to the fire that, as noted in Sect. 1.2 above, had already been heating up the drive in the EU for a right to delete.

Proper privacy by design would require businesses to make this data much more accessible, compact and user friendly—and should at least theoretically make it possible to delete that data easily. What is more, it could also allow for one of the further rights set out in the proposed Data Protection Regulation, the right of data portability,[45] to become a practical proposition. That right would allow users of social networking services to move their entire data from one provider to another. This right could potentially undermine the business models of the bigger social networking sites, and hence might generate considerable resistance. Moreover, it could be seen as an unjustified interference with free enterprise—telling a business how to organise its data is tantamount to telling it how to run its business.

4.5.2 Free Enterprise vs Free Speech: The Copyright Debate

The importance of free enterprise as opposed to free speech can be seen in how the issue has played out in relation to copyright. Though the idea that free speech is absolute is spoken about a great deal in the US, where copyright might be infringed it is clear which takes priority. In the 2012 US Supreme Court case of Golan vs Holder,[46] an extension of the copyright was given explicit precedence over free expression. Further, music and videos are regularly removed from YouTube even on suspicion of copyright infringement—the 'notice and takedown' regime is both powerful and effective.

That has two implications: firstly, that even to talk about free speech as being an absolute is misleading, and secondly that mechanisms can be and are in place to allow for items to be either removed or links to them hidden. If personal data were the subject of copyright, Google, Facebook and others would be both able and willing to remove it: if they can do so for the purposes of copyright, why not for the purposes of privacy? If they can do that to fulfil the 'rights' of the entertainment industry, why not to fulfil the 'rights' of individuals?

---

[43] Set out in Article 18 of the proposed Data Protection Regulation, the article immediately following the Right to be Forgotten and Erasure.

[44] Golan, et al., v. Holder (Attorney General), et al. 565 U.S. ___, 132 S.Ct. 873 (online at http://www.supremecourt.gov/opinions/11pdf/10-545.pdf).

[45] E.g. in October 2012, European regulators challenged Google's amalgamation of privacy policies. See http://www.bbc.co.uk/news/technology-19959306.

[46] See e.g. http://www.law.com/corporatecounsel/PubArticleCC.jsp?id=1202575007168&IBMs_New_Privacy_Chief_Calls_Data_Privacy_Cornerstone_of_Trust&slreturn=20120916094250.

4.5.3 *Free Enterprise vs Individual Rights*

From the European perspective this can be seen as a reflection of a general preference in the US for the rights of big corporations to those of individuals—and explicitly what the data protection regime in general and the right to be forgotten in particular are intended to fight. The European data protection regime is intended to support individual rights—and one of the most significant threats to individual rights is seen as coming from the actions of businesses.

In one way this comes down to the question of whose data it is anyway? Specifically who has rights over personal data—the person about whom the data has been gathered or derived, or the person who has gathered, derived or otherwise obtained that data. From a European perspective the answer to that question is being increasingly clearly elucidated: the individual should take priority. From the US perspective that is less clear—though even in the US the idea that individual privacy rights should be able to at least compete with the rights of businesses to exploit data is starting to gain currency.

As things currently stand, it is the businesses that have control—even over the issue of forgetting. Businesses can forget individuals—they can lose or delete their data whenever they feel like—but individuals find it hard to be forgotten if they would like to. There may even be a converse right—some kind of 'right to be remembered'—that allows individuals to prevent the deletion of their data or accounts—but that is something for a later debate. This issue for now, at least as far as the data deletion aspect of the right to be forgotten is concerned, is whether or how individuals should be able to exercise control over businesses.

From the US vs. EU standpoint, the debate polarises even further: the businesses most affected by the right would be US businesses. The biggest players of the internet—Google, Facebook, Apple, Microsoft etc—are primarily US businesses. An ability and willingness to 'take on' those businesses could be seen as part of a bigger conflict between the EU and the US as a whole: Google, in particular, seems to be increasingly in the sights of the European regulators.[47] Indeed, just as free speech might be seen in some ways as some kind of a cover for US protection of its own business interests, privacy could be seen as a cover for EU attacks on US businesses in order to promote or protect existing and future European businesses. Neither side has a monopoly on philosophical purity—and it all adds both tension and importance to the debate over the right to be forgotten.

4.6 Conclusions and Ways Forward

The first and most important thing to understand about the debate over the right to be forgotten is that both sides have valid points. From the EU perspective, the 'right to be forgotten' does attempt to address real problems—the excessive amounts of personal

[47] See e.g. http://news.cnet.com/8301-1023_3-57583022-93/googles-schmidt-the-internet-needs-a-delete-button/.

data being gathered, held and used by online businesses, the growing perception from people all over the world that 'their' data is increasingly out of control, and that online businesses such as Facebook and Google seem unwilling to address those concerns properly. From the US perspective, the right as currently drafted does have the potential to infringe on their properly cherished freedom of speech, and the exemption currently built into the regulation appears to be flawed and is likely to be ineffective. Neither side, in public at least, seems to be able to fully acknowledge the strength and importance of the other's position. That acknowledgement should be the starting point to finding a solution.

Finding that solution is not likely to be easy. As has been shown, there are issues between the EU and the US over the right to be forgotten at a number of levels—and the possible solutions have to work at all those levels. The conflict over the balancing between freedom of speech and privacy is the most obvious—but may not be as hard to resolve as it might seem. A better worded and more carefully couched 'exception' for freedom of speech could help—as could a renaming of the right as a right to erasure, dropping the emotive and misleading label 'right to be forgotten'. Ensuring that the focus is on the erasure rather than the forgetting, and on held data rather than published stories, could blunt the challenge of the free speech argument—and bring it, instead, to more of the same kind of level as the copyright vs. free speech argument.

As that debate has shown, the US can and does make what are to most intents and purposes 'compromises' in relation to the First Amendment when other interests are in play. Solicitor General Verrilli said, when commenting on Golan vs Holder (see above) in relation to compliance with the Berne Convention, "[Section] 514 is, in essence, the price of admission to the international system." Could an equivalent move be possible as the price of admission to the international system of privacy and individual rights over personal data? It would not be at all easy, but it might not be impossible either, if the terms of the right to be forgotten were expressed more appropriately.

Addressing the 'free enterprise' issue may not be so easy: in practice, business generally seems to win. Even when it does not, as for example in the recent apparent 'defeats' of the copyright-related bills SOPA and PIPA, it wasn't that individual rights triumphed over business rights so much as that one set of business rights (those of Google, Wikimedia etc.) triumphed over another. Is there an equivalent business interest that could compete on behalf of privacy? Currently there does not appear to be, but that might be changing. Ultimately, businesses depend on their customers, and if customers increasingly demand privacy, businesses may respond. In October 2012, IBM's new Chief Privacy Operator called data privacy the 'cornerstone of trust':[48] if more businesses follow that approach, if more businesses decide that privacy and individual control over personal data, is ultimately to their advantage, then the balance of power could start to shift.

That shift of power may be beginning to happen—and one sign has come from what might previously have been seen as a most unlikely source. Eric Schmidt, the

[48] See for example http://news.cnet.com/8301-1009_3-10036090-83.html.

Executive Chairman of Google, signalled what might be significant shift in May 2013 when he suggested that the Internet needed a 'delete button'. As he put it:

> In America, there's a sense of fairness that's culturally true for all of us. The lack of a delete button on the Internet is a significant issue. There is a time when erasure is a right thing.

Until that point, and as noted above, Google had been seen as one of the prime opponents of the right to be forgotten. If Google are now beginning to embrace the idea, it might be that some kind of a compromise is possible. A more limited right, relabelled as a *right to delete* or a *right to erasure* rather than the emotive and misleading *right to be forgotten*, could be the way forward for that kind of compromise. The emphasis would be on data rather than stories. It could focus on *held* rather than *published* data, on *gathered* more than *provided* data, and steer clear of the key areas where free speech in its more natural sense would apply.

There would still be very large barriers to overcome from a legal perspective—not least the kind of broadened interpretation of the First Amendment to cover more (or even all) data suggested by Bambauer and others—but it might be workable. There does, however, appear to be a long way to go in this debate. There are many possible amendments to the Regulation still on the table, there is a great deal of lobbying still going on—and that lobbying is likely to continue beyond the point at which the new regulation has been agreed in Europe. From the European perspective, too, compromise would not be easy: the Commission and indeed individual commissioners have invested a great deal in the idea of a 'right to be forgotten', and even to recast it would require a volte face that might seem embarrassing at best. As a consequence, it would be wise not to expect any kind of transatlantic consensus in the short term. The gap between the EU and the US remains substantial and it will take a lot of bridging, at least in a legal and technical sense.

Ultimately, however, that legal and technical agreement may not be what really matters. Google and other key players might choose to comply and allow people a 'delete button'. The words of Eric Schmidt suggest that such an attitude could be possible, and it would not be unprecedented for Google to make such a 'voluntary' shift. In 2008, under pressure from the Article 29 Working Party, Google cut its data retention periods for search logs in half, from 18 months to 9 months. What is more, they made that cut worldwide, when the Article 29 Working Party asked for compliance only within the EU.

If Google and others chose to do this then it would not matter whether any potential action brought by European regulators would be deemed unconstitutional or unenforceable in the US, because the regulators would not need to take such an action. Changes in policy in practice would satisfy both the legal requirements of the EU and the preference for self-regulatory actions in the US. The disagreements, profound though they might be, would largely be confined to legal scholars and commentators. From the perspective of the individual people whose rights the regulators seek to protect that would not be something of great concern.

References

Bell, C. G., and J. Gemmell. 2009. *Total recall: How the E-memory revolution will change everything*. New York: Dutton.

Bernal, Paul. 2011. A right to delete? *European Journal of Law and Technology* Vol. 2, No. 2. http://ejlt.org//article/view/75/144.

Bambauer, Jane. (forthcoming 2014). Is Data Speech? 66 Stan. L. Rev.

Brin, David. 1998. The transparent society: Will technology force us to choose between privacy and freedom? Reading: Addison-Wesley.

Haynes Stuart, Allyson. 2013. Search results—Buried if not forgotten, paper presented at the Privacy Law Scholars Conference, Berkeley, June 2013.

Hoven, Michael. 2012. Balancing privacy and speech in the right to be forgotten, Harvard Journal of Law & Technology 2012. http://jolt.law.harvard.edu/digest/privacy/balancing-privacy-and-speech-in-the-right-to-be-forgotten.

Rosen, Jeffrey. 2012. The right to be forgotten, Stanford Law Review Online. http://www.stanfordlawreview.org/online/privacy-paradox/right-to-be-forgotten.

Chapter 5
Stage ahoy! Deconstruction of the "Drunken Pirate" Case in the Light of Impression Management

Paulan Korenhof

5.1 Introduction

In the world, an increasing number of people make use of the Internet.[1] The Internet is a rich source of information and a medium that is widely used on a daily basis for information exchange. In a relatively short time, the quality and quantity of digital data storage and online accessible information have grown explosively. In his book *Delete: The Virtue of Forgetting in the Digital Age*, Viktor Mayer-Schönberger describes this qualitative and quantitative growth of digital data storage.[2] Compared to the analogue era, people have easier access to more information in the age of Web 2.0 and can more easily reach and store information. The Internet is also a very popular medium for the management of self-presentations and corresponding social relations. Websites like Facebook[3], MySpace[4], Google+[5] and LinkedIn[6] provide a platform for social interaction and information exchange (some are more focused on leisure interaction like Facebook, and some more on professional interaction like LinkendIn). This big flow of information has many benefits, but when it comes to personal data, it is also a reason for concern. The core concerns of personal information being accessible on the Internet are the lack of control that an individual has over this information and the possible consequences of that lack of control; for instance, people being unable to "escape" from past online information about them or people experiencing professional consequences due to their off-time behaviour that can be viewed on the Internet. Online information can severely affect the offline lives of individuals.

---

[1] Castells 2010, p. 382.
[2] Mayer-Schönberger 2009.
[3] www.facebook.com.
[4] www.myspace.com.
[5] plus.google.com.
[6] www.linkedin.com.

P. Korenhof (✉)
Privacy & Identity lab, Tilburg Institute for Law, Technology, and Society (TILT),
Tilburg University, P.O. Box 90153, 5000 LE, Tilburg, The Netherlands
e-mail: p.e.i.korenhof@tilburguniversity.edu

S. Gutwirth et al. (eds.), *Reloading Data Protection*,
DOI 10.1007/978-94-007-7540-4_5, © Springer Science+Business Media Dordrecht 2014

When one is interested in the manners in which offline life can be affected (negatively) by the Internet and starts digging through literature and articles concerning the matter, one is bound to stumble upon the so-called "drunken pirate" case sooner or later. This case received much media attention because it showed the possible destructive consequences of posting information on social media websites.[7] The data subject in this case—in this paper referred to as "S"[8]—became a news item because information on her MySpace website led to the end of her career as a teacher. S is denied her teaching diploma because she showed an apparently compromising photo of herself on her website.[9] The picture in question showed S with a pirate hat while drinking from a plastic cup. She captioned the photo "drunken pirate". The case has been repeatedly used to illustrate the need for a "right to be forgotten"[10] or need for deletion or erasure of 'expired' data.[11] Mayer-Schönberger writes:

> S(. . .) considered taking the photo offline. But the damage was done. Her page had been catalogued by search engines, and her photo archived by web crawlers. The Internet remembered what S(. . .) wanted to have forgotten.[12]

These approaches have put a lot of emphasis on the 'remembering' capacities of the Internet in the current debate on data protection. The question is whether the problems with regard to individual information control on the Internet and the solutions to these problems are (all) best approached from (only) a temporal framework of 'remembering the past,' since the Internet also affects the sharing of information over a spatial distance at a single point in time. In order to figure out how to cope with the problems that can arise due to information being online, I therefore believe it is necessary to get a clear picture first of the character of the problem(s) that can arise due to information being on the Internet. Because the "drunken pirate" case seems to be becoming an iconic case with regard to the offline problems that can be caused by people having access to online information, I believe it is worthwhile to explore this specific case in detail. Therefore, the role that the Internet played in the "drunken pirate" case will be examined in this paper. The main question is: which role did the Internet play in the downfall of S's career as a teacher?

To answer this question I will first give an outline of the case. Next, I will discuss the relation between impression management and the control of information and subsequently the manner in which the use of Internet affects an actor's ability to control his self-presentation. After that, I will consider the case in the light of the

[7] See e.g.: Rosen (2010); Stross (2007); Read (2007).

[8] This paper is written as response to a case that received a lot of media attention. In the media articles S is repeatedly named with her full name. In order to try to preserve some degree of privacy of the subject by not adding to the prevalence of her name online, I anonymized the data subject's name to "S".

[9] Mayer-Schönberger 2009, p. 1.

[10] http://ec.europa.eu/justice/data-protection/document/review2012/com_2012_11_en.pdf.

[11] Mayer-Schönberger 2009.

[12] Ibid., 1.

previously discussed elements. Finally, I will draw a conclusion as to what extent the Internet played a role in the downfall of S's teaching career.

5.2 The "Drunken Pirate" Case: What Happened?[13]

In reality, the "drunken pirate" case is a bit more complex than the straightforward dismissal of an individual because of a single photo on the Internet. It has been a combination of factors and decisions that to a greater or lesser degree all played a role in the turn of events.

S, who studied at the Millersville University (MU) wanted to obtain a degree as Bachelor of Science in Education (BSE). In order to receive this, she had to complete a student-teacher program successfully, part of this being an internship during which she had to fulfill the duties of a teacher for a certain period of time. During this internship, that S fulfilled at the Conegesta Valley High School (CVHS), the student-teachers had to adhere to the same professional standards as their professional colleagues and "fulfill as effectively as possible every role of the classroom teacher."[14] During the orientation for the teaching program, S was cautioned not to refer students to personal websites. In addition it was pointed out to her that student-teachers who ignored this warning, could be dismissed. Despite this warning and others from her supervisor S repeatedly communicated to her students that she had a website at the social network service 'MySpace.'[15] When one of the students approached a friend of S that was pictured on S's MySpace website, S became aware of the fact that at least one of her students visited her MySpace. She told this student that it was inappropriate for students to look at the MySpace website of a teacher since this had to be regarded as crossing a teacher-student boundary. However, on 4 May 2006 S posted the following message on her MySpace:

> First, [friend X] said that one of my students was on here looking at my page, which is fine. I have nothing to hide. I am over 21, and I don't say anything that will hurt me (in the long run). Plus, I don't think that they would stoop that low as to mess with my future. So, bring on the love! I figure a couple of students will actually send me a message when I am no longer their official teacher. They keep asking me why I won't apply there. Do you think it would hurt me to tell them the real reason (or who the problem was)?[16]

With 'they' S claims to refer to her students. Besides the above message, S also uploaded the "drunken pirate" picture. S stated that the photo had a personal meaning and that the message was only intended for her best friends.

A day later, on 5 May 2006, one of S's colleagues brought the message and the photo on her MySpace to the attention of her supervisor. Especially the message

[13] Summary of the events as described in S v. Millersville University et al., case 2:07-cv-01660-PD, document 47. In the documents prior to 47 one can find conflicting statements of the parties. Since piece 47 shows the ground for the court's ruling, it is held as being the closest approach of the facts.

[14] S v. Millersville University et al., case 2:07-cv-01660-PD, document 47, 5.

[15] See http://www.myspace.com.

[16] S v. Millersville University et al., case 2:07-cv-01660-PD, document 47, 10.

was condemned by CVHS, because it referred to S's work at the school. Next to that S already had a difficult understanding with one of her supervisors and the message disrupted this relationship even further. CVHS decided to bring S's teaching practicum to an early stop and bar her from campus. They gave three reasons for S's dismissal: S disobeyed her supervisors by communicating with her students about personal matters through her MySpace website, S had acted unprofessionally by criticizing her supervisor in the 4 May 2006 post and S was judged to have performed incompetently as a teacher. S's supervisors stated that S had problems with maintaining a formal teaching style and had difficulty adopting an appropriate role as a teacher in relation to both students and colleagues. She was considered too amicable towards her students and was accused of sharing too much information with them regarding her personal life.

As a result of this S had failed her internship and was graded as inadequate for the student-teacher program. She therefore did not meet the requirements to qualify for her BSE degree at MU.

This case shows that S's made a wrong impression on her colleagues and supervisors; in their eyes she was not up to the task of functioning as a teacher (in this paper I will leave aside whether this judgement was just). The impression that S made with her post and photo on MySpace was the straw that broke the camel's back and has been used by her supervisors to have her dismissed. Evidently something went wrong with S's impression management.

5.3 Impression Management

Before determining which role the Internet played in the "drunken pirate" case, it is important to explain first how information plays a role in social interactions.

5.3.1 The Theatre Metaphor: Performing for an Audience

Most people behave differently in different settings without perceiving their own identity as 'changed': despite being the same persons they show different aspects of their character depending on the context and setting that they find themselves in. For instance, a lot of people behave differently around their loved ones in the private spheres of their home than around colleagues at their work, sometimes they even speak in a higher or lower register of their voice. In different situations they share other information, including which 'part of themselves' they show. In his book *The Presentation of Self in Everyday Life* the sociologist Erving Goffman explains this phenomenon[17] and to make things clear, he uses theatrical terms: an actor plays a certain role and provides signals to the audience to inform it about the role that he

[17] Goffman 1959.

is playing. The performance is the 'front' of the actor.[18] The information that does not match the role is kept 'backstage' by the actor.[19] What counts as front stage and backstage is not a rigid distinction; the stages can swap roles depending on the performance that is regarded.

The audience receives information about the performance of the actor in various ways: by the actor's intentional communication, his appearance, his body language, his props and the stage of the interaction.[20] He may also unconsciously provide his audience with information[21] whilst the people around him (co-actors) also can provide the audience with important information.[22] The information to which the audience has access is crucial: the audience-members use the information to define the situation, to form a mental picture of the actor's identity and to get an idea what to expect from the actor and what the actor will expect from them in return.[23] Audience-members use the impressions that they have of an actor to ascribe certain social attributes and categories to him: his 'social identity.'[24] This interpretation of the actor's social identity forms the basis for the audience's assumptions about the actor's traits and behaviour and gives rise to the audience's normative expectations and demands.[25] These normative expectations depend on the social norms of the audience.

The audience members use the information they get to decide on the way in which they will respond to the actor's performance.[26] Therefore it is vital for an actor's performances that he controls the information to which his audiences has access. By sharing certain information with some people and not with others, an actor can give shape to his self-presentation and distinguish between different types of social relationships in order get to different types of responses.[27]

Making a distinction between the information one shares and the information one omits, based on the role that one is playing, is not only important to distinguish between roles. It can also be vital for a credible performance: information that is essential for a certain performance can be detrimental to another performance of the same actor. An audience that gets access to information that is detrimental to the performance it beholds, can become disillusioned. For an actor it will be difficult or even impossible to convince a disillusioned audience of the reality of the performance that he is giving.[28] Goffman states: "... the impression of reality

[18] Ibid., 32.

[19] Ibid., 114.

[20] Ibid., 14.

[21] Ibid., 14.

[22] Goffman 1963, p. 43.

[23] Goffman 1959, p. 13.

[24] Goffman 1963, p. 12.

[25] Ibid., 12.

[26] Goffman 1959, 21/22.

[27] Ibid., 17.

[28] Ibid., 136/137.

fostered by a performance is a delicate, fragile thing that can be shattered by very minor mishaps."[29] It is therefore necessary that an actor segregates his audiences to accomplish that the same audience will not see him in two inconsistent or conflicting performances.[30] This also is the case when an audience in the past has seen him in a performance that is inconsistent with his current one.[31] Information about the actor that harms a performance in any way, is "destructive information."[32] A disrupted performance can lead to a disturbed relationship between the parties on the level of the social interaction. To give an example: when the patients of a relationship therapist learn that the therapist himself is divorcing his own partner, this information has a high risk of affecting the trust of the patients in the skills of their therapist. If so, the performance of this therapist as an expert on mending troubled relations is disrupted, since his professional performance as an expert in mending relations is not credible to his patients, while the fact that the relationship therapist himself is divorcing his partner, does by no means necessarily mean that his skills as a relationship therapist are poor. The interaction on the level of the relation between patient and therapist is disturbed and the therapist will have problems doing his work properly because he lacks the trust of his patients. Goffman therefore states: "A basic problem for many performances, then, is that of information control; the audience must not acquire destructive information about the situation that is being defined for them."[33]

5.3.2 It is in the Eye of the Beholder

As pointed out in the previous section the control over personal information is of great importance to an actor's impression management. It is in the interest of the actor to decide for himself how he presents himself to others, so that he has maximum control over the image his audiences can form of him[34] and in this process informational privacy plays a crucial role. An actor can only present himself in different ways if he has sufficient privacy to control who has access to which information about him.

Privacy is often defined as a form of access control, wherein privacy means having control over the access that others have to something personal, in this case personal information. Alan Westin, for instance, defines privacy as "(. . .) the claim of individuals, groups, or institutions to determine for themselves when, how, and to what extent information about them is communicated to others."[35] Charles Fried states that privacy "is not simply an absence of information about us in the minds of others;

[29] Ibid., 63.

[30] Ibid., 137.

[31] Ibid., 138.

[32] Ibid., 141.

[33] Ibid., 141.

[34] Ibid., 15.

[35] Westin 1966, p. 7.

rather it is the *control* we have over information about ourselves."[36] Privacy as a form of access control over information regarding oneself is necessary for the construction of an identity of one individual between others individuals; "(self-identity) has to be routinely created and sustained in the reflexive activities of the individual."[37] A lack of privacy can hinder an actor to act autonomously by depriving his choices concerning his self-presentations and limiting him in the choice of the types of social relationships that he can establish.[38] I therefore adopt Floridi's view of the right to informational privacy as "a right to personal immunity from unknown, undesired or unintentional changes in one's own identity as an informational entity."[39]

Floridi points out that this interpretation of informational privacy "suggests that there is no difference between one's informational sphere and one's personal identity."[40] However, because I am interested in the individual as an informational entity *within* social interactions, I want to make a small nuance in this perspective and therefore I may deviate somewhat from what Floridi had in mind. In general an actor as an informational entity within a social interaction only shows a part of his information to a specific audience (the distinction between performing on the front stage and keeping certain information back stage) and this part does not necessarily have to coincide with his "actual identity".

To start with, an actor's own sense of self will always first be interpreted by the actor himself and translated into a performance before an audience can even perceive it. Also, an actor may not always perform conform his own sense of self (for instance, because he is afraid of the reactions of his environment) and as a result he may choose not only to play different roles, but also to play different characters for different audiences. Information concerning a choice to perform in correspondence with one's sense of oneself (or not) is a part of one's identity (when an actor's information is regarded as being his identity), but usually that information is not something to which audience members have access. This means that an audience has only a limited view of the information—and therefore identity—of an actor.

Furthermore, because an audience cannot look inside an actor's consciousness in order to perceive his actual identity, it cannot know the "identity-in-itself" (lending part of the term from Immanel Kant[41]) of the actor, but it can only perceive (part of) the informational entity and interpret the information in correspondence with its own knowledge (its experience with and knowledge of language, signals, attributes and norms). People are aware of feelings and experiences of other persons on the basis of their own empathic inferences.[42] The impression an audience has of an

[36] Fried 1984, p. 209.

[37] Giddens 1991, p. 52.

[38] Rössler 2001, p. 112.

[39] Floridi (2005, p. 195).

[40] Ibid., 195.

[41] Bluntly put, Kant stated in his work "The Critique of Pure Reason" that humans could never see the "thing-in-itself" because they would always see the thing in their own empirical perception of space and time, which are not necessary characteristics of the thing-in-itself (Kant 1781).

[42] Giddens 1991, 50/51.

actor, gets coloured by its own knowledge and experiences. Therefore the social norms of the audiences of the actor will be important for the way in which they will respond to certain information. Societies link different expectations to certain social characteristics as certain social identities are associated with specific stereotypes and lead to expectations about the actor's behaviour, regardless the specific situation. With regard to the social identity that an audience imprints on an actor based on the impressions that they have of him Goffman speaks of a *virtual social identity*.[43] The social identity that an audience imprints on an actor can deviate from the category and attributes that the actor actually possesses, which Goffman calls the actor's *actual social identity*.[44]

Summarizing, we may conclude that on the level of social interactions an actor's identity is perceived by his audiences as *their interpretation of the presented information*. Consequently, what an actor needs to share and what to omit in order to play a certain role without running the risk of a disrupted or faulty performance, depends for a great deal on the social norms of his audience. All societies create the norms for the way information is shared and interpreted. The social norms people inherit on a cultural and social level largely determine what is considered to be private information in which context,[45] and what information in what kind of relationship we are expected to share.[46] Such social conventions shape our expectations of what others know about us and how they deal with this knowledge. Especially social roles are associated with specific stereotypes and lead to expectations about the actor's behaviour, regardless the specific situation. Such roles are said to be 'institutionalized'[47] and the traits of character associated with an institutionalized role are culturally determined. Because of social conventions an actor is sometimes expected by society to keep certain information private in specific contexts.[48] This counts especially with regard to institutionalised roles. For instance, there generally is a difference in what an actor is expected to share in professional *intera*ctions and social interactions. The point about sharing information in a social interaction is therefore that it is an interaction: audiences respond to the performer on the basis of the information that they receive from the actor and other sources, combined with the knowledge that they already have. If an actor wants to get (or avoid) a certain response from an audience and wants to play certain roles successfully, he will need to act in correspondence with the norms of his audience. And for a great part what one is expected to share or to omit will also depend on the context. So even if an actor believes he has nothing to hide, he *does* have to abide by certain restrictions on the information that he shares (this covers the whole possible spectrum of information: content of the information, appearance, props, stage etc.) *in order to* perform certain roles in a socially recognizable and acceptable way.

[43] Goffman 1963, p. 12.

[44] Goffman 1963, p. 12.

[45] Cf. generally, Nissenbaum 2010.

[46] Rössler 2001, p. 118.

[47] Goffman 1959, p. 37.

[48] Schoeman 1992, p. 137.

5.3.3 Synchronous Audience Segregation

An actor who wants to be able to play different roles and to reduce the risk of any disruption of his performances, will need to segregate his audiences in such a way that audience members only have access to performances of roles that are intended for them. The stage on which a role is performed is an important factor in the audience segregation.

In general, different roles continue to exist over time; their performance is repeated on a daily/weekly/monthly/etc. basis. The roles that actors play depend on the setting and usually they adjust their performance accordingly. This works two ways; an actor adjusts his performance and the role that he is playing when he finds himself in a certain setting, but he can also actively seek a certain setting in order to play a specific role. To differentiate between roles and their corresponding audiences, an actor will usually swap (a part of) his appearance, props, co-actors, stage and audience. The quickest way to realize such a swap is by moving in space to another stage. Physically humans can only be in one place at a time, so by moving in space, they generally swap audiences and co-actors. By physically moving to another stage, an actor will not only move himself to another setting with different people (audiences and co-actors), but also to another stage and props. Since our physical world is divided in different "stages" and roles are generally performed on a certain stage—like the home, the school, the office and the supermarket—a role swap by changing stages is a very convenient and relatively clear method. However, roles are not fixed to a certain stage, since the role that will be played, will also depend on other aspects of the setting, likewho else is present (and who not).[49] When for instance a colleague of the actor will visit the actor at home in order to prepare a presentation for work, the actor will then play his role as employee at home.

Since an actor can physically be on one physical stage at a time, the audience segregation for a physical performance is based on the stage—the place in space—where the actor is performing (but of course an actor can also perform on one stage for two different audiences who interpret the roles differently based on their own knowledge). In order to have different stages and audiences, an actor will need to have a front stage and a back stage. Ergo, he needs to have the privacy to distinguish between his front stage and back stage information and control the access to these stages, so in fact he creates a different (front) stage for each audience. The control over the access to the performances on these different stages will differ depending on the nature of the stage. In *The Ontological Interpretation of Informational Privacy*[50] Floridi gives a fruitful account of privacy that I shall use to elaborate on the consequences that the nature of a stage can have for an actor's privacy.

In relation to the performance of an actor the setting of his performance, including its stage(s), props, actors, and audiences, would be what Floridi's calls

[49] van den Berg 2010.

[50] Floridi (2005, pp. 185–200).

the *infosphere*.[51] In the infosphere a certain amount of data is available for the audience to access. The larger the gap between the available information concerning the actor and the information the audience has, the larger the actor's privacy.[52] The accessibility of the information depends on "the ontological features of the infosphere,"[53] the features and characteristics of the actor, the audience-members, the props and—for this paper most importantly—the stage, so a performance given in a locked room with brick soundproof walls will be far less accessible for a would-be audience member who is not in the room, than if the same performance was given on a public square. A would-be audience member would be able to access the performance on the public square quite easily and become a real audience member, but features like a brick wall determine the degree of what Floridi calls "ontological friction:"[54]

> "Ontological friction" refers here to the forces that oppose the information flow within (a region of) the infosphere, and hence (as a coefficient) to the amount of work required for a certain kind of agent to obtain information (also, but not only) about other agents in a given environment.[55]

When performing on a stage with limited characteristics to stop or delay a flow of information, that therefore provides for a low or completely no degree of ontological friction, an actor has to keep in mind that he has almost no (if any) control over who has access to his performance. The features of the stage on which the performance is given, are therefore fundamental factors in the possibilities for an actor to effectively segregate his audiences. Part of controlling and managing one's impressions is therefore selectively choosing the stage for a certain performance based on the intended audience in combination with the amount of ontological friction provided by the stage. Technology that enables us to perform outside of our physical existence—like the Internet—turned that selection into a big challenge.

5.4 The Internet as Stage

Due to the interactive nature of the Internet and the fact that it is often used as a platform for the exchange of social information, Internet webpages become potential stages for the performance of roles. As a result of this we see that social network sites (SNS) in particular are transformed into important stages for the performance of various self-presentations, as was the case with the "drunken pirate". S made use of the SNS MySpace to share information with her audiences. However, Internet stages do not occupy a place in space and time in the same way as physical stages

---

[51] Ibid., 186.

[52] Ibid., 186.

[53] Ibid., 186.

[54] Ibid., 186.

[55] Ibid., 186.

and that creates a fundamentally different situation. In order to determine the role that the Internet has played in the "drunken pirate" case, it is necessary to get an idea of the manner in which the Internet forms a different sort of stage for an actor's self-presentations than a physical stage.

5.4.1 Layered Stages

As stated above, Internet stages do not occupy a place in space and time in the same way as physical stages: the Internet has a fundamentally different character than the physical world.

A performance on the Internet consists of digital information; the actor gives his performance in bits. An important characteristic of digital information is that it is aspatial.[56] It is not bound to any physical information carrier (like a newspaper or an actor that is giving a performance) and thus lacks certain ontological frictions that are typical for information that is 'fixed' to a certain physical form. Digital information can be easily transported.[57] Spatial ontological frictions (like distance or walls) are insignificant with regard to the sharing of digital information; the digital information can be distributed worldwide in a matter of seconds as long as one has access to the Internet. Also temporal ontological restrictions (like the opening times of libraries) are severely reduced too.

Another characteristic of digital information that distinguishes online stages from offline stages is the fact that digital information usually is a nonrival good.[58] This means that the consumption of the good by one person, does not diminish the usefulness of the good for others.[59] Information on a website can generally be viewed by a massive amount of people at the same time, without any of them preventing another person to see exactly the same content. This is a sharp contrast with physical performances, where no audience member can have exactly the same view of the performance as another (the audience members cannot be on the same spot with their eyes on exactly the same place) and where at the same time they can physically block each other's views. Although there is a limit to the maximum amount of people that can view a website at exactly the same time due to the capacity of the server that is hosting the website, this is only a small limitation compared to the limitation of the number of people that can access a physical performance at the same time, like a teaching performance in a classroom.

Because an online performance is not fixed to a physical form, it gives the actor of an online performance great freedom with regard to self-presentation: he can present himself as anyone or anything without any necessary resemblance to his own physical existence. In that sense the Internet provides an actor with a far-reaching

---

[56] Michalis Vafopoulos (2012), 412.

[57] van den Berg and Leenes (2010).

[58] Michalis Vafopoulos (2012), 411.

[59] Ibid., 9.

control over his self-presentation. However, his options for self-representation are limited and affected by the manner in which the online stage is programmed. If for instance he uses a SNS website that requires him to either tick "male" or "female" as part of his required personal information, one of those two categories will be attributed to the character that he is playing.

Additionally, because the online performance is detached from the actor's physical form, he can perform multiple roles on multiple online stages simultaneously, while in his physical form, he is restricted to one physical stage at a given point in time. Because potentially the Internet is always accessible from anywhere and depending on privacy settings, the online performance of the actor can be too. That means that an audience of an actor's performance in the physical world can attempt to get access to his online performance(s) as well. The detachment of an online performance from the actor in his physical form as being positioned in space and time, can lead to "layered" performances; because of the position that Internet stages can occupy in relation to physical stages—they provide a stage for multiple performances that is theoretically always present, but not necessary seen—the Internet stages can give an extra interpretative layer to a physical performance (or vice versa) by showing the actor in other performances and possibly other roles. The distinction between an actor's front stage and back stage will become vaguer due to the multiple performances (the back stage of one performance can be the front stage of the other) and may collapse. Performances on Internet stages—when accessible—can thus affect offline performances (and vice versa) by influencing the manner in which performances are interpreted by audiences. Because of the mutual influence that on- and offline performances can have on each other, the audience segregation in relation to multiple stages is vital for impression management.

5.4.2 *Performing on the Internet Stage: The General Challenges*

When an actor uses the Internet as a stage for performances, this stage can provide quite some challenges for him with regard to the control over his (on- and offline) performances and his corresponding audience segregation.

First of all, the amount of people that can populate an 'Internet space' (a website) is much higher than a physical space. Because the Internet is aspatial, it easily overcomes any spatial ontological frictions like distance and walls. Consequently, it also is not limited by the "distance between the walls;" it does not have a maximum physical mass that can occupy a certain space. For example, we can all watch our friend A perform her role as friend online without needing to be cramped up together in her house in order to see the performance. The amount of people that potentially have access to an online performance, can therefore be much higher than the maximal amount of people that can see a performance on a physical stage. Additionally, an online performance can continue unchanged and indefinitely over time, it can be more or less 'frozen' in time. In contrast, a physical performance is an action that actively happens in time and therefore is a series of moments that eventually ends.

The aspatial character and potential timelessness of an online performance infers that the access to online data could possibly involve a potentially infinite audience (depending on inter alia the privacy settings) through space and time (people from all over the world, future generations).

Secondly, the aspatial character of the Internet stage makes it difficult to keep an overview of the presence and composition of online audiences that are viewing a certain performance. Because an actor on an online stage has no physical presence in front (or between) physical audience members, he depends on 'signals' of his audience that they are watching the performance. An example of this is audience members on Facebook clicking the "like" button under a certain post. Due to this dependence on signals, actors that perform on such a stage have therefore a limited view of their audience.[60] Because of the limited view, it is hard—maybe even impossible—for an actor to timely register when an unintended audience has access to his Internet stage(s) and adjust his performance accordingly. The presence of unintended audience members will generally only come to an actor's attention when he receives a reaction from the unexpected audience member on his performance, and by then, most of the damage is already done.

Due to the aspatial character of the Internet—which nullifies any spatial ontological frictions- an actor runs the risk of performing on an all-encompassing online stage for the whole world if he cannot control who has access to his performance and who not. Controlling the access to a performance and being able to segregate audiences is therefore vital for an actor if he wants to be able to play different roles successfully, because this would not be possible if his audiences are able to regard him in all his roles. The control over this access depends on the architecture of the Internet stage is programmed. It depends on the features of a website whether an actor can limit access and can segregate his audiences by distinguishing between friends, colleagues, family etc. Most social network sites have limited options to differentiate between different sorts of relationships.[61]

Additionally, the control over the self-presentation and any inferences thereupon by others is problematic when performing on the Internet stage.[62] The online self-presentation consists of information that is added to the Internet by both the actor and his audience(s). Controlling such self-presentations is difficult because other parties can influence the interpretations of the performance. In this sense the Internet stage seems to allow more interaction with regard to the construction of a self-presentation than a physical stage, because the audience has more possibilities to add a 'comment' on the actor's performance that can 'stick' and be perceived by other audience members.

Furthermore, because the performance consists of digital information, the audience members can multiply and copy the performance information flawlessly without

---

[60] van den Berg and Leenes (2010).

[61] Ibid., 1111.

[62] Ibid., 1112.

any loss of quality or quantity of the original information. Digital information is infinitely expansible.[63] Online, the information can be stored for a long time and with the help of search engines it can usually be retrieved relatively easy. Due to these characteristics, the digital information can get a certain persistence.[64] And because digital information can be copied and reproduced anywhere on the web, it is hard to keep track (if that is even possible) of where all the copies are, let alone to exercise control over all the copies. Once the information is taken out of context, it runs the risk of being misinterpreted.

Because of the above discussed issues, an actor can generally segregate his audiences with far less nuances when performing on current Internet stages like MySpace and Facebook, in comparison to offline stages. Performances that can be viewed online have a higher risk of reaching an audience for whom certain information can be disillusioning. When performing online, it is therefore difficult to be sure that one is performing for the intended audience.

5.5 The "Drunken Pirate" on Stage

In the case of the "drunken pirate" the digital information that motivated S's supervisors to have S dismissed, were the message and to a lesser extent the "drunken pirate" photo that S had posted on her MySpace website. S had used her MySpace website as a stage to ventilate her dissatisfaction about her internship and more specifically to hint at the fact that a certain person was "the real problem."[65] The MySpace stage fulfilled a role as back stage with regard to her teaching role, and the CVHS campus ground formed her main front stage. According to S, the performance on the MySpace stage was intended for her best friends only—and as a result this was the front stage for them). However, in her message she assumes that a breach of audience segregation by her students will not be a problem. S believed that she "had nothing to hide"[66] and states: "... I don't say anything that will hurt me (in the long run). Plus, I don't think that they would stoop that low as to mess with my future."[67] Unfortunately S misjuded the situation on quite a few levels.

To start with, the "I have nothing to hide" position expressed by S is problematic, even more with regard to her role as teacher at CVHS. "I have nothing to hide" is a statement that tends to rear its head regularly in discussions regarding privacy.[68] Leaving aside the flaws of the "I have nothing to hide" notion in general[69] and assuming that an actor sincerely believes that he does have nothing to hide, he still

---

[63] Michalis Vafopoulos (2012), 411.

[64] van den Berg and Leenes (2010).

[65] S v. Millersville University et al., case 2:07-cv-01660-PD, document 47, 10.

[66] Ibid., 10.

[67] Ibid., 65.

[68] Solove (2007), pp. 745–772, p. 747.

[69] Ibid., 745–772, 747.

has to keep in mind that there are certain restrictions on the information that he can share (this covers the whole possible spectrum of information: content of the information, appearance, props, stage etc.) in order to perform a role in a socially recognized and accepted way. The success of a performance depends on the norms and knowledge of the audience and in this case in the eyes of S's supervisors a credible performance of her role as a teacher was dependent on their norms. The role of 'teacher' is generally associated with a number of requirements that people have to meet before they are found fit to educate the younger generations and is therefore an institutionalised role. In CVHS the view on the "script" that a teacher had to follow was quite clear and strict; as a teacher she should not share too much personal information with her students and she should not mention any issues regarding the school on personal webpages or let students access them. S's supervisors told S that she had to abide by these restrictions in order to complete her internship successfully. However, S disregarded the informational restrictions that her supervisors believed to be appropriate for a teacher and because she did not (want to) perform the role of teacher according to the "script" her supervisors believed to be important, she ran a risk of her performance being not credible for them with all due consequences.

Secondly S did not fully realize that her MySpace website could form a layered stage with regard to her performance as a teacher and could affect this performance. By pointing out to her students that she had a MySpace website, S even drew her professional front stage audience's attention to the existence of her MySpace back stage. When using a stage as a back stage for a certain front stage in order to ventilate feelings about the front stage performance, a collapse of the front stage with the back stage will very likely be disruptive for the front stage performance. The only manner in which an actor can prevent such a collapse is by strictly controlling the audiences' access to the back stage.

Because S's back stage was the Internet stage MySpace, it lacked the typical ontological frictions of a physical stage. The aspatial character of the MySpace stage turned the control over and view of the stage's audiences into a challenge. Any possibilities to cope with this challenge depended on the options that are offered by the programmers of this stage. When it comes to online stages, the design of the stage is determined by its programmers in a fundamental way: actions that are not part of the design, are excluded from performance[70] as all performances on the Internet stages are regulated by the technology underlying these stages (the so called techno-regulation[71]). In the offline world one can usually influence a stage in ways that are not part of its intended design, like demolishing and rebuilding parts (like adding an extra door for security), but in the online world one would just get an error notice when trying to do something that is not part of the design.[72] This design not only limits our choices, but it also affects the way in which we behave on that stage. Pariser writes: "we're contextual beings: how we behave is dictated in part by the

[70] Pariser 2011, p. 175.

[71] Cf. generally, Leenes (2011), pp. 143–169.

[72] Pariser 2011, p. 175.

shape of our environments."[73] Thus the design of MySpace plays at least a role of some importance in the "drunken pirate" case. However, the exact scope of this role will remain unclear since it is unknown what S's privacy settings were at the time of the case who exactly had access to her MySpace website. S has stated that she changed her profile name every few months in order to protect her privacy[74] and she believed that she was hard to find on MySpace; one had to own a MySpace account and had to take the trouble to find her. She even uses the word 'hacking' with regard to the effort that her colleague must have taken in order to be able to view her MySpace website.[75] However, the incident with the student showed that apparently at least one of S's students did not have any trouble with accessing S's MySpace website either. This suggests that S's profile was not properly shielded. Additionally we may assume that being in the safety of her home in front of a pc-monitor and adding messages to a stage called "my space," may very well have given the "drunken pirate" the illusion of a private and controlled setting. Would S for instance have thought twice about posting the message and the photo if the SNS she used was called "OurSpace"?

The design of MySpace obviously plays an important role with regards to an actor's impression management, when that actor performs on a MySpace stage. However, in the case of the "drunken pirate," the actor was confronted with the flaws of the stage long before S gave her "fatal" performance. Due to the incident with the student who viewed her MySpace website, S was confronted with the fact that her performances on her MySpace stage reached her professional audiences. Instead of taking this breach in her audience segregation as a warning and pause her MySpace use until her internship was over, she posted the 4 May 2006 posts. With these posts she seemed to ignore the possibility that next to students, also her colleagues and supervisors might be trying to access her MySpace. Because of the viewpoint of CVHS on personal webpages of teachers, combined with the fact that CVHS knew that S informed her students about her MySpace website, S could have expected that someone of CVHS would try to access her webpage. With the suspicion that an unintended audience may breach the segregation, an actor needs to be alert and adjust either the access to the stage or the performance itself.

5.6 Conclusion

The "drunken pirate" case received a lot of media attention because it was a clear example of a case where the use of Internet led to consequences for someone's professional career. But what role did the Internet play in the downfall of S her career as a teacher?

The problem in the "drunken pirate" case was that a part of S's performance for her best friends ended up with her professional audience. Her front and back stage

---

[73] Ibid., 174.

[74] S v. Millersville University et al., case 2:07-cv-01660-PD, document 45, 9.

[75] Ibid., 9.

with regard to her role as teacher collapsed and impaired her self-presentation. Her performance was disrupted.

The role of the Internet in the turn of events is significant, but at the same time limited. S's 4 May 2006 posts had almost immediate consequences and were seen by her professional audience on her own MySpace stage. The problems in this case did not arise due to the Internet having a 'perfect memory' or being a place where information can be easily copied and reproduced. The problems arose because S disregarded the script for the teacher role set by CVHS and thereby failed to segregate the audiences of her online performances properly.

The aspatial characteristics of the Internet make it a tricky stage to perform on and an Internet stage can become an ever-present layered stage overlapping a physical performance. The use of the MySpace stage as a back stage to ventilate about her performance on her physical professional front stage, was therefore risky. Additionally, S had been warned by CVHS that the use of a personal webpage could undermine her professional performance and if that happened, CVHS would react accordingly. More importantly, S knew that the audience of her professional front stage performance had access to her MySpace back stage performance. S reacted to this audience-breach not by taking it as a warning and pause her MySpace use, but by posting the "drunken pirate" photo and the 4 May 2006 message to ventilate her feelings to her friends. The "drunken pirate" case therefore could have been prevented if S used her MySpace stage with more discretion. We need to learn how to deal with a life that consists of performances on layered stages. However, not only the user is up for improvement, but also that which she used: the Internet stage. The manner in which the online world is programmed can severely decrease any ontological friction in the information flow, but because the design is the online world, it could also be programmed to increase the degree of ontological friction. And if we want to be able to differentiate in our relations and play different roles, we need to think about whether and how we need to design our online stages if we want to be able to have control over which audiences have access to which performances. This is not an easy task. Most current solutions that propose to cope with the impression management-undermining characteristics of the Internet, like the "right to be forgotten or erasure" in the proposal for the General Data Protection Regulation,[76] are focused on the remembering capacities of the Internet. They therefore propose solutions in time, like erasure, and are not be of any help for actors who want to be able to play different roles in the same timeframe. An actor that wants to be able to play different roles does not want her information forgotten or erased, but wants to keep her different audiences segregated from performances that are not intended for them.

However, despite the fact that the role of the Internet in the case of the "drunken pirate" is 'space'-related, it could also become time-related. As a result of the case the name of S and her "drunken pirate" picture can be found all over the Internet. Articles are written about it. Due to the characteristics of the Internet, this case could haunt S for a very long time. Hence, the discussion of the "drunken pirate" case

[76] European Commission, COM (2012).

leads to a new question: can S ever start with a clean slate, or will she always be S the "drunken pirate" as a result of the information storing and sharing characteristics of the Internet? If the last option turns out to be the case, we may need to find a way to draw the curtains on the stage. Thus, the paradoxical result of the "drunken pirate" case is that while the case in itself did not illustrate the need for a "right to be forgotten" (but rather the need of good methods for audience segregation on SNS), the role that the case is playing in the academic and media discussion on the Internet's 'iron' memory does give rise to a need for S to be forgotten as a "drunken pirate". But can the genie be put into the bottle again?

Acknowledgments This research is conducted within the Privacy and Identity Lab (PI.lab) and funded by SIDN.nl (http://www.sidn.nl). Additionally I would like to acknowledge the help of Ronald Leenes, Bert-Jaap Koops and the CPDP reviewers by providing me with comments and suggestions.

References

Castells, Manuel. 2010. *The information age: Economy, society, and culture volume I: The rise of the network society*. 2nd ed. Chicester: Wiley-Blackwell.

European Commission, COM. 2012. 11 final, "Proposal for a regulation of the European parliament and of the council on the protection of individuals with regard to the processing of personal data and on the free movement of such data (General Data Protection Regulation). (Brussels, 25 January 2012).

Floridi, Luciano. 2005. The ontological interpretation of informational privacy. *Ethics and Information Technology* 7:185–200.

Fried, Charles. 1984. Privacy (a moral analysis). In *Philosophical Dimensions of Privacy,* ed. Ferdinand D. Shoeman. Cambridge: Cambridge University Press.

Giddens, Anthony. 1991. *Modernity and self-identity*. Standford: Stanford University Press.

Goffman, Erving. 1959. *The presentation of self in everyday life*. London: Penguin Books. (used print: 1990).

Goffman, Erving. 1963. *Stigma: Notes on the management of spoiled identity*. London: Penguin Books. (used print: 1990).

Kant, Immanuel. 1781. *Kritik der reinen Vernunft*, Insel, Darmstadt, (used print: English translation, ed. and trans. P. Gruyer and A.W. Wood, Cambridge: Cambridge University Press, 1998).

Leenes, Ronald. 2011. Framing techno-regulation: An exploration of state and non-state regulation by technology. *Legisprudence* 5 (2): 143–169.

Mayer-Schönberger, Viktor. 2009. *Delete: The virtue of forgetting in the digital age*. Princeton: Princeton University Press. (used print: 2011).

Nissenbaum, Helen. 2010. *Privacy in contex: Technology, policy, and the integrity of social life*. Stanford: Stanford Law Books.

Pariser, Eli. 2011. *The filter bubble*. London: Viking, an imprint of Penguin Books.

Read, Brock. 2007. 'Drunken Pirate' learns costly lesson from her myspace posting. http://chronicle.com/article/Drunken-Pirate-Learns/38725. Accessed 10 Sept 2013.

Rosen, Jeffrey. 2010. The web means the end of forgetting. http://www.nytimes.com/2010/07/25/magazine/25privacy-t2.html?pagewanted=all&_r=0. Accessed 10 Sept 2013.

Rössler, Beate. 2001. *Der Wert des Privaten*. Frankfurt a. M.: Suhrkamp Verlag. (used print: English translation, Cambridge: Polity Press, 2005).

Schoeman, Ferdinand D. 1992. *Privacy and social freedom*. Cambridge: Cambridge University Press. (used print: 2008).

Solove, Daniel J. 2007. I've got nothing to hide' and other misunderstandings of privacy. *San Diego Law Review* 44:745–772.

Stross, Randall. 2007. How to loose your job on your own time. http://www.nytimes.com/2007/12/30/business/30digi.html?pagewanted=all. Accessed 10 Sept 2013.

Vafopoulos, Michalis. "Being, space, and time on the Web." *Metaphilosophy* 43.4 (2012): 405–425.

van den Berg, Bibi, and Ronald Leenes. 2010. Audience segregation in social network sites. Proceedings for Social Com2010/PASSAT2010 (Second IEEE International Conference on Social Computing/Second IEEE Internation al Conference on Privacy, Security, Risk and Trust), 1111–1117. Minneapolis: IEEE.

van den Berg, Bibi. 2010. *The situated self: Identity in a world of ambient intelligence*. Oisterwijk: Wolf Prodcutions.

Westin, Alan F. 1966. *Privacy and freedom*. New York: Atheneum. (used print: 1970).

Part III
Surveillance and Law Enforcement

Chapter 6
New Surveillance, New Penology and New Resistance: Towards the Criminalisation of Resistance?

Antonella Galetta

6.1 Introduction

The proliferation of pervasive surveillance technologies and practices confront us constantly with new challenges and concerns. On the one hand, they raise attention to the manifold implications and consequences linked to the implementation of new surveillance systems. On the other, they encourage surveillance scholars to fuel the on-going debate about new surveillance theories. Although the surveillance literature benefits from an array of different concepts and interpretations of contemporary surveillance, the Panopticon still represents the main term of comparison. Nonetheless, it is widely recognised that the panoptic model fails to grasp contemporary representations of surveillance. There are considerable differences between 'old' and 'new' surveillance and new patterns are emerging. One of them is represented by resistance to surveillance, a topic that was still in its infancy within the panoptic scheme but that deserves more consideration.

After having illustrated the main differences between old and new surveillance, this contribution will focus on resistance to surveillance highlighting how it operates in today's surveillance societies. In particular, it analyses how resistance is developing into new forms and patterns which might turn resistance into criminalisation. This circumstance is examined taking into account the 'Facebook case' presented in Section 4.1. New surveillance and New Penology contribute to a great extent to create new forms of resistance to surveillance in today's societies. Nonetheless, they can originate misinterpretations and/or misconceptions over the true meaning of resistance. The criminalisation of resistance can be considered as the degeneration of

This chapter is based on research undertaken in the framework of the European Commission FP7 Project IRISS: Increasing Resilience in Surveillance Societies.

A. Galetta (✉)
Law, Science, Technology and Society (LSTS), Faculty of Law
and Criminology, Vrije Universiteit Brussel (VUB), Pleinlaan 2, 1050 Brussels, Belgium
e-mail: antonella.galetta@vub.ac.be

S. Gutwirth et al. (eds.), *Reloading Data Protection*,
DOI 10.1007/978-94-007-7540-4_6, © Springer Science+Business Media Dordrecht 2014

surveillance and of the New Penology. Given its potential consequences, it represents a huge risk that should be avoided.

6.2 From 'Old' Surveillance to 'New' Surveillance[1]

Moving from the assumption that surveillance is a "key feature of modern life",[2] it comes as no surprise that that there is not a unanimous understanding over its meaning. Surveillance is as complex and multifaceted as modern life. One of its most intriguing features is represented by its two-faced nature which relates both to care (looking after) and control (looking over).[3] Surveillance implies control 'by default' and the focused and systematic gaze of surveillance as control opens up people to examination and scrutiny while interfering with individual autonomy.[4] As such, it reveals itself in many forms and interpretations which should all be encompassed in its definition. Contemporary surveillance can be referred to as the "focused, systematic, and routine attention to personal details for the purposes of influencing and protecting those whose data have been garnered".[5] This broad definition is usually further tailored on the basis of the specific circumstances at stake. Online surveillance for example can be defined as "any collection, and processing of personal data, whether identifiable or not, for the purpose of influencing or managing those whose data have been garnered".[6]

Surveillance is a changing phenomenon and today's surveillance technologies contribute to stratify existing knowledge about surveillance and surveillance theories. The shift from an old to a new surveillance can be better appreciated looking back at panoptic surveillance. The definition of surveillance one can still find in the Oxford Dictionary is definitely in line with the Panopticon scheme. Here surveillance is qualified as the "close observation, especially of a suspected spy or criminal".[7] As surveillance scholars would point out, this definition provides a limited representation of the phenomenon of surveillance which is far from contemporary reality.[8] Although surveillance theories are still heavily anchored

---

[1] For a more detailed analysis of differences between old and new surveillance, see Marx (2004, pp. 18–37).

[2] Lyon (2001, p. 2).

[3] Lyon (1994, 2001, p. 3).

[4] Monahan (2010, pp. 91–110); Haggerty and Samatas (2010).

[5] Lyon (2007, p. 14).

[6] Lyon (2001, p. 2).

[7] Oxford Dictionaries (2013).

[8] Surveillance scholars have identified several models of surveillance which oppose or differ from the idea of the Panopticon, such as the 'superpanopticon' (Poster); 'global panopticon' (Gill); 'banopticon' (Bigo); 'synopticon' (Mathiesen); 'neo-panopticon' (Mann), 'omnicon' (Goombridge); 'urban panopticon' (Koskela); etc. See Haggerty (2006, pp. 23–45).

to the Panopticon and to the triad crime-punishment-surveillance of *Surveiller et Punir*,[9] the surveillance experience we deal with in our everyday life goes well beyond this definition. Surveillance does not only consist in a close observation but also in a remote, faraway gaze whose level of detail can be identical or at least very similar to that of a close observation. Today's surveillance does not target suspected persons in particular but the whole society. Indeed, the idea of a surveillance restricted to the criminal and law enforcement spheres fails to capture contemporary surveillance practices. Finally, whereas the final aim of old surveillance was to redress and/or educate the inmate subjugating him to "a state of conscious and permanent visibility",[10] new surveillance can have many other purposes which may range from mere identification to social sorting and profiling. Nonetheless, this wide spectrum of purposes originates the so-called function creep.[11]

6.2.1 The Surveillant and the Surveilled: A Truly Imbalanced Relationship?

The shift from old to new surveillance is apparent when analysing the relation of power between the surveillant and the surveilled. Surveillance does always express a balance of power between opposing entities. Traditional surveillance was a system that reproduced a disproportionate and asymmetric relation of power which was exercised by a few individuals over a large number of people. This imbalanced relationship of power was framed around the technology availability of both parties at stake which made the surveilled feel helpless in the face of the overwhelming force of the surveillant.[12] Power asymmetry was reflected in the design of the panoptic machine that induced social compliance and facilitated the exercise of disciplinary power. Nevertheless, the panoptic logic served for Foucault to explain the technology discipline that was replicated in key social institutions such as prisons, schools, factories and hospitals.[13] Although the top-down surveillance paradigm tends to replicate itself in contemporary societies, it is changing and becoming more articulated. Today's surveillance is more dispersive, pervasive, fluid,[14] ubiquitous and invisible than traditional surveillance.[15] The technology gap that characterised the relationship between the surveillant and the surveilled is being narrowed and reframed. Nowadays

---

[9] Foucault (1975).

[10] Foucault (1975, p. 201).

[11] Function creep refers to the use of surveillance for purposes and targets beyond those originally envisaged. See for example Marx (1988).

[12] Simon (2005, p. 3).

[13] Elmer (2012, pp. 21–29).

[14] Lyon (2010, pp. 325–338); Bauman and Lyon (2013).

[15] Simon (2005).

surveillance societies are information societies in which the surveillant is not immune to forms of control himself.[16]

Yet, Foucault recognised that a certain evolution in the disciplinary power was already underway. He argued that disciplinary power was also organised as a multiple, automatic and anonymous power. As he pointed out "for although surveillance rests on the individuals, its functioning is that of a network of relations from top to bottom, but also to a certain extent from bottom to top and laterally; this network 'holds' the whole together and traverses it in its entirety with effects of power that derive from one another: supervisors perpetually supervised."[17] Even though Foucault perceived different nuances in the articulation of power between the surveillant and the surveilled, today this relationship is far more horizontal and oblique (and less vertical) than before. These dynamics are supported by the proliferation of surveillance in an array of 'opticons',[18] such as the superpanopticon[19] and synopticon[20] and by amplifications of the panoptic model to lateral surveillance.[21]

6.2.2 From Disciplinary Power to Social Sorting

As mentioned above, today's surveillance is not primarily aimed at exercising a disciplinary power but has a broader scope. The main purpose of today's surveillance is to sort out individuals and arrange social categories. It would be naïve to argue that the final purpose of surveillance as social sorting is simply to identify individuals and communities. Mere identification can be considered as the preliminary aim of surveillance as social sorting, while differentiation is at the core of this process. Indeed, social sorting is also referred to as a "mechanism of societal differentiation".[22] In fact, social sorting aims to cluster populations "in order to single out different groups for different kinds of treatment".[23]

Although contemporary surveillance implies the exercise of a horizontal gaze, it entails an unequal exposure to surveillance systems.[24] Torpey has been analysing the disproportionate effects of surveillance distinguishing between thin and thick surveillance.[25] Everyone is subjected to thin surveillance and it disproportionately

---

16 Surveillance of the surveilled over the surveillant goes under the name of 'sousveillance' (Mann et al. 2003, pp. 331–355).

17 Foucault (1975, p. 175–176).

18 Haggerty (2006, pp. 23–45).

19 Poster (1990, p. 93).

20 Mathiesen (1997, pp. 215–234; 1999, pp. 1–36).

21 Andrejevic (2005, pp. 479–497); Reeves (2012, pp. 235–248).

22 Monahan (2010, p. 97).

23 Lyon (2007, p. 98).

24 For instance, as Norris and Armstrong reported, black young men who are casually dressed have a higher chance of being the target of surveillance in our societies. Norris and Gary (1999, pp. 108–116).

25 Torpey argues that "thin surveillance monitors our movements, our business transactions, and our interactions with government, but generally without constraining our mobility per se. Thick

affects the non-poor. In contrast, thick surveillance disproportionately affects the poor.[26] Surveillance as social sorting clusters individuals and communities, while creating social categories. The differential deployment of surveillance systems causes marginalising and excluding effects which tend to discriminate particularly the poor, ethnic minorities and women. The result of this process is 'marginalising surveillance'[27] which operates by excluding individuals actively but invisibly. It follows that surveillance as social sorting tends to encourage existing socio-spatial inequalities while resulting in a higher sense of injustice.[28] In fat, social sorting is considered as a "powerful means of creating and reinforcing long-term social differences"[29] while emphasising the disturbingly antidemocratic character of surveillance.[30] As a consequence, surveillance contributes to social stratification[31] and amplifies existing social inequalities while reproducing conditions of social discrimination and marginalisation.[32] This confirms that, "rather than being neutral or impacting seemingly random on individuals, forms of surveillance may sustain or even create group-based harm, through for example, racial profiling".[33] The true nature of new surveillance reveals itself when looking at it not only as a horizontal and ubiquitous gaze but as an "intervention in the social world"[34] of individuals for the purpose of sorting them out. Thus, it is more than appropriate to claim that today's surveillance societies are societies of control instead of disciplinary societies.[35]

6.3 From 'Old' Penology to 'New' Penology

The great development of surveillance technologies in the last decades is closely linked to military and security applications which flourished during the Cold War. In fact, the invention and design of some of the surveillance systems we use nowadays

---

surveillance, on the other hand, involves confinement to delineated and often fortified spaces, in which observation is enhanced by a limitation of the range of mobility of those observed". Torpey (2007, pp. 116–119, p. 117).

26 Torpey (2007).

27 Monahan (2008, pp. 217–226, p. 220).

28 The marginalising and excluding features of surveillance are reflected in many of the urban securisation projects implemented in developing countries such as Brazil. See for example Kanashiro (2008, pp. 270–289); Melgaço (2001).

29 Lyon (2003).

30 As Monahan argues, "The dominant manifestations of surveillance-based control today are disturbingly antidemocratic because of the way they sort populations unequally, produce conditions and identities of marginality, impinge upon the life chances of marginalized populations, and normalize and fortify neoliberal word orders". Monahan, "Surveillance as governance. Social inequality and the pursuit of democratic surveillance", p. 100.

31 Lianos (2003, pp. 412–430).

32 Monahan (2008).

33 Lyon et al. (2012, p. 423) See also Lyon (2003).

34 Lyon (2001, pp. 171–181).

35 Deleuze (1990/2003, pp. 240–246).

(such as Global Positioning Systems) date back to that specific context.[36] Since then, many of the surveillance technologies and measures that were implemented in exceptional war-time conditions have become routine practice. This event has marked the massive introduction of surveillance systems in policing and criminal matters (criminal surveillance).

The exercise of state powers over criminals, suspects and prisoners is unavoidably made through surveillance measures and practices nowadays. From a theoretical perspective, the basic foundations of criminal surveillance are rooted in the Panopticon which was indeed a prison in its early design. However, criminal surveillance has undergone substantial changes in the last decades which are substantiated by differences between Old and New Penology. In 1992 Feeley and Simon argued that a paradigm shift was occurring in the discourse of criminal justice, prompted by "a new strategic formation in the penal field".[37] While the Old Penology was concerned with the identification of criminals for the purpose of ascribing blame and guilt, the New Penology focused on "techniques to identify classify and manage groupings sorted by dangerousness".[38] As the authors pointed out, the New Penology was highly dependent on surveillance as it sought to sort, classify and separate the less from the more dangerous, as well as to "deploy control strategies rationally".[39] Although there is not unanimous consensus over this thesis,[40] there are two main trends in policing and criminology that support it, namely: the expansion of powers of law enforcement authorities and intelligence forces (and the enhanced cooperation between the two)[41] and the shift to more proactive, preemptive, predictive and pre-crime patterns. This shift represents one of the main trends emerging in policing and criminology.[42] 'Pre-crime' implies that the "possibility of forestalling risks competes with and even takes precedence over responding to wrongs done".[43] Accordingly, pre-crime societies are based on calculation, risk, uncertainty, surveillance, precaution, prudentialism, moral hazard, prevention and the pursuit of security.[44] These dynamics are also associated with similar trends in criminal law which can be summarised as follows: greater us of diversion; greater use of fixed penalties; greater use of summary trials; greater us of hybrid civil-criminal processes; greater use of strict liability; greater incentives to plea guilty; and greater use of preventive orders.[45] Surveillance is influenced by pre-crime patterns and adapts to them but it also contributes to trigger them. On the one hand, surveillance is considered as an effective

[36] Surveillance Studies Network (2006, pp. 13–15).

[37] Feeley and Simon (1992, p. 451).

[38] Feeley and Simon (1992, p. 452).

[39] Feeley and Simon (1992, p. 452).

[40] Cheliotis (2006).

[41] McCulloch and Pickering (2009, pp. 628–645).

[42] Van Brakel and Hert (2011, pp. 163–192, p. 3.); De Goede (2008).

[43] Zedner (2007, pp. 261–281, pp. 262).

[44] Zedner (2003); Hudson (2003); Ericson and Haggerty (1997).

[45] Ashworth and Zedner (2008 pp. 21–51).

antidote to combat and prevent crime. On the other, criminals and delinquents, it is said, can be found anywhere and this legitimises the broad resort to surveillance.[46]

Nowadays surveillance technologies are introduced not only to detect but also to prevent, deter crime and avoid criminal deviance. New Penology reaches these goals by implementing specific surveillance practices such as profiling, dataveillance and biometrics. In a proactive, predictive and pre-crime society every single person is considered as a potential target of surveillance systems and practices. As mentioned earlier, the horizontal gaze of surveillance narrows the gap between the surveillant and the surveilled, and so between criminals and non-criminals. This creates the so-called 'correctional continuum' or 'correctional spectrum' between criminals and the community in which they live.[47] In addition, the blurred difference between criminals and non-criminals is coupled with the disappearance of the ideal line between prisons and the rest of society, so that "it is by no means easy to answer such questions as to where the prison ends and the community begins or just why any deviant is to be found at any particular point".[48] From a criminological point of view, today's surveillance societies appear as an "undifferentiated open space",[49] which conflicts flatly with the idea of *clôture* emphasised by Foucault.[50] In Lianos' words "the boundary between normal and deviant has largely been erased" in contemporary dangerised societies.[51]

6.4 From 'Old Resistance' to 'New Resistance'

Resistance often ensues as a result of the normalisation of surveillance in today's societies. Like surveillance, it is not an offspring of modernity but one of the distinctive features of the modern world.[52] It is not an epiphenomenon of surveillance but "a basic and necessary co-development of surveillance, existing in many forms that often go unrecognised".[53] Resistance operates within certain surveillance schemes for the purpose of causing detrimental effects on surveillance. It is considered as a way to counter or at least mitigate surveillance and its negative side-effects. Resistance includes actions and forms of inaction which have different degrees of intensity. In fact, it may range from passive actions (such as avoidance) to active actions (such

---

[46] Sewell (2006, pp. 934–961).

[47] Cohen (1979, p. 344).

[48] Cohen (1979, p. 344).

[49] Cohen (1979, p. 344).

[50] In fact, Foucault argued that "discipline requires enclosure". Michel Foucault, *Surveiller et punir*, p. 166.

[51] Lianos and Douglas (2000, pp. 261–278). Lianos describes 'dangerization' as "the tendency to perceive and analyse the world through categories of menace. It leads to continuous detection of threats and assessment of adverse probabilities, to the prevalence of defensive perceptions over optimistic ones and o the dominance of fear and anxieties over ambition and desire". Lianos, and Douglas, "Dangerization and the end of deviance", p. 276.

[52] Lyon (2003, p. 161); Misa et al. (2003).

[53] Martin et al. (2009, pp. 231–232, p. 216).

as counter-surveillance). They all consist in differential responses to the application of surveillance practices and are modulated according to the degree and intensity of surveillance at stake. Individuals tend to implement and develop differential forms of acceptance of surveillance. They depend upon several variables which are usually linked to the personal, social and economic status of the surveillance target. Nonetheless, there are "varieties of overt and covert responses to surveillance both within a given form and across forms of surveillance"[54] which result in differential degrees of acceptance and resistance to surveillance. These different attitudes are usually associated with conditions of ignorance, manipulation, deception or seduction[55] which vary across peoples, places and times.

Thus, there is a tight relationship between surveillance and resistance which is framed alongside the pair action-reaction. This idea was clearly illustrated by Foucault who highlighted the dialectical nature of the relationship between surveillance and resistance. In his thinking, resistance rose from the relation of power between the surveillant and the surveilled and was an unavoidable component of this relationship. As he pointed out, "in the relations of power there is necessarily the possibility of resistance, for if there were no possibility of resistance—of violent resistance, of escape, of ruse, of strategies that reverse the situation—there would be no relations of power".[56] Furthermore, he stressed that in order to exercise a relation of power, there must be "at least a certain form of liberty" from the side of the surveillant and the surveilled.[57] As a consequence, liberty is a condition for resistance and resistance does always express a certain form of liberty. However, in the framework of the disciplinary surveillance described by Foucault resistance did not emerge as a concrete option and did not bring the surveilled to perform active moves. In fact, in the framework of the panoptic machine, resistance appeared helpless in the face of surveillance.[58]

The changing nature of today's surveillance is followed by the emergence of a 'new resistance' which is exercised through new and/or innovative forms of resistance. This process is triggered by a more balanced relation of power between the surveillant and the surveilled, as well as by the proliferation of new surveillance technologies and practices. Given that technology is within the surveilled's reach, resistance appears more like a concrete option than a potential one. New resistance has a higher intensity than 'old' resistance and can be expressed through multiple forms. Marx has identified 11 prominent types of response to surveillance, namely: discovery moves; avoidance moves; piggybacking moves; switching moves; distorting moves; blocking moves; masking (identification) moves; breaking moves; refusal moves; cooperative moves; and counter-surveillance moves.[59] The wide array of forms of resistance includes also privacy which in this context becomes "the legal recognition

[54] Marx (2005, p. 377).

[55] Marx (2005, p. 342).

[56] Foucault (1994, pp. 1–20, p. 12).

[57] Ibid.

[58] Simon (2005).

[59] Marx (2003, pp. 369–390, pp. 374–384).

of the resistance or reticence to behaviour steered or induced by power".[60] From a Foucauldian perspective, it is apparent that new surveillance legitimises the exercise of several and wide forms of liberty.

Nonetheless, resistance will keep on evolving in the future alongside new surveillance patterns. As Haggerty et al. argue, new resistance is an "inevitable by-product of the increased level of social monitoring. To the extent that surveillance is perceived to be unjust or stands in the way of desirable ends, more individuals will likely find themselves resisting surveillance in new and innovative ways".[61] Yet, given today's advances in technology, it is reasonable to claim that resistance will be highly dependent on these developments in the future and that forms of counter-resistance will become more manifest. Thus, the relationship between the surveilled and the surveilled will turn out to be even more circular or spiral. In Leistert words, "once one side re-empowers itself by technical measures", the other will find "new ways to gather information or hide better their activities".[62] The following section will throw light on resistance to online surveillance focusing on avoidance and making reference to the 'Facebook case'. This analysis will be useful to show how resistance patterns are evolving nowadays and will emphasise their main challenges, threats and controversial aspects.

6.4.1 Facebook Resistance: Turning Resistance Into Criminalisation?

With more than a billion monthly active users, Facebook is the fastest growing and most used social media nowadays.[63] It is the most popular network which serves as a tool for the presentation of the self. Facebook users insert bits of their lives into the system, posting pictures, videos and may other items, so making them 'visible' to their actual and potential friends. On the one hand, Facebook lets users share information with friends in an easy way. On the other, it is an electronic monitoring system. Apart from all information that can be accessed and monitored by users, each piece of data can be tracked by the makers of Facebook (or by hackers).[64] Furthermore, there are specific functions on Facebook that are considered particularly detrimental to privacy and can be used for surveillance purposes, such as the Like button.[65] Profiling is a common practice on Facebook, considered that users create profiles themselves and make them accessible into the system.

Resistance to Facebook is expressed mainly by avoidance moves. As mentioned above, avoidance is one of the forms of resistance to surveillance. It involves withdrawal and consists in making passive moves in order to avoid confronting

---

[60] Hert and Gutwirth (2006, pp. 61–104).

[61] Haggerty and Ericson p. 21.

[62] Leistert (2012, pp. 441–456).

[63] Facebook (2013).

[64] Andrejevic (2007); Rachel (2011, pp. 111–129).

[65] Roosendaal (2012, pp. 3–19); Roosendaal (2013).

surveillance. In this event the surveilled prefers to elude surveillance ex-ante rather than to deal with its consequences ex-post. He does not make any effort to engage in surveillance. Instead, he is extremely cautious in disclosing information and concerned over leakage. Displacement to times, places and means characterises avoidance moves.[66] The surveilled avoids places and settings in which surveillance is present or may be present and reduces the exposure to surveillance taking decisions that have the least impact on his privacy. In the case of Facebook, resistance results from the choice of having no account, not using the system and/or using it seldom if ever. Thus, this conscious choice denotes the margin of liberty that characterises resistance. However, resistance to Facebook is sometimes misinterpreted and misunderstood. Given the huge number of Facebook users, the mere fact of not appearing on Facebook is sometimes associated with hiding something to the external world. According to this logic, suspicious conducts are linked to not having any Facebook account, having deactivated any prior account or not using Facebook.

A few months ago a piece of news about Facebook avoidance hit the headlines and appeared in several news media such as the German newspaper *Der Tagesspiegel*,[67] the USA magazines *Forbes*[68] and *Time*[69] and the British *The Guardian*[70] and *Daily Mail.*[71] Based on research conducted by the German psychologist Christoph Möller, journalists reported that the fact of not having a Facebook profile or not using Facebook could indicate mental illness or social dangerousness. It could be "the first sign that you are a mass murderer".[72] Initially *Der Tagesspiegel* flagged the news noting that mass murderers like Anders Breivik[73] and James Holmes[74] lacked social media presence and showed Facebook resistance. Although there is not any official scientific study supporting this claim, this case is paradigmatic of how resistance can be misinterpreted and how pre-crime logics may impact on resistance. Even though we are far from giving criminal labels to Facebook abstainers, this mere hypothesis warns us against the risk of turning resistance into criminalisation. In addition, it shows how resistance is modulated alongside the development of new surveillance technologies and practices. Given that resistance is an expression of liberty, its criminalisation is a degeneration of that resistance into crime.

---

[66] Marx (2003, pp. 375–377).

[67] Schulze (2012).

[68] Hill (2012).

[69] White (2012).

[70] Bennett (2012).

[71] "Is not joining Facebook a sign you're a psychopath? Some employers and psycologists say staying away from social media is 'suspicious'", Daily Mail Online, 6 August 2012, http://www.dailymail.co.uk/news/article-2184658/Is-joining-Facebook-sign-youre-psychopath-Some-employers-psychologists-say-suspicious.html (last accessed 1 March 2013).

[72] "Facebook abstainers could be labeled suspicious" *Slashdot*, 29 July 2012, http://tech.slashdot.org/story/12/07/29/1627203/facebook-abstainers-could-be-labeled-suspicious (last accessed 1 March 2013).

[73] Anders Behring Breivik was convicted of mass murderer and terrorism in 2012, further to the 2011 Norway attacks which killed 77 people.

[74] James Eagan Holmes is the suspected perpetrator of a mass shooting that occurred in July 2012 in Colorado, which killed 12 people and injured 58 others.

6.5 Conclusions

Resistance is changing nowadays alongside new surveillance technologies and practices and the New Penology. New surveillance is more horizontal and lateral (and less vertical) than traditional surveillance and a more balanced power exercise characterises the renewed relationship between the surveillant and the surveilled. The changing nature of resistance and surveillance represents an interesting subject matter which deserves in-depth analysis.

Similarly, criminal surveillance is undergoing significant changes nowadays, influenced by new trends in surveillance studies, criminology and policing as well by the raise of New Penology. The gap between criminals and non-criminals is narrowing within a "growing culture of dangerisation, wherein others constitute by default a source of threat, unless one has good reasons to think otherwise".[75] As a consequence, paradigms of social deviance and dangerousness are subject to tensions nowadays. Individuals tend to modulate their forms of resistance to surveillance alongside surveillance developments. Deviance has always been referred to forms of divergence from social and/or moral values which, translated into the legal norms, originated categories and species of criminality. Social dangerousness denoted the possibility of a person to subvert the established legal order or to contravene social and/or moral norms. Although these patterns are still in place and are regularly applied nowadays, they tend to be adjusted to new surveillance practices. As the Facebook case illustrated, deviance seems to be increasingly associated with forms of resistance to surveillance. Still, dangerousness tends to indicate the possibility of a person to avoid or reject surveillance.

It is hard to say if resistance and criminalisation will converge in the future. Despite the inclusiveness of new surveillance, is important to ensure that the surveilled could always be given the possibility exercise forms of resistance to surveillance freely. Nonetheless, it is crystal clear that clustering individuals according to their Facebook affiliation would neither help prevent and counter crime, nor detect social deviance. Finally, the decision not to be subject to surveillance and not to engage in any social media activity should be respected like any other consumer choice.

References

Andrejevic, Mark. 2005. The work of watching one another: Lateral surveillance, risk, and governance. *Surveillance & Society* 2 (4): 479–497.

Andrejevic, Mark. 2007. *iSpy: Surveillance and power in the interactive era*. Lawrence: University Press of Kansas.

Ashworth, Andrew, and Lucia Zedner. 2008. Defending the criminal law: Reflections on the changing character of crime, procedure and sanctions. *Criminal Law and Philosophy* 2 (1): 21–51.

Bauman, Zygmunt, and David Lyon. 2013. *Liquid surveillance. A conversation*. Cambridge Polity Press.

[75] Lianos (2003, p. 421–422).

Bennett, Catherine. 2012. Not on Facebook? What kind of sad sicko are you?, *The Guardian*, 12 August 2012, http://www.guardian.co.uk/commentisfree/2012/aug/12/catherine-bennett-facebook-psycopaths. Accessed 1 March 2013.

Cheliotis, Leonidas K. 2006. How iron is the iron cage of New Penology? The role of human agency in the implementation of criminal justice policy. *Punishment & Society* 8:313.

Cohen, Stanley. 1979. The punitive city: notes on the dispersal of social control. *Contemporary Crisis* 3:344.

De Goede, Marieke. 2008. The politics of preemption and the war on terror in Europe. *European Journal of International Relations*. 14 (1).

De Hert, Paul, and Serge Gutwirth. 2006. Privacy, data protection and law enforcement. Opacity of the individual and transparency of power. In *Privacy and the criminal law*, In Anthony Duff Erik Claes, and Serge Gutwirth. ed. Oxford: Intersentia.

Deleuze, Gilles. 1990/2003. *Pourparlers 1972–1990*, 240–246. Paris: Les éditions de minuit.

Dubrofsky, Rachel E. 2011. Surveillance on reality television and Facebook: From authenticity to flowing data. *Communication Theory* 2 (2): 111–129.

Elmer, Greg. 2012. Panopticon-discipline-control. In Routledge Handbook of Surveillance Studies, ed. David Lyon, Kirstie Ball and Kevin Haggerty, 21–29. Routledge.

Ericson, Richard V., and Kevin D. Haggerty. 1997. *Policing the risk society*. Oxford: Oxford University Press.

Facebook. 2013. *Facebook Newsroom, key facts*. http://newsroom.fb.com/Key-Facts. Accessed 1 March 2013.

Feeley, Malcolm M., and Jonathan Simon. 1992. The New Penology: notes on the emerging strategy of corrections and its implications. *Criminology* 30 (4): 451.

Feeley, Malcolm, and Jonathan Simon. 1994. Actuarial Justice: The Emerging New Criminal Law. In *The Futures of Criminology*, In D. Nelkin. ed. London: Sage.

Foucault, Michel. 1975. *Surveiller et Punir. Naissance de la prison*, Gallimard.

Foucault, Michael. 1994. The ethic of care of the self as a practice of freedom. In *The final Foucault*, In James Bernauer, and David Rasmussen, ed. 1–20. Cambridge: MIT Press.

Haggerty, Kevin D. 2006. Tear down the walls: on demolishing the panopticon. In *Theorizing surveillance. The panopticon and beyond*, In Davind Lyon, ed. 23–45. USA: Willan Publishing.

Haggerty, Kevin D., and Minas Samatas. 2010. *Surveillance and Democracy*. New York: Routledge-Cavendish Publishing.

Haggerty, Kevin D., and Richard V. Ericson. *The new politics of surveillance and visibility*.

Hill, Kashmir. 2012. Beware, tech abandoners. People without Facebook accounts are 'suspicious', *Forbes*, 8 June 2012, http://www.forbes.com/sites/kashmirhill/2012/08/06/beware-tech-abandoners-people-without-facebook-accounts-are-suspicious/. Accessed 1 March 2013.

Hudson, Barbara. 2003. *Justice in the risk society*. London: Sage.

Kanashiro, Marta Mourão. 2008. Surveillance cameras in Brazil: Exclusion, mobility, regulation, and the new meanings of security. *Surveillance & Society* 5 (3): 270–289.

Leistert, Oliver. 2012. Resistance against cyber-surveillance within social movements and how surveillance adapts. *Surveillance & Society* 9 (4): 441–456.

Lianos, Michalis, and Mary Douglas. 2000. Dangerization and the end of deviance. *British Journal of Criminology* 40 (2): 261–278.

Lianos, Michalis. 2003. Social control after Foucault. *Surveillance & Society* 1 (3): 412–430.

Lyon, David. 1994. *The electronic eye. The rise of surveillance society*. Cambridge: Cambridge Polity Press.

Lyon, David. 2001a. Facing the future. Seeking ethics for everyday surveillance. *Ethics and Information Technology* 3:171–181.

Lyon, David. 2001b. *Surveillance society. Monitoring everyday life, (Issues in Society)*. Buckingham: Open University Press.

Lyon, David. 2003a. *Surveillance after September 11*. Polity Press.

Lyon, David. 2003b. *Surveillance as social sorting. Privacy, risk, and digital discrimination*. New York: Routledge.

Lyon, David. 2007. *Surveillance Studies, An overview*. Polity Press.

Lyon, David. 2010. Liquid surveillance: The contribution of Zygmunt Bauman to surveillance studies. *International Political Sociology* 4:325–338.

Lyon, David, Kirstie Ball, and Kevin Haggerty, eds. 2012. *Routledge handbook of surveillance studies*. Routledge.

Mann, Steve, Jason Nolan, and Barry Wellman. 2003. Sousveillance: Inventing and using wereable computing devices for data collection in surveillance environments. *Surveillance & Society* 1 (3): 331–355.

Martin, Aaron K., Rosamunde E. Van Brakel, and Daniel J. Bernhard. 2009. Understanding resistance to digital surveillance. Towards a multi-disciplinary, multi-actor framework. *Surveillance & Society* 6 (3): 231–232.

Marx, Gary T. 1988. *Undercover: Police surveillance in America*. Barkley: University Press.

Marx, Gary T. 2003. A tack in the shoe: neutralizing and resisting the new surveillance. *Journal of Social Issues* 59 (2): 369–390.

Marx, Gary T. 2004. What's new about the "new surveillance"?: Classifying for change and continuity. *Knowledge, Technology and Policy* 17 (1): 18–37.

Marx, Gary T. 2005. Seeing hazily (but not darkly) through the lens: some recent empirical studies of surveillance technologies. *Law & Social Enquiry* 30 (2): 377 (Spring).

Mathiesen, Thomas. 1997. The Viewer society: Michael Foucault's "Panopticon" resivited. *Theoretical Criminology* 1 (2): 215–234.

Mathiesen, Thomas. 1999. *On globalisation of control: Towards an integrated surveillance system in Europe*. A Statewatch Publication.

McCulloch, Jude, and Sharon Pickering. 2009. Pre-crime and counter-terrorism: Imagining future crime in the 'War on Terror'. *British Journal of Criminology* 49:628–645.

Melgaço, Lucas. 2001. The injustices of urban securization in the Brazilian city of Campinas. *Spatial Justice* 5. available at http://jssj.org/media/dossier_focus_vt7.pdf. Accessed 1 Mar 2013.

Misa, Thomas, Philip Brey, and Andrew Feenberg, eds. 2003. *Modernity and Technology*. Cambridge: The MIT Press.

Monahan, Torin. 2008. Editorial: surveillance and inequality (eds. Torin Monahan and Jill Fisher). *Surveillance & Society* 5 (3): 217–226.

Monahan, Torin. 2010. Surveillance as governance. Social inequality and the pursuit of democratic surveillance. In *Surveillance and Democracy*, In Kevin D. Haggerty, and Minas Samatas, ed. 91–110. New York: Routledge-Cavendish Publishing.

Norris, Clive, and Armstrong Gary. 1999. *The maximum surveillance society. The rise of CCTV, 108–116*. Oxford: Berg.

Oxford Dictionaries. 2013. Surveillance. http://oxforddictionaries.com/definition/english/surveillance?q=surveillance. Accessed 1 March 2013.

Poster, Mark. 1990. *The mode of information: poststructuralism and social context*. Chicago: University of Chicago Press.

Reeves, Joshua. 2012. If you see something, say something: Lateral surveillance and the use of responsibility. *Surveillance & Society* 10 (3/4): 235–248.

Roosendaal, Arnold. 2012. We are all connected to Facebook ... by Facebook. In *European Data Protection in Good Health?*, ed. Serge Gutwirth, Ronald Leenes, Paul De Hert and Yves Poullet, 3–19. Dordrecht: Springer.

Roosendaal, Arnold. 2013. Facebook tracks and traces everyone: Like this! Tilburg Law School, Research Paper No. 03/2011. http://papers.ssrn.com/sol3/papers.cfm?abstract_id=1717563. Accessed 1 March 2013.

Schulze, Katrin. 2012. Machen sich Facebook-verweigerer verdächtig? *Der Tagesspiegel*, 24 July 2012. http://www.tagesspiegel.de/weltspiegel/nach-dem-attentat-von-denver-kein-facebook-profil-kein-job-angebot/6911648-2.html. Accessed 1 March 2013.

Sewell, Graham. 2006. Coercion versus care: Using irony to make sense of organizational surveillance. *Academy of Management Review* 31 (4): 934–961.

Simon, Bart. 2005. The return of Panopticism: Supervision, subjection and the new surveillance. *Surveillance & Society* 3 (1): 3.

Surveillance Studies Network. 2006. *A Report on the Surveillance Society*, Information Commissioner's Office. Wilmslow: UK.

Torpey, John. 2007. Through thick and thin: surveillance after 9/11. *Contemporary Sociology* 36 (2): 116–119.

Van Brakel, Rosamunde, and Paul De Hert. 2011. Policing, surveillance and law in a pre-crime society: Understanding the consequences of technology based strategies. *Journal of Police Studies* 20 (3): 163–192.

White, Martha C. 2012. Does not having a Facebook page make you 'suspicious' to employers?, Time, 8 August 2012. http://business.time.com/2012/08/08/does-not-having-a-facebook-page-make-you-suspicious-to-employers/. Accessed 1 March 2013.

Zedner, Lucia. 2003. *Pre-crime and post-crime criminology?, ibid. Barbara Hudson, Justice in the risk society*. London: Sage.

Zedner, Lucia. 2007. Pre-crime and post-crime criminology? *Theoretical Criminology* 11 (2): 261–281.

Chapter 7
Surveillance and Criminal Investigation: Blurring of Thresholds and Boundaries in the Criminal Justice System?

John A. E. Vervaele

7.1 Introduction

The classic objective of criminal justice is an ex-post and reactive determination of guilt for criminal behaviour and imposing the related criminal punishment. The crime control function of criminal justice (the sword) has to go hand in hand with notions of due process and fair trial (the shield) in accordance with the rule of law and related constitutional and human rights standards. As long as the defendant is a suspect and is not convicted, then he/she has to be protected by the presumption of innocence (*presumptio innocentiae, in dubio pro reo principle*). This presumption is a long-standing principle which lies at the heart of the criminal justice system and can be traced back to the Enlightenment, but is also enshrined in human rights conventions.[1] In the European context it is protected as one of the fair trial rights under article 6(2) European Convention on Human Rights (ECHR) and repeated verbatim by article 48(1) of the EU Charter of Fundamental Rights:

"everyone charged with a criminal offence shall be presumed innocent until proved guilty according to law".

The wording of this right is somewhat misleading, as the combination of "charged" and "proved guilty" could be understood as applying to the trial stage. However, thanks to the interpretation by the European Court of Human Rights (ECtHR) we know that article 6(2) does apply during the procedure as a whole, which means from the formal opening of a criminal judicial investigation onwards or from the first investigative acts from which a person can deduce that he/she is suspected of having committed an offence.[2] When the authorities use investigative measures within the framework of criminal proceedings they have to comply with the standards of fair trial, including respect for the presumption of innocence.

---

[1] Ashworth (2006).

[2] *Adolf v. Austria,* application no. 8269/78, 26 March 1982, para. 30.

J. A. E. Vervaele (✉)
Willem Pompe Institute for Criminal Law and Criminology, Utrecht University,
Boothstraat 6, 3512 BW Utrecht, Netherlands
e-mail: J.A.E.Vervaele@uu.nl

S. Gutwirth et al. (eds.), *Reloading Data Protection,*
DOI 10.1007/978-94-007-7540-4_7, © Springer Science+Business Media Dordrecht 2014

Surveillance is increasingly used as an investigative technique and implies different modus operandi: behavioural surveillance, communication surveillance, data surveillance, location and tracing, body surveillance, attitude surveillance or combinations thereof. The constantly renewal of technical devices and the digitalisation of society result in constantly new modus operandi. Surveillance technologies such as satellite or wireless bugging or remote computer surveillance are quite recent. Although the mentioned technologies are quite recent new trends can already be identified: through mobile devices surveillance is delocalizing and is becoming global. Cloud surveillance and cloud tapping are the most recent examples.

When looking for a definition of surveillance, we encounter very different ones From a classic view of criminal procedure surveillance is defined as electronic monitoring:

> Electronic monitoring is a general term referring to forms of surveillance with which to monitor the location, movement and specific behaviour of persons in the framework of the criminal justice process.[3]

Researchers dealing with surveillance, such as for instance within the framework of the 7th EU framework programme (Surveille/IRISS), have deliberately opted for a wider definition of surveillance:

> Targeted or systematic monitoring of persons, places, items, means of transport or flows of information, in order to detect specific, usually criminal, forms of conduct, or other hazards, and enable, typically, a preventive or reactive response or the collection of data for preparing such a response in the future.[4]

This definition includes anticipative ex-ante (judicial) investigations,[5] thus investigations before the commission or the preparation of an offence in order to deal with situations of risks and threats to security. In some countries these investigations are submitted to ex ante judicial review, in others they are not. The "war" against the drugs trade, organized crime and terrorism justifies the use of intrusive measures, also in situations in which there is no suspicion that an offence has been committed and thus there can legally be no suspect yet In situations of ex-ante risk assessment or threat assessment the only reference is dangerousness (relating to behaviour and/or the mind). The pre-emptive investigation and surveillance is part of a new security paradigm that redefines not only the classic concepts of criminal justice, but also the applicable human rights standards.

The human rights dimension of these surveillance measures is mostly related to article 8 ECHR, as in many cases they interfere with privacy, the protection of the home and, family life and data protection. Secret surveillance in private places is in many states considered to be a highly intrusive measure, but systematic or targeted surveillance in public places can also interfere with privacy. Privacy protection under

---

[3] Council of Europe, European Committee on Crime Problems (CDPC), Council for Penological Cooperation (PCCP), Scope and Definitions—Electronic Monitoring, PC-CP (2012) 7 rev 2.

[4] http://cordis.europa.eu/search/index.cfm?fuseaction=proj.document&PJ_LANG=EN&PJ_RCN=12717722&pid=0&q=A6F3555B2B910072A30F6A42438C154D&type=adv.

[5] McCulloch and Pickering (2009); Brakel and De Hert (2011).

human rights standards applies to all interferences, whether within the framework of criminal justice or not.

The way in which the presumption of innocence comes legally into play is however very different, as the presumption is only triggered once a person is suspected of having committed a criminal offence or other irregularities for which punitive penalties can be imposed.[6] Legally speaking, there can be no presumption of innocence without a suspect. Moreover, even when we focus on the surveillance of suspects, there is very little case law on the presumption of innocence in relation to surveillance. The use of surveillance measures as (coercive) investigative techniques is as such not a violation of this presumption. The status of suspect include that there must be reasonable grounds of suspicion against him/her. The use of surveillance techniques as investigative measure is only possible if certain thresholds are met and these thresholds include standards of suspicion. Even if surveillance is used for identification (naming) and for profiling, the results are rarely made public and do not result in the voicing of suspicion (shaming). If the measures are challenged as being disproportionate the test is mostly related to the infringement of privacy. Surveillance can of course stigmatise persons and violate their dignity. This does not mean, however, that their right to the presumption of innocence is legally infringed. Recently the ECtHR has seemed to widen the concept and to build a link between article 6(2) and article 8 of the ECHR in the case of *S. and Marper v. United Kingdom*[7], dealing with the indefinite retention of applicants' fingerprints, cellular samples and DNA profiles after their acquittal:

> 122. Of particular concern in the present context is the risk of stigmatisation, stemming from the fact that persons in the position of the applicants, who have not been convicted of any offence and are entitled to the presumption of innocence, are treated in the same way as convicted persons. In this respect, the Court must bear in mind that the right of every person under the Convention to be presumed innocent includes the general rule that no suspicion regarding an accused's innocence may be voiced after his acquittal (see Asan Rushiti v. Austria, no. 28389/95, § 31, 21 March 2000, with further references). It is true that the retention of the applicants' private data cannot be equated with the voicing of suspicions. Nonetheless, their perception that they are not being treated as innocent is heightened by the fact that their data are retained indefinitely in the same way as the data of convicted persons, while the data of those who have never been suspected of an offence are required to be destroyed.

This is an interesting reasoning in the framework of article 8, but not an *expressis verbis* admission of a violation of the presumption of innocence in this context. The legal reasoning could of course also be applied to the ex-ante situations in which pre-emptive surveillance results in disproportionate storage and processing of data. It remains to be seen if and to which extent the ECtHR is willing to expand the presumption of innocence outside the framework of article 6(2) in that direction. For the moment, there is no case law in that direction and for that reason not much to say about it, at least not from a legal point of view. Moreover, it is more likely

[6] I am referring to the Engel criteria as elaborated in ECtHR 21 February, 1984, *Óztürk v. the Federal Republic of Germany*, no. 8544/79.

[7] Bellanova and De Hert (2009).

that the results of ex-ante surveillance could be addressed under human rights standards when dealing with the principle of collection limitation and the principle of purpose specification under article 8 than under a widened concept of presumption of innocence. Interesting might be also the application of article 14 ECHR, since there can be an un unjustified differentiation between those innocent who have been suspected of a crime during their lifetime, and those who have not, an aspect that was also invoked in *S. and Marper v. United Kingdom,* but not further elaboration on the discrimination after having established a violation of Article 8.

Anyway, investigative surveillance is an important issue from the human rights dimension of fair trial. Considering the secret character of many surveillance measures (and their modus operandi), many human rights cases deal with restricted access to the file, the limited disclosure of evidence between parties, ex-parte proceedings in the pre-trial or trial phase and anonymous witnesses. The aim of my contribution is however not to elaborate on surveillance and fair trial, as there is abundant doctrine[8] and case law thereon, but on the questions to which extent and by which conceptual changes surveillance has invalidated the classic thresholds and boundaries of criminal justice and the related presumption of innocence. The conceptual changes are strongly related to the information society and to transformations in the criminal justice system under the security paradigm.

7.2 Post-industrial Information Society and Transformations in the Criminal Justice System[9]

7.2.1 Post-industrial Information Society

The processes of globalization have been combined in the past decades with the transformation of our societies into post-industrial information societies, by which our social behaviour and social structure have been fully reshaped. A single information society concept does not exist. Scientists are struggling about definitions and values of the concept and focus on economic, technical, sociological and cultural patterns. Postmodern society is often characterized as an information society, because of the widely spread availability and usage of Information and Communication Technology (ICT). The most common definition of information society indeed emphasizes the technological innovation. Information processing, storage and transmission have led to the application of ICT, and related biotechnology and nanotechnology, in virtually all corners of society. The information society is a post-industrial society in which information and knowledge are key-resources and are playing a pivotal role.[10] However, information societies are not solely defined by the technological infrastructure in place, but rather as multidimensional phenomena. Any information society is a

[8] De Hert (2012); Gutwirth (2002); Gutwirth et al. (2013).

[9] IRISS (2013)

[10] Bell (1976).

complex web, not only of technological infrastructure, but also as an economic structure, a pattern of social relations, organizational patterns, and other facets of social organization. Therefore, it is important to focus not only on the technological side, but also on the social attributes of the information society, which include the social impact of the information revolution on social organizations, such as the criminal justice system.

Moreover, the postmodern age of information technology transforms the content, accessibility and utilization of information and knowledge in the social organizations, including the criminal justice system. The relationship between knowledge and order has fundamentally changed. The transformation of communications into instantaneous information-making technology has changed the way society values knowledge. In this rapidly changing age, the structure of traditional authority is being undermined and replaced by an alternative method of societal control. The emergence of a new technological paradigm based on ICT has resulted in a network society,[11] in which the key social structures and activities are organized around electronically processed information networks. There is an even deeper transformation of political institutions in the network society: the rise of a new form of state (network state) that gradually replaces the nation-states of the industrial era. In this rapidly changing age, the structure of traditional authority is being undermined and replaced by an alternative method of societal control (surveillance society). The transition from the nation-state to the network state is an organizational and political process prompted by the transformation of political management, representation and domination in the conditions of the network society. All these transformations require the diffusion of interactive, multi-layered networking as the organizational form of the public sector. Information and knowledge (Information Power) are key-resources of the information society, affecting the social and political structure of society and state[12] and affecting the function, structure and content of the criminal justice system. The increased possibilities to gather information, to store and process the data have substantially changed the way in which law enforcement is designed and functions.[13]

7.2.2 *Transformations in the Criminal Justice System*

The classical rationale for the use of criminal justice (starting with the primary criminalization by the definition of offences), based upon *ultimum remedium*, strict conditions of harmful conduct that violates legally protected interests and concepts derived from the Enlightenment and Kantian philosophy, has been replaced in the past decades by a globalizing criminal policy concept, translated into criminal policy paradigms: combat/war against drugs, combat/war against organized crime, combat/war against terrorism. I call them paradigms, because they function as a frame of

[11] Castells (2000).

[12] Lyon (1994).

[13] Lyon (2007) and Brown (2006).

reference for the perception of reality and thus for the definition of social constructs as crime, danger, risk and insecurity. These criminal policy paradigms have been used both at the domestic and at the international level in order to justify substantial changes in the relation between state-society and criminal justice and within the criminal justice system itself.

Although there has been a substantial shift (from drugs to organized crime and to terrorism)[14], the three paradigms have transformed, through a common and cumulative security-orientated approach (security paradigm) the objectives, nature and instruments of the criminal justice system. The objective of the criminal justice system has changed from punishment of guilty perpetrators of committed offences (with general and special preventive aims, including rehabilitation) towards a broader field of social control of danger and risk.[15] The net widening and function creep has affected general criminal law, special criminal law (the definition of the offences), criminal procedure and mutual legal assistance (MLA) in criminal matters. The substantive application has been widened in order to include preparatory acts and incrimination of criminal organizations and terrorist organizations (or conspiracy variants of it).[16] As a result, the commission of criminal conduct by a suspect is no longer the triggering threshold for the *ius puniendi* of the state. The threat of organized crime or terrorist crime by setting up organizations (with a very low threshold definition) is sufficient for criminalization. The criminalization of apology of terrorism or other apologies (xenophobia) demonstrates a similar trend.[17] Such offenses concern the criminalization of the mind (and may touch upon the freedom of expression) of a person, instead of the criminalization of a criminal act, based upon conduct. By redefining the objective of criminal justice, its very nature has been converted. The greater the risk or the danger, which is based on a social construction and certainly not on empirical facts, the lower the threshold for using the *ius puniendi,* which means that criminal law turns into security law. Security law is not so much based on a legal definition of suspect and criminal conduct, linked to serious harm to a legal interest, but is based upon a pre-set definition of an enemy[18] that is associated with risk, danger and insecurity. The security approach in criminal law has led to an expansion of substantive criminal law (general part and special part) beyond the traditional boundaries and limits as defined by the Enlightenment. The growth of modern surveillance technologies have facilitated this shift, as pre-emptive identification and profiling of potential perpetrators or potential dangerousness has become

[14] van Duyne (1996); Fijnaut et al. (2004).

[15] In continental theories of criminal law, a basic distinction is made between the effects of punishment on the man being punished, individual prevention or special prevention and the effects of punishment upon the members of society by general prevention. The characteristics of special prevention are termed: deterrence, reformation and incapacitation. General prevention, on the other hand, may be described as the restraining influences emanating from the criminal law and the legal machinery. See: Andenaes (1965–1965).

[16] Pelser (2008).

[17] de la Cuesta (2007), http://scholar.google.nl/scholar?start=20&q=apology+of+terrorism&hl=en &as_sdt=0,5 and van Noorloos (2011).

[18] Jakobs (2004).

possible through information and intelligence[19] obtained by large-scale surveillance operations, and high-tech storing and processing. Surveillance is however not only a net widening device and technique but also a new function and objective of criminal justice: from punishment to social control.[20]

The transformation of the criminal justice system, especially in the era of counter-terrorism, has had even more far-going consequences, especially for the field of criminal procedure.[21] Proactive criminal investigation includes the situation in which there is not yet any reasonable suspicion that a crime has been committed, is about to be committed or that specific preparatory acts have taken place and in which, of course, there can be no suspect(s) legally speaking. The objective of proactive investigations is to reveal the organizational aspects in order to prevent the preparation or commission of a serious crime and to enable the initiation of criminal investigation against the organization and/or its members. This use of coercive measures for crime prevention can be realised by intelligence agencies, police authorities or judicial authorities. When doing so, they belong to the intelligence community, even if they are normally authorities belonging to the law enforcement community. In that time frame they might collect information and use certain coercive measures of criminal procedure in order to prevent the preparation or commission of the crime. In this area of criminal law without suspects we see a new combination between proactive or anticipative enforcement and coercive investigation (*Vorbeugende Verbrechensbekämpfung, Vorfeldaufklärung* and *Vorermittlung*).[28] These function creep has affected in the criminal justice system:

a. the type of players/authorities;
b. their powers and investigation techniques (the sword dimension);
c. the safeguards and constitutional and human rights to be respected (the shield dimension).

7.2.2.1 Redefinition of Players (Authorities)

In the first place, traditionally, criminal investigation is supervised by judicial authorities and coercive measures are authorized and/or are executed by members of the judiciary (investigating judges or pre-trial judges or trial judges). In many countries we can see a shift in the pre-trial phase from judicial investigation to prosecutorial and police investigation. We can clearly speak of a reshuffling of responsibilities in the law enforcement community. Magistrates are less and less involved in the pre-trial phase as such; there is a clear shift to the executive or to semi-executive branches of state power.[22]

Secondly, there is not only a shift between the classic players; new actors, such as administrative enforcement agencies also play an increasing role in the field of

[19] Bureau of Justice Assistance (2005).

[20] Lianos and Douglas (2000).

[21] For a more elaborated version, see Vervaele (2009).

[22] Ost and van de Kerchove (2002).

fighting serious crime. The intelligence community is also gaining ground in the criminal justice system, both as specialized police units dealing with police intelligence and as security agencies. These intelligence entities are responsible as the forerunners of police and intelligence-led investigations, and in some countries they have even obtained coercive and/or judicial competence. Furthermore, classic law enforcement agencies convert into intelligence agencies and change their culture and behaviour. This shift goes hand in hand with the increase of surveillance, as many of these administrative agencies have specialised surveillance tasks (f.i. the intelligence community) or are specialised in storage and processing of data surveillance. The Financial Intelligence Unit's (FIU's), dealing with money laundering and Terrorist Finance Tracking Programs (TFTP) is an excellent example. Contemporary financial intelligence[23] consist mostly of a set of surveillance measures applied by law enforcement agencies in the fight against financial crime and terrorism. All current strategies for combating terrorism and financial crime include financial measures increasing the surveillance of capital movements.

Thirdly, many countries have increased the use of private service providers (telecom operators, business operators, financial service providers) and professions with information privileges (such as lawyers and journalists) as gatekeepers and as the long-arm collectors of enforcement information. In these move towards privatisation of law enforcement, journalistic and legal privileges are no longer safe havens and key players in the private information society (producers, service providers, key consumers) are endorsed with law enforcement obligations. Data retention and data and communication surveillance by private players has become a key tool of criminal law enforcement.[24]

7.2.2.2 Redefinition of Players' Competences and Techniques (the Sword)

Firstly, the information society has substantially changed the ways in which law enforcement authorities can obtain information and evidence. The building up of information positions is as important as the use of investigative powers. This means that judicial authorities (police, prosecutors, administrative agencies with judicial powers, investigating magistrates) have own specialised databases at their disposal, but also that they have access to huge amounts of data in all types of public and private databases. The storage of these data is steered by intelligence led policing and by data retention obligations for public and private players. The storage, processing and use of these data have substantially changed in the last decades. Law enforcement authorities are not only checking certain facts, but are elaborating techniques of profiling in order to steer their investigation work.

Second, in most countries the paradigms of the drugs trade, organized crime and terrorism are not only used to redefine investigative, coercive instruments, but also to introduce new special investigative techniques, such as wiretapping, infiltration

[23] Biersteker and Eckert (2007).

[24] De Busser (2009) and de Busser (2010).

and surveillance, which can only be applied to investigate serious crimes. The result is a set of coercive measures with a double use (for serious and less serious offences) and a set of coercive measures with a single use for certain serious crimes.

Thirdly, in many countries the classic measures dealing with securing evidence and the confiscation of dangerous instruments or products in relation to crime have become an autonomous field of security measures concerning goods and persons (e.g., seizure and confiscation, detention orders and security orders). Related to that, investigations into the financial flows from the drugs trade, organized crime (financing, money laundering) and terrorism (financing) have been converted from a classic investigation for gathering evidence into an autonomous financial investigation, dealing with extensive seizure and confiscation of the proceeds of crime (asset recovery) and/or into autonomous financial surveillance and investigations into the financing of serious crime.

Fourthly, the triggering mechanisms or minimum thresholds for the use of coercive measures to combat serious crimes are changing. Criminal investigation no longer starts with a reasonable suspicion that a crime or an offence has been committed or attempted, or with a reasonable suspicion that a preparatory act for committing a serious crime has been committed or attempted. Investigative techniques and coercive measures are also used in a proactive or anticipative way to investigate, *anti-delictum,* the existence and behaviour of potentially dangerous persons and organizations in order to prevent serious crimes or dangerousness. The conversion from a reactive punishment of crime into a proactive prevention of crime has far-reaching consequences. The distinction between police investigation and judicial investigation is under pressure. Coercive proactive enforcement becomes important for serious crimes. The intelligence community becomes a main actor in the law enforcement field. Preventive criminal law is not about suspects and suspicion, but about information gathering (information and criminal intelligence investigation) and procedures of exclusion against potentially dangerous persons. The criminal justice system is increasingly used as an instrument to regulate the present and the future and not to punish for behaviour in the past. The criminal process is becoming a procedure in which the pre-trial investigation is not about truth-finding related to committed crime, but about construction and de-construction of social dangerousness.

Fifthly, the sword of criminal justice has changed substantially by the use of digital-led investigation (online criminal searches, the monitoring of data flow, data processing) and the use of advanced technology in judicial investigations (digital surveillance, detection devices, etc.). Information-led investigation replaces mere suspicions. The expansion of the judicial investigation into a proactive investigation and the increasing overlap between the law enforcement community and the intelligence community has been further increased by the technological developments in investigative devices: the sword of technology with far-reaching eyes and razor-sharp edges. Thanks to new technology, the methods of surveillance for communication, the physical surveillance of persons and their movements and activities and for transactional surveillance (of their services) have changed dramatically. Technology has completely changed not only the behaviour of citizens, but also, through the use of wiretapping, video surveillance, tracking devices, detection devices and see-through devices, data mining, remote digital searches, Trojan horses, and so forth, the environment of enforcement and proactive enforcement.[25]

7.2.2.3 Redefining the Safeguards and the Constitutional and Human Rights Dimension (the Shield)

In many countries, the legislator considered some procedural guarantees as burdens to the efficiency of serious crime prevention, serious crime investigation and serious crime prosecution. First of all, the use of existing instruments such as search and seizure and police detention is submitted to other parameters for serious offences than for less serious offences. Moreover, judicial authorization (in the form of warrants) is weakened or abolished for some coercive measures (warrantless coercive measures). The role of the defence and of the judge as procedural guarantees is reduced. This means in practice that the police and prosecutors have more autonomy and are subjected to diminished supervision by the judiciary on their investigative work. We could speak of a two-fold expansion of the existing coercive measures: a general expansion of the powers of the police and prosecutors with relaxed safeguards, which trend is even stronger for the investigation of serious crimes because of the presence of a security interest. Generally speaking, we can say that the seriousness of the crimes under the aforementioned paradigms is used to justify raising the sword and lowering the shield. In many countries, in the case of serious crimes, the relationship between the intrusiveness of the measures and judicial control has changed: the greater the security interest, the less the judicial control and the procedural safeguards.

Secondly, by lowering the thresholds (reasonable suspicion or serious indications to simple indications, reversed burden of proof, legal presumptions of guilt) for triggering the criminal investigation and for imposing coercive measures, the presumption of innocence is undermined and replaced by objective security measures. The shields protecting the citizen against the *ius puniendi* of the state are put at the back of the stage in the theatre of criminal justice. This has, of course, direct consequences for habeas corpus, habeas data, fair trial rights, redefinition of evidence rules, public proceedings, etcetera.

In the third place, in many countries there is also a need to secure the functioning of the criminal justice system and its players. The protection of witnesses has also been converted into the protection of anonymous witnesses, including those from the police authorities and intelligence agencies involved in infiltration. The criminal justice system is increasingly shielding its surveillance agents against the defence through *ex parte* proceedings, forms of secret evidence-gathering and the use of secret evidence in the pre-trial and trial setting.

Fourthly, several countries have amended their mandate for intelligence forces and their powers. Their investigative competences now include coercive powers, parallel to the ones in the Code of Criminal procedure, and their objective also includes

[25] Casey (2011) and Pradillo (2011).

the prevention of serious crime, as this constitutes a threat to national security. In some countries they need the authorisation for the use of these powers by a public prosecutor or by the executive branch of government. *De facto*, the intelligence community is using judicial coercive powers without being a judicial authority and without the guarantees of some form of judicial warrant and/or judicial supervision. We can see an overlapping competence between the intelligence agencies and the police authorities acting as intelligence community in the preliminary proactive or anticipative investigation. Intelligence led enforcement has blurred the conceptual boundaries and thresholds.

In the fifth place, we see an increasing use of intelligence in the criminal justice system. As long as it is used as steering information or as data sharing or as triggering information for the opening of a judicial investigation it does not affect or infect the criminal justice system. However, when intelligence is used as triggering information, establishing probable cause for using coercive measures, or as evidence in criminal proceedings it does infect the classic rules of fair trial and equality of arms, as most of this type of intelligence can only be used in shielded and secret *in camera* and *ex parte* proceedings.

It goes without saying that all these transformations affect the position of the defence lawyer in the criminal process. His legal privilege is under pressure. In certain countries, when dealing with secret evidence in cases of organized crime and terrorism, the defence lawyer has no full access to the file (limited disclosure) or only special security screened bar lawyers can act on behalf of the suspect. The defence lawyer's role and his duties and responsibilities are redefined.

The transformation have resulted in a clear expansion of the punitive state,[26] thereby disfavouring the rule of law. The focus on public security and preventive coercive investigation is clearly undermining the criminal justice system and its balances between the sword and the shield. Administrative and preventive forms of punitive justice are expanding. The result is also that the equilibrium between the three branches of the trias politica is under great pressure in favour of the executive.

In the majority of European countries transformations have resulted in a distinction within the ordinary criminal justice system between a criminal procedural regime for serious offences and one for 'petty' offences or in special legislation replacing substantial parts of the ordinary criminal justice system. In fact, criminal procedure is no longer organized in line with the general part of criminal law, but in line with the dual use in the special part of criminal law. The exceptional features for organized crime and terrorism changed from the exception into the main and common procedure for serious crimes, for which reason we can speak of the normalization of the exception.

---

[26] Frost (2006).

7.3 Conclusion

We are living in a setting of time in which many reforms of the criminal justice system are the result of a political instrumentalisation and mediatisation of crime and the fear of crime. These reforms are being justified by the criminal policy paradigms of combating drugs, organized crime and terrorism. The result is that the *ius puniendi* of the state (being one of the most repressive interferences in liberty on behalf of the state), is being instrumentalised and put at service of danger and risk management. When prevention of dangerousness becomes the triggering mechanism for the use of very intrusive investigative techniques, as secrete surveillance or systematic targeted surveillance and criminal punishment, the criminal justice system is risking perverting into a security system. These developments result in a substantial expansion of the criminal justice system, through substantive and procedural criminal law, and thus of expanded interference with the liberty of citizens. The expansion of criminal justice goes hand in hand with the erosion of its basic principles (*nullum crimen sine iniuria, nulla poena sine culpa, ultimum remedium*, fair trial, presumption of innocence, etcetera). At the same time, criminal repression becomes a passe partout formula for solving societal problems. The expectations about the problem-solving capacity of criminal justice are however in sharp contrast with the real performance. The expansion of criminal justice is very real in terms of social control, but very symbolic in terms of societal problem solving capacity.

The criminal policy paradigms (drugs, organized crime, and terrorism) are used as political justifications at the domestic, European and International levels. We can certainly not conclude that the European and/or international dimensions have unilaterally caused these shifts. The three levels are strongly interacting under the same paradigms and aiming at integrating further the security approach into the criminal justice system. It is clear that the basic concepts of modern criminal justice, as elaborated in the Enlightenment, and further substantiated in codifications, constitutions and human rights instruments, have come under strong pressure by the security paradigm. This is reflected by shifting responsibilities within the criminal justice system (between the public authorities and between the parties), but also by the expansion of the criminal justice system itself.

This net widening and function creeping towards the proactive prevention and pre-emptive use of coercive measures has been laid down in new concepts such as intelligence-led policing and information-led policing. Surveillance is a key tool of these forms of policing, but one that effects the whole conceptual design of criminal justice. So we could confidently say that surveillance-led enforcing has become a dominant feature of criminal justice and security law. Surveillance as a coercive measure is used in a pre-suspect setting, in which the legal presumption of innocence can play no role at all. Given the potential intrusive impact of surveillance and the coercive character of some surveillance techniques, also in the pre-emptive setting, it is logical to build in guarantees against disproportionate infringements of privacy, human dignity and the presumption of innocence. The latter could then be related not to the commission of offences, but also to the definition of dangerousness.

When used in a suspect setting evidential thresholds are lowered to justify the measure and prior authorization by the judiciary is either delegated to the prosecutor or is disappearing. Due to the secrecy of the modus operandi of surveillance techniques the basics of natural justice, such as equality of arms, disclosure between the parties and open confrontation are being adapted to shield surveillance agents, their modus operandi and part of the evidence. Investigative surveillance does contribute to inquisitorial secret proceedings. Equality of arms and fair trial are not absolute human rights. Legitimate aims (such as the protection of security) can justify restrictions. However, there is a bottom line for fairness: the procedure must be fair as a whole. This means that the defendant must be able to prepare his defence and challenge the evidence at trial. It also means that the judiciary must have full access to the file in order to balance the rights of the defence and security. Without judicial supervision (justiciability) surveillance is a potential undermining factor of the thresholds and guarantees in the criminal justice system.

It needs to be acknowledged that these basic concepts of criminal justice have a certain degree of flexibility, but also that they always have the function to limit the *ius puniendi* of the state. Only within a balanced approach between the sword and the shield function of criminal justice can the *ius puniendi* of the state become justice as we know it.

References

Andenaes, Johannes. 1965–1966. General preventive effects of punishment. *University of Pennsylvania Law Review* 114:949–983.

Ashworth, Andrew. 2006. Four threats to the presumption of innocence. *The International Journal of Evidence & Proof* 10:241–278.

Bell, Daniel. 1976. *The Coming of Post-Industrial Society*. New York: Basic Books.

Bellanova, Rocco, and Paul De Hert. 2009. Le cas S. et Marper et les données personnelles: l'horloge de la stigmatisation stoppée par un arrêt Européen. *Cultures & Conflicts* 76:101–114.

Biersteker, Thomas J., and Sue E. Eckert, eds. 2007. *Countering the financing of terrorism*. London: Routledge.

Brakel, Rosamunde van, and Paul De Hert. 2011. Policing, surveillance and law in a pre-crime society: Understanding the consequences of technology based strategies. *Journal of Police Studies* 20:163–192.

Brown, Sheila. 2006. The criminology of hybrids: Rethinking crime and law in technosocial networks. *Theoretical Criminology* 10:223–244.

Bureau of Justice Assistance. 2005. *Intelligence-led policing: The new intelligence architecture*, VII. Washington, D.C.: US Department of Justice. https://www.ncjrs.gov/pdffiles1/bja/210681.pdf.

Casey, E. 2011. *Digital evidence and computer crime*. Academic Press.

Castells, Manuel. 2000. *The rise of the network society. The information age: Economy, society and culture. 2nd ed. vol. 1*. Malden: Blackwell.

De Busser, E. 2009. *Data protection in EU-US criminal cooperation*. Maklu.

de Busser, E. 2010. EU data protection in transatlantic cooperation in criminal matters. Will the EU be serving its citizens an American meal? *Utrecht Law Review* 6 (1). (January 2010).

de la Cuesta, J. L. 2007. Anti-terrorist penal legislation and the rule of law: Spanish experience, Revue Internationale de Droit Pénal (RIDP).

De Hert, Paul, ed. 2012. *Privacy impact assessment*. Dordrecht: Springer.

Fijnaut, C., J. Wouters, and F. Naert, eds. 2004. *Legal instruments in the fight against international terrorism. A Transatlantic dialogue*. Martinus Nijhoff Publishers.

Frost, Natasha A. 2006. *The punitive state: Crime, punishment and imprisonment across the United States*. LFB Scholarly Publishing LLC.

Gutwirth, Serge. 2002. *Privacy and the information age*. Lanham: Rowman and Littlefield.

Gutwirth, Serge, Ronald Leenes, and Paul De Hert, et al. 2013. *European data protection: coming of age?* Dordrecht: Springer.

Increasing Resilience In Surveillance Societies (IRISS). 2013. Deliverable D1.1, surveillance, fighting crime and violence. http://irissproject.eu/wp-content/uploads/2012/02/IRISS_D1_MASTER_DOCUMENT_17Dec20121.pdf.

Jakobs, Günther. 2004. Bürgerstrafrecht und Feindstrafrecht. *HRRS* 3:88–95.

Lianos, Michaelis, and Mary Douglas. 2000. Dangerization and the end of deviance. The Institutional Environment. *British Journal of Criminology* 40:261–278.

Lyon, David. 1994. *The electronic eye: The rise of surveillance society*. Minneapolis: University of Minnesota Press.

Lyon, David. 2007. *Surveillance studies: An overview*. Cambridge: Polity Press.

McCulloch, Jude, and Sharon Pickering. 2009. Pre-crime and counter-terrorism: Imagining future crime in the war on terror. *British Journal of Criminology* 49:634–663.

Ost, F., et M. van de Kerchove. 2002. *De la pyramide au réseau? Pour une théorie dialectique du droit*. Bruxelles: Publications des Facultés universitaires Saint-Louis.

Pelser, C. 2008. Preparations to commit a crime. The Dutch approach to inchoate offences. *Utrecht Law Review* 4 (3).

Pradillo, O. 2011. Fighting against cybercrime in Europe: the admissibility of remote searches in Spain. European. *Journal of Crime, Criminal Law and Criminal Justice, núm* 19:363–395.

van Duyne, P. 1996. The phantom and threat of organized crime, Crime. *Law and Social Change* 24:341–371.

van Noorloos, M. 2011. Hate speech revisited. A comparative and historical perspective on hate speech law in the Netherlands and England & Wales. Intersentia.

Vervaele, J. A. E. 2009. Special procedural measures and respect of human rights, general report for the International Association of Criminal Law (AIDP). *Utrecht Law Review* :66–109.

Chapter 8
Privatization of Information and the Data Protection Reform

Els De Busser

The need for law enforcement authorities to use personal data originating from private entities has been questioned and discussed on several occasions, e.g. the 2010 EU-US Agreement on the processing and transfer of financial messaging data for the purposes of the terrorist finance tracking program (the TFTP Agreement)[1] and the 2006 Data Retention Directive.[2] As necessity is one of the key requirements for personal data to be used for a purpose that is different from or even incompatible with the purpose they were gathered for, this condition also forms the core of the debate.

The central theme of this contribution deals with the need for law enforcement authorities to receive personal data from private entities.[3] Even though the proposed directive on data protection in criminal matters is not applicable to Europol or Interpol, both agencies will also be covered by this analysis due to their involvement in cross-border criminal investigations and the correlated use of personal data from private entities. Directive 95/46/EC on the protection of individuals with regard to the processing of personal data and on the free movement of such data[4]—and the proposed regulation that should replace it in future—is the basic legal instrument

[1] Agreement on between the European Union and the United States of America on the processing and transfer of Financial Messaging Data from the European Union to the United States for the purposes of the Terrorist Finance Tracking Program, O.J. L 195, July 27, 2010, 5.

[2] Directive 2006/24/EC on the retention of data generated or processed in connection with the provision of publicly available electronic communications services or of public communications networks and amending Directive 2002/58/EC, O.J. L 105, April 13, 2006, 54.

[3] Both Europol and Interpol use a definition of "private parties", respectively "private entities". Apart from minor differences, both definitions are largely the same. The similarities in both is that a private entity should be a legal person that is governed by private law such as a business, company, commercial association or a not-for-profit organization and that is not categorized as an international organization. For the purpose of this research, this definition will be used.

[4] O.J. L 281/95, November 23, 1995, 31.

E. De Busser (✉)
Max Planck Institute for Foreign and International Criminal Law, Günterstalstraße 73, 79100 Freiburg im Breisgau, Germany
e-mail: e.busser@mpicc.de

S. Gutwirth et al. (eds.), *Reloading Data Protection*,
DOI 10.1007/978-94-007-7540-4_8, © Springer Science+Business Media Dordrecht 2014

covering the protection of personal data processed for commercial activities. Personal data processed for the purpose of criminal investigations and prosecutions are dealt with in other legal instruments due to the limited competences of the EU in this area and due to the specific sensitivity of data being processed for such purposes. Framework Decision 2008/977/JHA on the protection of personal data processed within the framework of police and judicial cooperation in criminal matters[5] should soon be replaced by the proposed directive. The first question this contribution thus focuses on is which legal instrument should regulate a transfer of personal data from private entities to law enforcement authorities: the proposed directive or the proposed regulation? To answer this question, it is necessary to analyze the effect of private-to-law-enforcement transfers on data protection. In other words, does the fact that data are transferred from a private to a public entity make them fall into a gap between two legal instruments, essentially affecting the protection of these data? The effects that the transfer will have on the quality of the data, their security and the purpose limitation principle will be analyzed due to their specific vulnerability when data are transferred from a private entity to law enforcement. It would be beyond the scope of this article to also include the risks regarding the right to a fair trial and the admissibility of the data as evidence in criminal proceedings, although this should be the subject of a separate research paper. What has been included as a second question in the present research is the situation in which personal data held by private entities are requested by a third state for the purpose of a criminal investigation or prosecution on the territory of that third state. Due to the potentially different data protection approaches of third states and possible conflicts of jurisdiction, the analysis should be made as to how to regulate the transfer of personal data. Such an analysis includes the question of whether the transfer of personal data should be regulated in the current data protection reform package or in another legal instrument such as a bilateral agreement with the third state.

8.1 How to Regulate Private-to-Law-Enforcement Transfers?

When a private entity processes personal data collected during commercial activities in the EU, these data fall within the scope of the data protection regime of Directive 95/46/EC. This means that they should be accurate, adequate and not excessive in relation to the legitimate commercial activity they were collected for. Furthermore, they should not be processed for purposes that are incompatible with this commercial activity and also should not be stored for longer than is necessary for the purpose of their collection. The principle of informed consent plays an important role in Directive 95/46/EC as consent is one of the legal grounds for processing of personal data.

When a law enforcement authority conducts a criminal investigation into corruption, for example the bank accounts of the persons involved are particularly interesting to investigate and cooperation of the bank(s) in question will be needed. In accordance with Framework Decision 2008/977/JHA the aforementioned data protection

[5] O.J. L 350, December 30, 2008, 60.

principles also apply here. The informed consent rule obviously does not apply as this is a principle that does not work when dealing with suspects of a criminal offence.

Processing personal data for commercial activities and processing personal data for law enforcement purposes are two distinct operations that are governed by two distinct types of legislation in the EU legal framework on data protection. The reasons for this separation of legal instruments are the particular context of criminal investigations and prosecutions, on the one hand, and the competences of the EU, on the other hand. Transferring personal data from a private entity to a law enforcement authority, however, is a transfer that links both legal instruments. This particular type of transfer and what the applicable legal instrument should be is analyzed in the first part of this chapter. Transferring personal data from a private entity to a law enforcement authority involves a change of data controller, who is responsible for compliance with the data protection principles. In the second part of this chapter, the risks concerning data accuracy and data security during this type of transfers will be assessed.

8.1.1 Applicable Legal Instrument

Directive 95/46/EC is applicable to the processing of personal data wholly or partly by automated means. It is also applicable to the processing of personal data other than by automated means which are part of a filing system or are intended to become part of a filing system in the course of an activity that falls within the scope of Union law. This roughly translates into the processing of personal data for the purpose of commercial activities. The proposed regulation does not change this scope. Framework Decision 2008/977/JHA is applicable to the transmitting of personal data that a Member State receives from another Member State[6] for the purpose of the prevention, investigation, detection, or prosecution of criminal offences or the execution of criminal penalties. The proposed directive expands its scope to include domestically gathered personal data. There are several reasons for using two distinct legal instruments.

First, processing personal data for the purpose of a criminal investigation activates the fair trial rights laid down in Article 6 ECHR from the moment a criminal charge is made. This means that the presumption of innocence, the right to defend oneself against the charges made, etc. should be safeguarded when personal data are processed for the purpose of a criminal investigation and a criminal charge is made. This is not the case for data processing for the purpose of a commercial activity. An important distinction between both purposes is also the principle of informed consent as a ground for lawful processing in commercial matters. Including consent as a ground for lawful processing in criminal matters would not work.

[6] Additionally, the scope of the Framework Decision includes personal data that have been transmitted or made available by Member States to authorities or to information systems established on the basis of Title VI of the Treaty on European Union; or are or have been transmitted or made available to the competent authorities of the Member States by authorities or information systems established on the basis of the Treaty on European Union or the Treaty establishing the European Community.

Second, the competences of the EU to make laws are limited. Due to the creation of the internal market, commercial activity is one of the main areas in which the EU enacts legislation. In the field of criminal matters, its competences are restricted. In the traditional pillar structure, it was the third pillar dedicated to judicial and police cooperation in criminal matters that was intergovernmental while the first pillar on Union law was organized in a supranational fashion. Even though the Lisbon Treaty has merged the three pillars, there is still a difference in the legal instrument to be used for judicial cooperation in criminal matters and police cooperation in comparison to commercial matters. Regulations are the legal instruments governing commercial matters, binding in their entirety and directly applicable in all Member States. Directives are only binding with regard to the result leaving the Member States some measure of flexibility with regard to the means they choose to achieve that particular result. This makes directives an appropriate tool for harmonizing national criminal law. Since the legal instruments currently known as directives were called framework decisions up to the entry into force of the Lisbon Treaty, the legal instrument currently still applicable for data protection in criminal matters is Framework Decision 2008/977/JHA, which will be replaced by the proposed directive in the future.

With data processing in commercial matters governed by Directive 95/46/EC and data processing in criminal matters governed by Framework Decision 2008/977/JHA, where is the transfer of data from a commercial entity to a law enforcement authority regulated? This question cannot be answered by naming a legal instrument that governs these types of transfers in general. The EU legal instruments that regulate private-to-law-enforcement transfers all have a scope that is restricted to a transfer between specific private entities and law enforcement authorities. They include Directive 2006/24/EC on the retention of data generated or processed in connection with the provision of publicly available electronic communications services or of public communications networks and amending Directive 2002/58/EC (the Data Retention Directive), the 2012 Agreement between the EU and the US on the processing and transfer of passenger name record data (the PNR Agreement) and the aforementioned 2010 TFTP Agreement. Directive 2005/60/EC on the prevention of the use of the financial system for the purpose of money laundering and terrorist financing (the Third AML Directive) should also be mentioned here.[7] Additionally, both Europol[8] and Interpol have laid down rules for processing the data they receive from private entities or persons.[9]

No legal instrument adopted on the EU level lays down standards on private-to-law-enforcement personal data transfers in general. The question then arises as to

[7] Article 22 of Directive 2005/60/EC on the prevention of the use of the financial system for the purpose of money laundering and terrorist financing, O.J. L 309, November 25, 2005, 27.

[8] Recently a new proposal was published that further strengthens the data protection regime that Europol has in place: Regulation on the European Union Agency for Law Enforcement Cooperation and Training (Europol) and repealing Decisions 2009/371/JHA and 2005/681/JHA, COM(2013) 173 final, March 27, 2013. This proposal is currently in its preparatory phase in the European Parliament.

[9] Article 25 Europol Decision and Article 28 Interpol Rules on the Processing of Data.

how this should be done. In order to answer this question, it is first necessary to define the precise nature of this transfer and whether it is necessary to regulate it on the EU level or leave it up to the Member States.

When the provisions of the currently applicable legal instruments are consulted, a definition of this particular transfer is found lacking. Not having a definition of what constitutes a private-to-law-enforcement transfer, we can nonetheless conclude what it is not. It is certainly not "processing under the authority of the controller and processor"[10] since there is no reporting or supervision between both parties. It can also not be qualified as "processing on behalf of a controller"[11] because that would mean that the law enforcement authority would be processing the data for a commercial purpose or vice versa. It is also not a transfer to a third state or international organization. Obviously, it could constitute a transfer to a third state's law enforcement authority but it can also be a transfer within the EU or even within one Member State.

Because both the private entity and the law enforcement authority are data controllers but both process the data for the performance of different activities unrelated to each other, the transfer of personal data from a private entity to a law enforcement authority can be defined as a transfer from a data controller to another data controller where the purpose of the data processing changes from a commercial purpose to that of a criminal investigation or prosecution.

To answer the question of why this transfer should be regulated on an EU level, we need to go back to the data protection principle of purpose limitation. As mentioned earlier, the processing of personal data is protected in a different manner based on the purpose of the processing. When the processing serves the purpose of commercial activities, different rules apply than when the processing serves the purpose of a criminal investigation or prosecution. Transferring data from one purpose to the other is restricted by the purpose limitation principle, meaning that the processing of data for a purpose that is incompatible with the purpose they were collected for is not allowed. Derogating from this principle is foreseen on the condition that it has been laid down in law and on the condition that processing for the "incompatible" purpose is necessary and proportionate.

What exactly constitutes an incompatible purpose has recently been explained by the Article 29 Data Protection Working Party.[12] In an elaborate opinion, key factors are identified to facilitate deciding upon the compatibility of a purpose. Earlier, scholars stated that the processing of personal data for a criminal investigation or prosecution after they were originally collected for a commercial purpose can be considered incompatible due to the lack of functional equivalence and foreseeability.[13] The Article 29 Working Party identifies the relationship between the purposes for

[10] Article 16 of Directive 95/46/EC (Article 27 of the proposed regulation) and Article 22 of the proposed directive.

[11] Article 17 of Directive 95/46/EC (Article 26 of the proposed regulation) and Article 21 of the proposed directive.

[12] Article 29 Working Party (2013).

[13] Bygrave (2002, p. 340). See also De Busser (2009, p. 68).

which the data have been collected and the purposes of further processing as one of the key factors. It is considered important to also take the factual context into account. The context in which the data were collected and the reasonable expectations of the data subject as to the further use of the data is a second key factor to be considered according to the Article 29 Working Party. This is very close to the concept of foreseeability previously formulated by aforementioned authors. The third and fourth key factors identified by the Article 29 Working Party have not been highlighted in this context before but are nonetheless of great significance. They stress the impact that processing for incompatible purposes can have on the data subject and which safeguards are in place to ensure fair processing and prevent negative impact. This shows a certain extent of flexibility with regard to processing for other purposes while at the same time safeguarding the rights of the data subject.

The requirement of legality can be fulfilled by providing for data protection rules on this type of transfer in national legislation. Nonetheless, a set of rules adopted on the EU level would be more efficient for the following reasons.

The protection of personal data as such is regulated on the EU level and the question being dealt with here is which one of the existing legal instruments applies or should apply. Due to the current debate on reforming the data protection legal framework, additional provisions could be included in either the proposed regulation or the proposed directive. It would be more efficient to regulate the private-to-law-enforcement now than to wait for an ECJ ruling on the matter.

8.1.2 Data Protection Related Risks

8.1.2.1 Data Quality

In accordance with Directive 95/46/EC, the data controller must ensure that personal data are accurate and, where necessary, kept up to date. Framework Decision 2008/977/JHA also states that personal data shall be rectified when inaccurate as well as completed or updated when possible and necessary. The proposed directive made this provision more precise by explicitly making it the competent authority's responsibility as a data controller to adopt policies and implement appropriate measures to ensure that the processing of personal data is performed in compliance with the provisions adopted pursuant to the directive. This includes the right to rectification of inaccurate or incomplete data and the right to deletion. More importantly, in accordance with the proposed directive, law enforcement authorities are also obliged to indicate the degree of accuracy and reliability of the personal data they process.

Thus, when personal data are transferred from a private entity that is a data controller to a law enforcement authority, which then becomes the data controller, the accuracy of the data should be safeguarded. With regard to data accuracy, what is then the risk of this transfer? The risk lies in the assessments that law enforcement authorities make based on personal data received from private entities. The personal data as such can be accurate but that does not make the assessments or conclusions drawn from them accurate. For instance, the data indicating that a certain individual buys a large amount of artificial fertilizer could lead to the conclusion either that this individual may be using the chemicals to produce explosives or that he legitimately needs the chemicals for agricultural activities.[14]

As mentioned above, the proposed directive has therefore included an obligation for law enforcement authorities to distinguish different degrees of accuracy and reliability when processing different categories of personal data: in particular, the distinction between personal data based on facts, on the one hand, and personal data based on personal assessments, on the other hand. This is not a new idea. It was already provided for by principle 3 of CoE Recommendation (87)15 regulating the use of personal data in the police sector. During the negotiations on Framework Decision 2008/977/JHA, the Austrian presidency proposed inserting a clarification of the term "accuracy" in the preamble.[15] This proposal resulted in recital 12. The Commission's first proposal of the framework decision, however, included the distinction of degrees of accuracy and reliability in the text of the legal instrument, more specifically in the principles relating to data quality.[16] The proposed provision did not survive the negotiations.[17]

In its rules on analysis work files, Europol has included the stipulation that data stored in these files for analysis purposes shall be distinguished according to the assessment grading of the source (see below) and the degree of accuracy or reliability of the information. Data based on facts are distinguished from data based on opinions or personal assessments.[18] Information is evaluated by Europol using a 4 × 4 system that awards a code to the source of the information and a code to the information itself. Based on these codes, decisions are made regarding the accuracy of the information or the reliability of the source.[19] The responsibility for data processed at Europol, particularly as regards transmission to Europol and the input of data, as well as their accuracy and their up-to-date nature, lies with the Member State that has communicated the data. However, with respect to data communicated to Europol by third parties, including data communicated by private parties, this responsibility lies with Europol.[20]

---

[14] Such purchasing behavior, while being the owner of an agricultural firm, was reportedly one of the preparatory acts of Anders Behring Breivik using the chemicals to produce and detonate a bomb in the centre of Oslo, Norway in July 2011.

[15] Council, 6450/3/06, May 11, 2006, 15.

[16] COM(2005) 475 final, October 4, 2005, 16. See also the European Data Protection Supervisor's second opinion on the proposal, O.J. C 91, April 26, 2007, 11.

[17] See also De Busser (2009, pp. 131–134).

[18] Council Decision on adopting the implementing rules for Europol analysis work files, O.J. L 325, December 11, 2009.

[19] Europol Information Management Booklet, File no: 2510–271. See also "Europol: '4 × 4' intelligence handling codes includes 'dodgy data'", Statewatch, accessed January 7, 2013, http://www.statewatch.org/news/2013/jan/03europol-dodgy-data.htm.

[20] Article 28 Europol Decision.

With regard to information processed in the Interpol Information System, the national central bureaus, national and international entities are responsible for ensuring that these data are still accurate and relevant before using them. Data that have been received from private entities in accordance with the Interpol rules of data processing can also be processed in the Interpol Information System.[21]

Therefore, where the quality of personal data that law enforcement authorities of the Member States have received from private entities is concerned, the currently applicable rules do not provide for the necessary safeguards. However, the proposed directive improves this situation. As long as the provision on distinguishing degrees of accuracy and reliability remains in the proposed directive, it is not necessary to provide for further rules on ensuring the quality of data transferred from private entities to law enforcement authorities.

8.1.2.2 Data Security

In accordance with Directive 95/46/EC, the data controller is responsible for implementing appropriate technical and organizational measures to protect personal data against accidental or unlawful destruction or accidental loss, alteration, unauthorized disclosure or access, and against all other forms of unlawful processing. The level of security should be appropriate for the risks presented by the processing and the nature of the data in question. The private entities that transfer personal data to law enforcement authorities should thus secure the data until the moment of transfer.

Law enforcement authorities have a similar obligation of ensuring data security under Framework Decision 2008/977/JHA. The provisions are more specific, also including equipment access control, data media control, storage control, communication control, transport control, etc.

Building on the personal data breach notification in Article 4(3) of e-privacy Directive 2002/58/EC, the proposed regulation introduces the obligation for the data controller to notify the supervisory authority of a personal data breach. This new obligation has been inserted in both the proposed regulation and the proposed directive; hence both private entities and law enforcement authorities are bound by it as data controllers. The provision states that the controller needs to document any personal data breaches, comprising the facts surrounding the breach, its effects, and the remedial action taken. This documentation must enable the supervisory authority to verify compliance with the regulation. It is neither laid down in the proposed regulation nor in the proposed directive, but if personal data is sent to law enforcement authorities that were the subject of a data breach when under the control of the private entity as data controller, the above-mentioned documentation should also be transferred to the law enforcement authority in question. The purpose is not to verify compliance with the regulation but to be informed of possible manipulation of personal data that can be used in a criminal investigation or prosecution at a later stage. In view of the accuracy and reliability of personal data processed by law enforcement authorities,

[21] Article 7, § 2 and Article 63 Interpol Rules of Processing Data.

the fact that a security breach may have affected or disclosed these data at an earlier stage could be vital information.

Both Interpol and Europol have data security rules in place that are based on risk assessment and proportionality. In the case of Interpol, the General Secretariat develops control mechanisms and the National Central bureaus are also responsible for adopting appropriate security measures that are at least equivalent to the minimum level of security laid down in the security rules established by the General Secretariat. In the case of Europol, Europol itself takes the necessary technical and organizational steps to ensure data security and that each Member State and Europol implement measures to ensure controls regarding data access, data media, etc.

Regarding the personal data supplied by private entities, the Interpol General Secretariat is obliged to ensure that the means used by private entities to supply data processed in the Interpol Information System allow those entities to access only the data authorized in accordance with the particular agreement that has been concluded to that effect.[22] The Interpol rules on the processing of data do not mention security breaches of personal data before they are transmitted from a private entity to Interpol. Nevertheless, cooperation with a private entity must respect Interpol's Constitution, and an agreement is concluded between Interpol and the private entity. In accordance with the Europol Decision, direct contact with private entities is not allowed. Europol may only process personal data transmitted by private entities via the National Unit of the Member State under whose law the entity was established, and the transfer should be in accordance with the national law of that Member State.[23] Thus, for the security of the personal data in the hands of the private entity, the national law, which needs to comply with Directive 95/46/EC and, in the future, with the proposed regulation, will be applicable.

The introduction of data breach notifications in the proposed data protection legal framework of the EU is highly important to both data processing for commercial purposes and data processing for the purposes of a criminal investigation or prosecution. In order to fulfill the requirement of distinguishing different degrees of accuracy and reliability, a mandatory transmission of the notification by the private entity to the receiving law enforcement authority should also be provided for.

8.1.2.3 Purpose Limitation

The transfer of personal data from a private entity to a law enforcement authority is a clear breach of the data protection principle of purpose limitation, which only allows processing for purposes compatible with the purpose the data were collected for. Key factors to decide upon the compatibility of the purpose have recently been defined by the Article 29 Working Party.[24] In order to make the processing of relevant personal data for the purpose of criminal investigations and prosecutions feasible but consistent

---

[22] Article 28, §§ 4 and 10 Interpol Rules on the Processing of data.

[23] Article 24 Europol Decision.

[24] Article 29 Working Party (2013).

with the data protection principles, a compatible purpose should have a link with the original purpose the data were collected for and the reasonable expectation of the data subject regarding the further processing should be considered. As mentioned before, these key factors correspond to theories on the compatible purpose that were formulated in academic publications. The concept of "functional equivalence" refers to the link or similarity between the original purpose and the data subject. The latter should be able to reasonably foresee the processing of his data for that purpose.[25] Functional equivalence is more than just relevance. It means that both purposes serve a function that is similar, e.g. personal data given to the bank to open a bank account can be used by the bank to contact clients with offers of better service with regard to their bank account. Processing for a different but compatible purpose should be reasonably foreseeable when any person in these circumstances can also predict that the data will be used for another—compatible—purpose. The fact that data can be retained or stored for longer periods of time could lower the level of foreseeability for the data subject. In this respect, the processing for other purposes is closely related to the duration of data retention.

An incompatible purpose could be defined as a purpose that is not reasonably foreseeable[26] and does not have functional equivalence with the original purpose. In order to process personal data for incompatible purposes, e.g. the processing of personal data that were collected for commercial purposes for the purpose of a criminal investigation or prosecution, the legality and necessity requirements should be fulfilled. Additionally, the Article 29 Working Party correctly attaches great importance to the impact that processing for a different purpose has on the data subject and the safeguarding of the data subject's rights.

Directive 95/46/EC and the proposed regulation both enshrine the traditional purpose limitation principle but allow restrictions of it when necessary to safeguard the prevention, investigation, detection and prosecution of criminal offences. This raises the question of how to ensure that only the necessary personal data are transferred from private entities to law enforcement authorities. What it really means is that no bulk transfer of personal data is allowed, which was the central issue with respect to the controversial Data Retention Directive but also with respect to the transfer of financial messaging data from the Belgian based company SWIFT to the US Department of the Treasury (UST). It is essential to maintain the link or nexus between the data that are transferred by a law enforcement authority and the criminal investigation or prosecution that they should be processed for. For example, when highly sophisticated printers are found and confiscated during the investigation into large scale counterfeiting of the euro currency, receiving data on the buyer of these machines from the manufacturer can be crucial to tracing the offender(s). In this case there is a clear nexus between the personal data and the ongoing investigation. If a manufacturer of printers were to be asked to transfer his client database to law

[25] Bygrave (2002, p. 340). See also De Busser (2009, p. 68).

[26] ECtHR, The Sunday Times v. United Kingdom (1979), 49; Malone v. United Kingdom (1984), 67–68; Rotaru v. Romania (2000), 55 and Amman v. Switzerland (2000), 56.

enforcement authorities for them to "comb" through, however, there would not be such nexus.

The addition should be made to the provisions of the proposed regulation as well as the proposed directive that the necessity requirement means that a nexus should be present between the personal data requested and the criminal investigation or prosecution for which their transfer and processing will be carried out.

In the 2012 TFTP Agreement, Europol was assigned a special task after the first version of the Agreement was rejected by the European Parliament.[27] The new role for Europol that is laid down in Article 4 of the TFTP Agreement gives Europol the power to give binding force to the requests from the UST. Europol was thus put in the unexpected key position as the authority that decides upon the legitimacy of the requests to obtain data from a private entity. Besides checking that the requests do not include SEPA-related data, Europol should check the requests formulated by the UST on three aspects. The request should identify as clearly as possible the categories of data requested, the necessity of the data should be demonstrated and the request should be tailored as narrowly as possible to minimize the amount of data. Europol receives a copy of each request that the UST sends to SWIFT and SWIFT must wait for Europol's authorization before carrying out the request.[28] At the moment of the first joint review of the TFTP Agreement in 2011, an inspection report by the Europol JSB concluded that the requests that had been sent made a proper verification by Europol within the terms of Article 4 of the agreement, impossible. In addition, the JSB revealed that UST staff gave Europol staff oral instructions—with the stipulation that no written notes would be made—that influenced Europol's decisions regarding the requests. As the content of these instructions is not known, the JSB as well as Europol's own data protection officer were unable to carry out an effective inspection.

The second joint review of the TFTP Agreement has a more positive view on Europol's task. Highlighting that this verification role is based on an operational assessment of the validity of the request, the reviewers concluded that Europol is best placed for deciding on the requirement of tailoring the requests as narrow as possible while enjoying a certain margin of discretion.[29] Involving also the Europol data protection officer is a positive innovation.[30] Nonetheless, verifying requests for data on compliance with data protection rules is not a task that belongs to the Europol tasks laid down in the Europol Decision. So far, the Europol Decision has not been amended regarding this additional task. Furthermore, the Europol JSB still has concerns regarding the amount of data being transferred since subsequent requests—that have all been positively verified by Europol—with an average of one per month

[27] In the first version of the Agreement, Article 4 referred to the 2003 EU-US MLA Agreement as the legal basis for the UST's requests to obtain SWIFT data.

[28] Article 4 of the TFTP Agreement.

[29] Report on the second joint review of the implementation of the Agreement between the European Union and the United States of America on the processing and transfer of Financial Messaging data from the European Union to the United States for the purposes of the TFTP, SWD(2012) 454 final, 14.12.2012, 6–7.

[30] JSB Report 2012, Public Statement, 2.

essentially cover a continuous time-period. Another concern expressed by the JSB is the continuing role that oral information provided by the UST to Europol plays in the verification process.[31] On 18 March 2013, a brief statement was made by the JSB concerning a third inspection. Again, the JSB repeats its unease with the amount of data that is being transferred in accordance with the TFTP Agreement. The JSB recognizes that this is a political issue and that it is "up to the legislators to balance the massive transfer of data sets—mostly of non-suspects—with proportionality".[32]

8.2 Where to Regulate Private-to-Law-Enforcement Transfers?

With a proposed directive and regulation on the EU institutions' negotiation table, the opportunity is there to incorporate clear provisions on how to properly organize transfers of personal data from private entities to law enforcement authorities and ensure that the data protection principles are respected. The question is whether these provisions belong in the proposed directive or in the proposed regulation.

The proposed directive is limited to the processing of personal data by competent authorities—i.e. law enforcement authorities—for the purposes of the prevention, investigation, detection or prosecution of criminal offences or the execution of criminal penalties. This type of processing is explicitly excluded from the scope of the proposed regulation. The latter is limited to the processing of personal data in the course of an activity which falls within the scope of Union law. Yet the processing that is envisaged is the processing of personal data that were collected in the course of an activity which falls within the scope of Union law but are afterwards processed for the purposes of the prevention, investigation, detection or prosecution of criminal offences or the execution of criminal penalties. Drawing up a third legal instrument for this particular type of processing would not be efficient. What is meant is a transfer from what is covered by the proposed regulation to what is covered by the proposed directive. Thus most of the applicable data protection provisions already exist; only the transfer of the data is not regulated yet.

To decide which of the two proposed legal instruments should regulate these transfers, the jurisprudence of the ECJ should be consulted. The ECJ used the element of *essential* objective of data processing to determine the correct legal basis for the first EU-US PNR Agreement. Limited by the action brought by the European Parliament, the Court did not elaborate on this point but decided that Article 95 TEC was not the appropriate legal basis for the agreement.[33] In its judgment on the Data Retention Directive, the ECJ was equally faced with the question whether the purpose of use for the investigation, detection and prosecution of serious crime should be the determining factor in deciding upon a third pillar instrument rather

[31] JSB Report 2012, Public Statement, 2–3.

[32] JSB of Europol, Implementation of the TFTP Agreement: assessment of the follow-up of the JSB recommendations, Ref. 13–01, March 18, 2013.

[33] ECJ C-317/04 and C-318/04, Parliament v. Council (2006).

than a directive.[34] However, the situation in the case of the Data Retention Directive is different from the PNR Agreement. According to the Court, the directive only pursues the objective of safeguarding the internal market and excludes activities under Title VI TEU. Thus, the directive remained a first pillar instrument as the action for annulment was dismissed.

The element of safeguarding the internal market could not be used in the case of transfers of personal data from private entities to law enforcement authorities. The element of essential objective or the final purpose of the data processing would lead to the conclusion of regulating these transfers in the proposed directive. Not only would this be in line with the jurisprudence of the ECJ, it would also respect the scope of the proposed directive that is limited to processing of data by law enforcement authorities. Thus the proposed directive should lay down that its provisions also apply to the personal data a Member State's competent authority receives from a private entity. This would avoid future confusion as to which legal instrument applies to a transfer from what is covered by the proposed regulation to what is covered by the proposed directive.

8.3 Transfer to Third States

Considering the difficulties in transferring personal data from a private entity to a law enforcement authority described above, additional issues can arise when the requesting law enforcement authority is located in a state that is not an EU Member State. When a third states' authority cooperates with a Member State, there are two possible scenarios: the third state has also ratified the CoE Data Protection Convention or it has not. In case of ratification of the Data Protection Convention, the Member State can safely assume that the personal data will be adequately processed, i.e., in compliance with the data protection principles that it adheres to. In case of a third state that has not ratified the Data Protection Convention, several scenarios are possible. The third state could have a data protection regime that is similar but not equal to a regime that complies with the Data Protection Convention, or it could have a legal framework that is based on fundamentally different data protection principles. It could even have no data protection rules at all. This is where the adequacy requirement comes in, i.e., the requirement for a Member State or the European Commission to assess the level of data protection in a third state to which personal data should be transferred. Besides the issue of transfer of personal data to a private entity in a third state, the adequacy requirement as such is not without problems.

8.3.1 Adequacy Requirement

The EU was the first to introduce the requirement of an adequate level of data protection. In Directive 95/46/EC, the EU laid down rules for exchanging personal data

[34] ECJ C-301/06, Ireland v. Council and Parliament (2009).

within the scope of European Community activities, including commercial trade. Due to this particular scope which essentially extends beyond the external borders of the EU, the directive also provided for rules on data protection when doing business with third states. Article 25 of Directive 95/46/EC requires an assessment of the adequacy of the level of data protection in the third state. As the first provision of its kind in the EU, the adequacy requirement drew attention among third states' authorities. Before the directive was even adopted, reactions surfaced on how this would affect the trans-border flow of data such as electronic payments in international trade.[35] The first effect of this requirement for an adequate level of data protection was felt in the US, resulting in the so-called Safe Harbor compromise. This compromise was a set of data protection rules that companies in the US promised to apply when receiving data from an EU Member State's company. In 2001 the CoE included the adequacy requirement for all automatic data processing in its Additional Protocol to the Data Protection Convention, but this Additional Protocol has not yet been ratified by all Member States.

Thus, Directive 95/46/EC includes the adequacy requirement for personal data being transferred to a third state for commercial purposes. In the area of judicial and law enforcement cooperation in criminal matters, the transfer of data to third states has been protected by the adequacy requirement in the Europol Decision and Framework 2008/977/JHA,[36] but they are not relevant here since the sending authority under these legal instruments is not necessarily a private entity. It is, however, essential to mention the specific agreements that have been made concerning transfers of data from EU-based private companies to public authorities in the US. These include the 2012 PNR Agreements between the EU and the US[37] and between the EU and Australia.[38] The 2005 PNR Agreement between the EU and Canada[39] is currently being renegotiated due to the expiry of the adequacy decision. The aforementioned 2010 TFTP Agreement between the EU and the US also belongs in this list.

The PNR Agreements with the US and Australia as well as Canada followed a decision by the Commission on the adequate level of data protection by the receiving authority in the third state. The agreements that have been concluded between the EU and Europol on the one hand and the US on the other hand, should have been based on an assessment of the level of data protection in the US as well. The receiving authority on US territory is different from the receiving authority in the case of the PNR Agreement. Nevertheless, this condition has not always been fulfilled. In accordance with Article 18, § 1, 2) of the Europol Convention—which was applicable in 2002—and in accordance with the rules governing the transmission of personal data by Europol to third States and third bodies, the level of data protection of the US

[35] Boehmer and Palmer (1993); Bennet and Raab (1997) and Long and Pang Quek (2002).

[36] The same goes for Eurojust: Council Decision 2009/426/JHA on the strengthening of Eurojust and amending Decision 2002/187/JHA setting up Eurojust with a view to reinforcing the fight against serious crime, O.J L 138, June 4, 2009, 14.

[37] O.J. L 215, August 11, 2012, 5.

[38] O.J. L 186, July 14, 2012, 4.

[39] O.J. L 86, March 24, 2006, 15.

should have been assessed before concluding the 2002 Supplemental Europol-US Agreement on the exchange of personal data and related information (2002 Europol-US Agreement).[40] Considering inter alia the fact that in an exchange of notes, the scope of this agreement is widened considerably,[41] comprehensive proof of the US endorsing a data protection regime that fulfils the conditions of the adequacy requirement in accordance with the Europol Convention has not been produced yet. On the contrary, the later concluded agreements with the US show how the adequacy requirement is pushed aside in favor of a smooth flow of information.

An informal explanatory note reflecting Europol's view on the 2002 Europol-US Agreement states that the Agreement is "generally in line with the major principles incorporated in Europol's legal framework". The note goes even one step further and states that the provisions of Article 5 of the Agreement on general terms and conditions "would not be used as a legal basis for generic restrictions, but only in specific cases where there was a real necessity."[42] The phrase "generic restrictions" can clearly be understood as the requirement involving an adequate level of data protection and as such appears also in other transatlantic agreements. The 2003 EU-US MLA Agreement[43] and the Eurojust-US Agreement[44] both include a provision on "limitations on use to protect personal and other data," which explicitly states that generic restrictions with respect to the legal standards of the requesting state or party in the processing of personal data may not be imposed by the requested state or party as a condition for providing evidence or information.[45] Where, in the 2002 Europol-US Agreement, the US was labelled an adequate partner with regard to its data protection regime, even though this was unjustified, in the 2003 EU-US Agreement and the 2006 Eurojust-US Agreement, the adequacy requirement as such was disregarded.[46]

8.3.2 The Problem with Private-to-Law-Enforcement Transfers to Third States

In addition to the inherent difficulties with the adequacy requirement, the principle that personal data should only be processed when they are necessary for the purpose that they are processed for is one of the more problematic areas in the cooperation with third states. The experiences with the aforementioned TFTP Agreement—even with Europol in the role of verification authority—and the history of the subsequent

[40] O.J. C 88, March 30, 1999, 1.

[41] See De Busser (2009, pp. 322–334).

[42] Council, 13696/1/02, November 28, 2002, 10.

[43] Agreement 25 June 2003 on mutual legal assistance between the European Union and the United States of America, O.J. L 181, 19 July 2003, 41.

[44] Agreement between Eurojust and the United States of America, November 6, 2006.

[45] Article 9 of the 2003 EU-US Agreement and Article 9 of the 2006 Eurojust-US Agreement.

[46] De Busser (2009, p. 343).

PNR Agreements between the EU and the US illustrate how challenging it is to strike a balance between the prevention, investigation and prosecution of criminal offences on the one hand, and data protection on the other.

The European Parliament has attempted to rule out bulk transfers of personal data in the TFTP Agreement. This is however connected to the setting up of an EU-system that is equivalent to the TFTP, an ambitious project that takes time to be developed.[47] Also, as mentioned before the joint reviews and the report by the Europol JSB show that in practice the ban on bulk transfers of data can be circumvented by making successive requests for specific data.

It is important to point out that the concept of law enforcement authorities is differently defined in the EU and the US. When defining the scope of the principles under consideration for a general EU-US agreement on data exchange in criminal matters, the differences were considered but unfortunately no compromise was made as to how this should work in practice. In the EU, the term law enforcement covers the use of data for the prevention, detection, investigation, or prosecution of any criminal offense. In the US, this encompasses the prevention, detection, suppression, investigation, or prosecution of any criminal offense or violation of law related to border enforcement, public security, and national security as well as non-criminal judicial or administrative proceedings related directly to such offenses or violations.[48]

The problem lies in the fact that personal data from EU origin can be used in the US as intelligence so the more data are transferred, the more will be used for intelligence analysis, possibly even for profiling, e.g. profiling potentially dangerous passengers before they board a flight. This is possible due to the US' structure of state and federal authorities and of authorities involved in both law enforcement and intelligence such as the FBI and the CIA.[49] Together with other elements of information, personal data are compared and analyzed to serve the intelligence purpose of following a lead to further reveal evidence of criminal activity. In such cases there is no link with a specific investigation or prosecution at the stage of processing the personal data. Striking examples of this have recently been revealed in the media after in December 2012 the US Department of Homeland Security granted the National Counterterrorism Center access to a number of databases including databases containing information about foreign-exchange students and visa applications.[50] These are not actual transfers of personal data in the sense of Directive 95/46/EC or Framework Decision 2008/977/JHA, however they can and most likely will involve personal data on EU citizens.

The necessity requirement is thus the Achilles heel of data exchange in the transatlantic cooperation in criminal matters and should be carefully dealt with when laying

---

[47] COM(2011) 429 final, July 13, 2011.

[48] Council, 9831/08, EU US Summit, June 12, 2008—Final Report by EU-US High Level Contact Group on information sharing and privacy and personal data protection, May 28, 2008, 2. See also European Data Protection Supervisor (2008).

[49] See inter Fijnaut (2004); Vervaele (2005) and Manget (2006).

[50] Angwin (2012).

down provisions on data transfers to third states. On this particular point, the proposed directive and the proposed regulation do not offer a solution.

8.3.3 Reforming the Adequacy Procedure

Improving the adequacy procedure was one of the main points of the reform of the data protection legal framework.[51] The proposed directive and the proposed regulation contain similar provisions when it comes to third state transfers and the assessment of a third state's level of data protection, however none mentions or covers the transfer from a private entity to a law enforcement authority in a third state.

8.3.3.1 Adequacy Assessments

Directive 95/46/EC, Framework Decision 2008/977/JHA and the Europol Decision provide in a list of example criteria to be included in an adequacy assessment.[52] No limited list was ever part of a legally binding instrument. Thus, assessments could differ as to the criteria being used.[53] In the context of reforming the data protection legal framework, the Commission introduced a list of three criteria that need to be taken into account: a legal component containing the rule of law, relevant legislation, and effective and enforceable rights, such as administrative and judicial redress; a supervision component containing not only the existence, but also, the effective functioning of an independent supervisory authority; and an international component covering all international commitments to which the third state is bound. It is new that the right to redress and the functioning of a supervisory authority are also explicitly mentioned as criteria to consider when assessing a third state's data protection level.

The list is inspired by policy documents from the Article 29 Working Party on Data Protection in the context of Directive 95/46/EC. In addition, the list refers to the umbrella CoE Data Protection Convention. 17 third states have ratified this Convention and are thus bound by the same data protection standards that inspired the EU legislation. This should make an adequacy assessment fairly easy; however it is not sufficient to rubberstamp the third state as adequate for the following reason. Framework Decision 2008/977/JHA does not provide for the exception set forth in the Additional Protocol to the Data Protection Convention that requires assessments *only* for states that did not ratify the Data Protection Convention, and neither does

[51] COM(2012) 9 final, January 25, 2012, 11–12.

[52] The same goes for Eurojust: Council Decision 2009/426/JHA on the strengthening of Eurojust and amending Decision 2002/187/JHA setting up Eurojust with a view to reinforcing the fight against serious crime, O.J L 138, June 4, 2009, 14.

[53] De facto the criteria being used are the same with regard to Directive 95/46/EC since the Article 29 Working Party published a working document in 1998 on the application of articles 25 and 26: Article 29 Working Party (1998).

the proposed directive. In fact, both legal instruments provide for more precise data protection rules in comparison to the Convention. A third state's ratification of the Data Protection Convention is thus not sufficient for establishing that it has an adequate level of data protection. The implementation at the national level also needs to be considered.

Directive 95/46/EC allows both the Commission and Member States' authorities to decide upon the adequacy of a third state's level of data protection. Framework Decision 2008/977/JHA only provides in the Member States making this decision. If the Commission makes the assessments, the result is obviously one decision on which all Member States could rely. Granting only Member States the authority to make these assessments could theoretically lead to different conclusions, thereby creating confusion and 'data shopping'. In this case the door is open for third states to abuse such a situation and request their data from the Member State that is most likely to rubberstamp their data protection system as adequate. In the proposed directive as well as the regulation, the decision on adequacy is to be taken by the Commission only.

8.3.3.2 Alternatives

Both the proposed directive and the regulation have formulated alternative solutions to the adequacy procedure but the provisions create some confusion as to what happens if the Commission decides that a third state does not provide in an adequate level of data protection. The text of the provisions could be interpreted as if a negative decision would block all possible data exchange with the third state in question. It could also be read as another way to activate the alternative solutions. This should be made clear in order to rule out all confusion.[54]

In case the Commission does not take any decision, two alternative solutions are offered. First, adequate safeguards could be enshrined in a legally binding instrument or the data controller or processor assesses the adequacy of the offered safeguards and thus allows for the data exchange. A bilateral agreement would be the most logical way of providing in a legally binding instrument. In the transatlantic cooperation in criminal matters, the European Commission started negotiations in 2010 with US representatives for drafting a general agreement on the protection of personal data transferred and processed for the purpose of preventing, detecting, investigating and prosecuting crime, including terrorism. After a Council Decision authorizing the opening of the negotiations in November 2011, the talks have not resulted in an agreement yet.[55] Since a positive adequacy decision on the US' level of data protection is rather unlikely[56] at the moment, these negotiations could lead to the development of an alternative solution. Possibly after the adoption of the proposed directive, both the EU and the US will find the described provisions to be a new incentive for their negotiations.

---

[54] Article 29 Data Protection Working Party (2012).

[55] 16908/11, November 15, 2011.

[56] De Busser (2009, pp. 293–303).

A second alternative amounts to placing the data controller or processor in the driver's seat and giving them the authority to label a third state as offering an adequate level of data protection. The European Data Protection Supervisor rightfully points out that for the proposed directive this is not only an inappropriate but also an insufficient safeguard for the protection of personal data.[57] For transfers of personal data from private entities to law enforcement authorities in third states, this is equally inappropriate. Due to *inter alia* the significant risk of private entities being pressured to deliver personal data to a third state's authorities for the purpose of a criminal investigation or prosecution, the option of having the data controller or processor decide on adequacy should not be available or at least be covered by the supervisory authority's verification. Derogating from the adequacy requirement could be seen as a third alternative. This is not new. It is also not new that derogations must be interpreted restrictively and that frequent, massive and structural transfers of personal data should not be the result.[58]

Considering the experience with the TFTP Agreement and the bulk transfer of personal data to the US Department of the Treasury as well as the questions regarding the necessity of transfers of PNR data, there is a need to stress the necessity requirement more in the adequacy assessment procedure. The provisions that would specify that the adequacy procedure should also be applied for the transfers of personal data from private entities to law enforcement authorities in third states, would thus have to restrict these transfers to only those data that are necessary. Article 33 of the proposed directive already contains such safeguard restricting the transfers to those that are necessary for the prevention, investigation, detection or prosecution of criminal offences or the execution of criminal penalties.

8.3.4 *Where to Regulate Private-to-Law-Enforcement Transfers to Third States?*

If the provisions governing the transfers of personal data from private entities to law enforcement authorities within the EU should be included in the proposed directive, should the same conclusion be made for such transfers to law enforcement authorities in a third state? The question cannot be answered with a simple yes or no. Due to the particularities of third states' criminal justice systems, e.g. the structure and competence of US' law enforcement authorities, bilateral agreements will still have to be concluded after a positive decision has been made on the third state's level of data protection. The aforementioned examples of EU-US agreements and the PNR Agreements with other third states illustrate this.

[57] Opinion of the European Data Protection Supervisor on the data protection reform package, March 7, 2012, 64.

[58] Opinion of the European Data Protection Supervisor on the data protection reform package, March 7, 2012, 65.

Also, the proposed directive is only applicable to processing by law enforcement authorities. In the case of a transfer to a third state, the data are transmitted from a private entity on EU territory to a law enforcement authority in a third state. The receiving law enforcement authority is thus not bound by the provisions of the proposed directive.

Therefore, the procedure for reaching a decision on the level of data protection of a third state should be included in the proposed regulation. This should not be incorporated in the proposed directive since Article 33 limits its scope to transfers by competent authorities. A better solution is to add a specification to the adequacy procedure in the proposed regulation that it is also applicable for transfers of personal data from private entities to competent authorities in a third state for the prevention, investigation, detection or prosecution of criminal offences or the execution of criminal penalties. Bilateral agreements should be concluded for laying down further provisions on the processing of the data by the receiving authority. With regard to third states such as the US, where data sharing between government agencies is a legally organized practice, provisions on purpose limitation and data retention should be carefully considered.

8.4 Conclusion

The topic of transfers of personal data from private entities to law enforcement authorities within and outside the territory of the EU raises many questions. Several of those are covered by this contribution, especially the question as to which legal instrument should contain the provisions governing both types of transfers.

On the level of data quality, attention should be paid to the different degrees of accuracy and reliability of the data before and after the transfer. Introducing an obligation for the private entities to inform the receiving law enforcement authority of any data protection breaches, is a tool that would assist determining the accuracy and reliability of the data in question.

Fulfilling the necessity requirement when processing personal data, especially when transferring them to another data controller, has been proven to be difficult in practice. It is crucial in this respect that respecting the necessity of the data should be a task to be taken seriously both by the private entity that sends the data and the law enforcement authority that receives and processes the data. It is the proposed directive that should regulate that its provisions also apply to the personal data a Member State's competent authority receives from a private entity.

Ensuring a clear nexus between data that are transferred and a particular criminal investigation or prosecution becomes especially complicated when the receiving law enforcement authority is located in a third state that has a different definition of law enforcement and where data sharing among government agencies is common. When data are then received to be used as intelligence, the nexus with a specific criminal investigation or prosecution is not present. This is however an issue that cannot be covered by the proposed directive or regulation. For such third state transfers, bilateral agreements should be concluded. The assessment whether the third state offers an adequate level of data protection however should be laid down in the proposed regulation. Its provisions on adequacy decisions should thus be expanded with the transfers from private entities to law enforcement authorities for the purpose of the prevention, investigation, detection or prosecution of criminal offences or the execution of criminal penalties.

Providing for such set of rules on the transfers from private entities to law enforcement authorities both within and outside the EU, would remove legal uncertainty regarding a practice that is not only useful but also necessary. While developing a new legal framework on data protection for the EU, it is the perfect timing to decide upon such provisions.

References

Angwin, Julia. 2012. U.S. terrorism agency to tap a vast database of citizens. *The Wall Street Journal*, 13. December. http://online.wsj.com. Accessed 20 Feb 2013.

Article 29 Data Protection Working Party. 2012. Opinion 01/2012 on the data protection reform proposals, WP 191, March 23, 2012, 30 and Opinion of the European Data Protection Supervisor on the data protection reform package, March 7, 2012, 65.

Article 29 Working Party. 1998. Working Document, WP12 Transfers of personal data to third countries: Applying Articles 25 and 26 of the EU data protection directive, July 24, 1998.

Article 29 Working Party. 2013. Opinion 03/2013 on purpose limitation, WP 203, April 2, 2013, 20–27.

Bennet, Colin, and Charles Raab. 1997. The adequacy of privacy: The European union data protection directive and the north American response. *The Information Society* 13:245–263.

Boehmer, Robert, and Todd Palmer. 1993. The 1992 EC data protection proposal: an examination of it implications for the US business and the US privacy law. *American Business Law Journal* 31:265–311.

Bygrave, Lee. 2002. *Data protection law. Approaching its rationale, logic and limits, 340*. The Hague: Kluwer law International.

De Busser, Els. 2009. *Data protection in EU and US criminal cooperation*. Antwerp: Maklu.

European Data Protection Supervisor. 2008. Press Release November 11, 2008, Opinion on transatlantic information sharing for law enforcement purposes: Progress is welcomed, but additional work is needed, 13.

Fijnaut, Cyrille. 2004. Inlichtingendiensten in Europa en Amerika: de heroriëntatie sinds de val van de Muur en 11 september 2001. *Justitiële Verkenningen* 3:10–42.

Long, William, and Marc Pang Quek. 2002. Personal data privacy protection in an age of globalization: the US-EU safe harbor compromise. *Journal of European Public Policy* 9:325–344.

Manget, Fred. 2006. Intelligence and the criminal law system. *Stanford Law & Policy Review* 17:415–435.

Vervaele, John. 2005. Terrorism and information sharing between the intelligence and law enforcement communities in the US and the Netherlands: Emergency criminal law? *Utrecht Law Review* 1:1–27.

Chapter 9
Quo Vadis Smart Surveillance? How Smart Technologies Combine and Challenge Democratic Oversight

Marc Langheinrich, Rachel Finn, Vlad Coroama and David Wright

9.1 Introduction

Surveillance is undergoing profound changes. Formerly distinct surveillance systems are converging and being combined.[1] Both data collection and data processing are becoming more complex, more varied, and are being automated.[2] Surveillance technologies, which used to be the prerogative of government agencies, are now in the reach of companies and citizens.[3] Taken together, these trends lead to what is increasingly called "smart surveillance".

We believe the rapid evolution of smart surveillance to be driven by five main trends: (i) a qualitative broadening of the types of data that can be collected, (ii) the quantitative increase for most of these data source types, driven by increased automation, (iii) more and better analytic and processing tools, (iv) the increasing ubiquity

[1] Haggerty and Ericson 2000, pp. 605–622.
[2] An example for the automation of data collection is given by Diffie and Landau (2009) for the surveillance of communications. The pervasiveness and automation of data analysis can be observed, for example, by means of the profiling technique, which is enabled by data mining: Hildebrandt (2008).
[3] Wright et al. (2010).

M. Langheinrich (✉)
Faculty of Informatics, Università della Svizzera italiana (USI),
Via Buffi 13, 6904 Lugano, Switzerland
e-mail: langheinrich@acm.org

R. Finn · D. Wright
Trilateral Research & Consulting, Crown House, 72 Hammersmith Road,
W14 8TH London, UK
e-mail: rachel.finn@trilateralresearch.com;

D. Wright
david.wright@trilateralresearch.com

V. Coroama
Center for Industrial Ecology, University of Coimbra,
Rua Luís Reis Santos, 3030-788 Coimbra, Portugal
e-mail: vlad.coroama@dem.uc.pt

S. Gutwirth et al. (eds.), *Reloading Data Protection*,
DOI 10.1007/978-94-007-7540-4_9, © Springer Science+Business Media Dordrecht 2014

of surveillance as an everyday tool, and (v) the convergence of surveillance systems and assemblages.

Surveillance technology has moved far beyond video and audio recordings. Recent types of data sources include: positioning data from the triangulation of mobile phones between GSM cell towers[4] or from the neighbouring relation to WiFi access points;[5] remote temperature readings from infrared cameras;[6] extensive data on vehicle whereabouts from traffic cameras or electronic toll collection systems; the content of e-mails and instant messaging chats from so-called "parental control" programmes.[7]

As new types of data sources appear, the quantity of existing data sources is continuously expanding—mostly driven by the rapidly falling cost of surveillance hardware. According to some estimates, around 0.5 million CCTV cameras were deployed in the UK in 1999;[8] in 2007, there were some 4 million.[9] Airports, especially in the US, are quickly increasing the number of body scanners: in 2011, a total of 486 machines were installed at 78 airports.[10] In many countries, Internet service providers record comprehensive data about the Internet usage of their customers, as required by law. Fingerprint databases are becoming commonplace.[11] DNA databases store ever more DNA profiles—in the UK, for example, the national DNA database contained around 6 million profiles in 2012.[12] And India is in the process of building the world's largest biometric database, which will comprise the fingerprints and iris scans of each of its 1.2 billion inhabitants.[13]

An emerging feature of new surveillance technologies is that both the data collection and the data processing are increasingly automated. In this aspect, smart surveillance extends the concept of dataveillance, which, as defined by Roger Clarke in 1988,[14] consists of two steps: gathering data from various databases and then "mining" this data. When Clarke coined the term in 1988, the main data sources

[4] Figueiras and Frattasi (2010).

[5] Skyhook.

[6] Lee (2010).

[7] Naraine (2007).

[8] Haggerty and Ericson (2000).

[9] Goodchild, Sophie, "Britain becoming a Big Brother society, says data watchdog", *The Independent*, 29 April 2007. These numbers have been widely and frequently quoted, however, there has been some controversy about just how many CCTV cameras there are in the UK and how many times a day on average a person in London is caught by CCTV cameras (see Aaronovitch 2009). The source of these "statistics" appears to be Norris and Armstrong (1999). However, Norris and Armstrong say that their numbers are "guesstimates".

[10] Transportation Security Administration, "Advanced Imaging Technology (AIT)", 2011. http://www.tsa.gov/approach/tech/ait/index.shtm. US airports have installed both millimetre wave and backscatter x-ray scanners. In early 2013, the TSA decided to remove the backscatter scanners, which some had described as equivalent to a strip search. Plungis (2013)

[11] Lyon (2008).

[12] GeneWatch UK.

[13] Polgreen (2011).

[14] Clarke (1988).

were financial transactions (credit card usage, ATM withdrawals, electronic transfers) and official records (court orders, criminal records, fingerprints). At the dawn of the new millennium, these sources had already grown to include more mundane tasks such as cellular phone usage, driving patterns (recorded by electronic toll collection systems), Internet usage, and shopping behaviour (as recorded by Internet shops and loyalty schemes).[15] Nowadays, with the emergence of ubiquitous computing,[16] the distinction between the online and the offline life is continuously fading, and increasing parts of what used to be the analogue offline world are now available in digital format. Sensors record and digitise information on our location, our encounters, what we see and hear,[17] or the way we drive.[18] CCTV cameras still sweep the public space but now complex activity recognition algorithms discovering distinct behaviour automatically trigger alarms.[19] Visual recognition algorithms help police to identify vehicle licence plates and photo sharing websites to "tag" friends in a photo. When local computation capabilities do not suffice, the data is sent over wired and wireless networks to be processed elsewhere.

Consequentially, while in the past surveillance has often been confined to high security applications driven by institutional actors, today's widespread availability of sensing and data processing capabilities leads to much more pervasive surveillance: surveillance becomes commonplace, a routine that can be performed with increasing easiness by ever larger circles, but is also less evident. The true power of future smart surveillance systems lies in the combination of all of these data collection and processing capabilities into an ever expanding array of interconnected surveillance tools, where the output of one system is the input to another. Haggerty and Ericson proposed the term "surveillant assemblage" to describe this increasing intertwining and almost discretionary interconnection of individual surveillance systems: "We are witnessing a convergence of what were once discrete surveillance systems to the point that we can now speak of an emerging surveillant assemblage."[20] Bauman and Lyon have gone even further by suggesting that surveillance is "liquid". Lyon explains what this means as follows:

> 'Liquid surveillance' is less a complete way of specifying surveillance and more an orientation, a way of situating surveillance developments in the fluid and unsettling modernity of today... Surveillance spreads in hitherto unimaginable ways, responding to and reproducing liquidity... A number of theorists have noted the ways in which surveillance, once seemingly solid and fixed, has become much more flexible and mobile, seeping and spreading into many areas where once it had only marginal sway.[21]

---

[15] Clarke (2003).

[16] Want (2009).

[17] Langheinrich (2009).

[18] Coroama (2006).

[19] Wright et al., op. cit., 2010.

[20] Cited from Haggerty and Ericson, op. cit., 2000, p. 606.

[21] Bauman (2013).

Fig. 9.1 A simple example for smart surveillance. A single data source (CCTV) is the input for a single data processing stage—in this case, a tool to automatically detect human shapes (e.g., in a parking lot). Ultimately, this tool allows the system to track people in a parking lot, highlighting them to a human operator and keeping statistics on movements in the parking lot over time

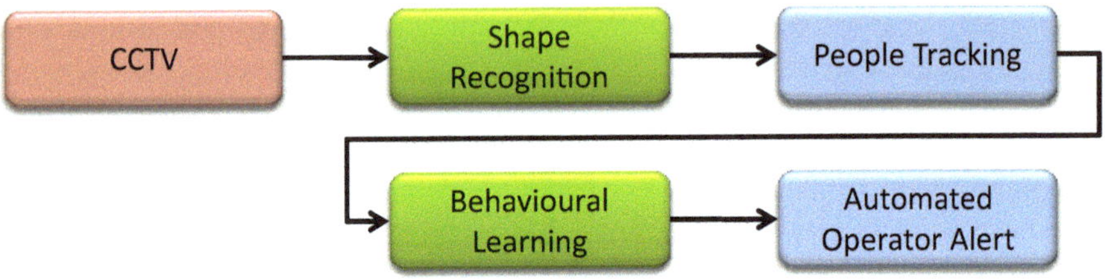

Fig. 9.2 Cascading smart surveillance elements into systems. Once the system is able to track people in a video, a learning algorithm can be used to classify the detected human behaviour in the image into "normal" and "suspicious", thus allowing an operator to receive automated alerts whenever a potential theft is in progress

Liquid surveillance is perhaps another way of expounding the notion that surveillance is becoming increasing pervasive, of saying that we live in a surveillance society, that we are entering a world of ambient intelligence, where all manufactured things will have some bit of "smart dust", which will enable all things to network, to communicate and to contribute to the ubiquity of surveillance.

9.2 Defining Smart Surveillance

To define smart surveillance, we first differentiate three types of building blocks of smart surveillance systems: sources, tools and functions (or in other words: data collection, data processing and data use). In the simplest case, one or more sources provide data to a tool in order to perform a particular function (cf. Fig. 9.1). By processing a digital video stream, a shape recognition processor is able to extract human shapes moving through the image, allowing an operator to better identify people in the video, and keep simple statistics about the number of people over time and their movements.

Multiple tools might be combined for a more complex function, as illustrated in Fig. 9.2. Once people can be tracked, a behavioural learning system can classify the movements into "suspicious" and "normal", thus supporting the new function of automatically alerting the operator of a potential theft in progress. Such a system greatly expands the number of cameras a single operator can control, as it frees the human from close monitoring.

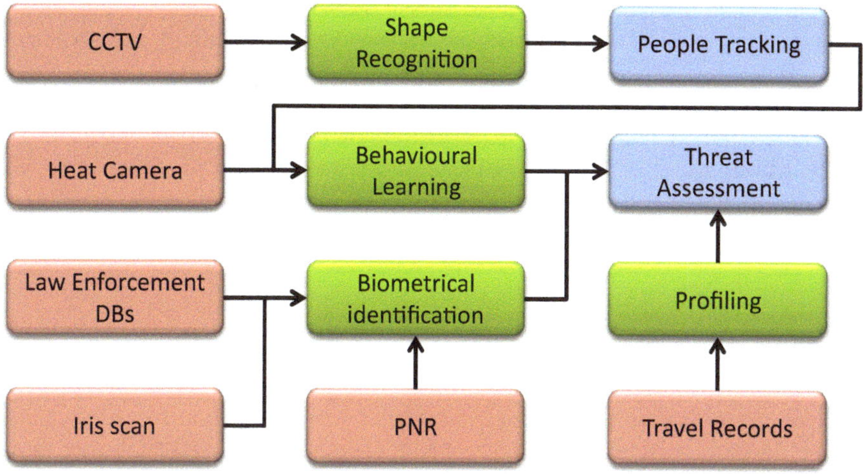

Fig. 9.3 Airport passenger screening as an example of a smart surveillance assemblage. Several independent surveillance systems contribute to the assessment of the threat posed by a passenger: The behavioural learning system is just one of the several inputs to a comprehensive threat assessment function

Tools may become sources for other tools, as may functions, leading—in principle—to systems of arbitrary complexity (cf. Fig. 9.3). A smart airport surveillance system may combine not only movements but also other physical signals such as body temperature into the behavioural alert system. This information may be joined with background information from passenger name records (PNRs) and other law enforcement databases to provide a comprehensive threat assessment of each traveller.

These complex combinations of sources, tools and functions lie at the heart of smart surveillance systems. We thus define smart surveillance as follows:

> **Smart surveillance systems** are those capable of extracting *application-specific information* from captured information (be it digital images, call logs or electronic travel records) in order to generate *high-level event descriptions* that can ultimately be used to make *automated or semi-automated decisions.*

Smart surveillance systems inherently offer a high level of scalability, as they in turn can act as input to other surveillance systems. Smart surveillance systems contribute to social reconfigurations in ways that essentially differ from previous surveillance techniques, especially by introducing new folding processes of the spatial and temporal dimensions with the purpose to go beyond "mere" re-action. For example, Gary Marx offered a definition of "new surveillance" as the use of technologies to extract personal information, and stated that they were characterised by a breakdown between the distinction between the watcher and the watched, a greater focus on monitoring large populations rather than closely monitoring particular individuals and the related distantiation of surveillance.[22] However, this does not account for the increasing interlinking of surveillance systems, nor the scalability and automated decision-making which characterises "smart" surveillance.

In order to better understand the potential of this development, we will look at the current research landscape of surveillance technology, offering a brief examination of the different types of surveillance technologies that are currently being deployed, the types of data they collect and the applications associated with each technology.

9.3 Technologies and Applications

We distinguish six families of surveillance technologies: visual surveillance, sensors, location determination technologies, biometrics, communication surveillance and dataveillance, and discuss individual technologies within each. Taken together, these different families of surveillance technologies demonstrate the extent to which the types and sources of surveillance data that can be collected has been broadened, and how surveillance has become increasingly ubiquitous and integrated into the everyday lives of many Europeans.

9.3.1 Visual Surveillance

The use of visual surveillance devices for surveillance purposes has expanded from the use of basic photography equipment to high-tech imaging scanners. Photography via the portable camera was the original form of visual surveillance technology and was used to link individuals to particular places at particular times. Portable devices such as mobile phones with picture and video recording capabilities perform a similar function today. However, although state or other authorities use this equipment to target the less powerful, these systems may be more democratic than other surveillance systems. Relatively non-powerful individuals can use this equipment to capture images of powerful individuals such as police, celebrities and/or state officials in a synoptic surveillance framework[23], as the Ian Tomlinson case in the G20 protests illustrated. CCTV represents another common form of visual surveillance and generally refers to "all semi-permanently installed video equipment… [and includes cameras that are] primarily used to monitor places or behaviour" usually by the police or other state or public authorities.[24] Such surveillance, according to Webster, is "considered ubiquitous, a normal part of everyday life, with citizens … happy to forego some personal privacy in return for greater levels of personal safety and

[22] Marx (2002).

[23] Mathiesen (1997).

[24] Nouwt et al. (2005).

security."[25] Cameras may be actively monitored in real time, where those monitoring the cameras can provide a response to incidents, or they may be passively monitored, in that they may only record data which can be later referred to if an incident occurs. Some examples of applications of CCTV systems include, but are not limited to, the protection of private property, national security, counter-terrorism, road traffic monitoring (associated with automatic number plate recognition), identification of individuals, monitoring for criminal or anti-social behaviour, behaviour or pattern recognition, border control and employee monitoring. Examples of places in which camera systems have been deployed are public spaces such as streets and town centres, motorways, casinos, housing association houses/estates, workplaces (including the home as a workplace as in "nanny cams"), shopping malls, convenience stores, banks, transport systems, airports and schools. CCTV cameras have also been fitted to mobile vehicles, including unmanned aerial vehicles (UAVs), which can generally be defined as a "device used or intended to be used for flight in the air that has no onboard pilot"[26] that include "multiple pieces of ancillary equipment, such as vehicle control equipment, communications systems, and potentially even launch and recovery platforms".[27] Here, the range of applications of visual surveillance technology, the ubiquity of surveillance technology and its increasing integration with other surveillance data signals the extent to which it is becoming "smart" in the theoretical sense of the term.

9.3.2 Communications Surveillance

Since ancient times, remote communications have been prone to interception, and modern forms are no exception; efforts towards their interception are as old as the communication technologies themselves. In the context of surveillance, the following technologies have been used to intercept communications. *Electronic eavesdropping* is "the act of electronically intercepting conversations without the knowledge or consent of at least one of the participants".[28] *Wiretapping* defines a specific subset of electronic eavesdropping, where an actual wire is involved in the communication. Electronic eavesdropping can be used to intercept communications along landline telephony, mobile telephony and calls using the Voice over Internet Protocol (VoIP). For a classic landline phone call, for example, interception can occur inside one of the telephones themselves, in a junction box, a phone closet, on a telephone pole, or in the telephone company's central office.[29] Digital communication data can also be collected via more recent digital wiretaps that work remotely, do not alter

[25] Webster and William (2009).

[26] Quoted from Aviation Safety Unmanned Aircraft Programme Office, 2008, in McBride (2009).

[27] McBride, op. cit., 2009, p. 629. See also Directorate of Airspace Policy (2010).

[28] Britannica (2011).

[29] Diffie and Landau, op. cit., 2009.

the communication stream, and are thus virtually undetectable.[30] As the voice travels in digitised form, though, users can use end-to-end encryption devices.[31] When strong end-to-end cryptography is used, the conversation cannot be wiretapped along the line—the only possibility lies in wiretapping one of the telephones directly. In addition, digital mobile phone signals can also be intercepted by accessing the unencrypted portion of the signal that travels unencrypted through the mobile provider's core network, to be encrypted again between the other telephone and its respective base station.[32] Voice over Internet Protocol (VoIP) is a collection of communication protocols that define how audio or audio-video conversations can use the Internet as a communication medium instead of telephone lines. Interception of communication within these technologies can be accomplished with a man-in-the-middle attack at the VoIP provider—this is the mechanism foreseen for lawful interception. The most popular VoIP software, nonetheless, works differently. Skype uses a proprietary, decentralised protocol that integrates the Advanced Encryption Standard (AES), and it is, by today's known technological standards, virtually impossible to eavesdrop on a Skype call along the line. The only possibility of wiretapping is before the voice signal has been encrypted by the Skype software, that is, on one of the communication partner's devices (computers, smartphones). Finally, other communication surveillance practices include call logging that records the time and duration of the conversation, as well as the identities of the communicating parties, albeit not the content[33] and the interception of text messages, e-mail and other digital communications, principally via spyware on smartphones and personal computers. Thus, despite the proliferation of communication devices, ubiquitous surveillance technologies can be deployed to intercept this varied range of communication data.

9.3.3 Sensors

Sensors are technological components that can recognise certain physical or chemical properties of their environment. They represent another type of surveillance technology, with a growing market in relation to security, law enforcement and commercial applications and which generate a range of different types of data. Sensors can range from traditional metal detectors or retail security systems at store entrances to complex, recently developed explosives "sniffing" or behavioural sensors. "Chemical sniffers" or "electronic noses"[34] are designed to detect and identify residual traces

---

[30] Diffie and Landau, op. cit., 2009.

[31] These come in numerous flavours, from rather home-brewed devices that can only be used in pairs by both parties and that use unknown, possibly unsafe algorithms (e.g., http://www.pimall.com/nais/voicekeeper.html) to enterprise-scale devices that use state-of-the-art encryption algorithms with a new key for every conversation (e.g., http://www.cisco.com/en/US/products/ps5853/index.html).

[32] Prevelakis and Spinellis (2007).

[33] Petersen (2007).

[34] Gardner and Bartlett (1999).

that indicate either the presence of, or someone's recent contact with, certain chemicals, such as drugs or explosives. Metal detectors are electromagnetic devices able to detect the presence of metals in their vicinity. They come either as portable units or walk-through gates, e.g., handheld metal detectors used by security staff or walk-through portals that are typically installed in points of access to zones where an increased level of security is needed.

In addition to measuring the presence or absence of particular materials, sensors such as heat sensors can provide data which is comparable and relative. There are two main types of heat sensors: passive infrared sensors and infrared cameras. Passive infrared sensors are small devices that create a small electrical potential when its temperature changes, while infrared (or thermographic) cameras sense the levels of infrared radiation (invisible to the human eye) in their field of sight and to transform them into a visual representation. Both types of heat sensors represent common security tools. Passive sensors are often used to sense human heat in relation to burglar alarms, and helicopter-mounted infrared cameras, for example, are often used to support ground forces in searches for suspects, especially at night.

Although each type of sensor often performs only one specific task, these sensing systems can be combined to consolidate a comprehensive, multi-modal system. For example, automated "behavioural profiling" aims to use multi-modal behavioural sensing to replace TSA agents in US airports who watch for suspicious behaviour among passengers (e.g., nervousness) and single out suspects for more detailed screening.[35] These systems use a large number of sensors (e.g., chemical, biometric, but also CCTV, licence-plate recognisers, retina scanners) to detect real-time behaviour to provide a centralised, real-time classification of travellers.[36] Thus, while these sensors are found in a number of quotidian security applications, and they collect a range of different types of data, they are furthermore becoming automated an integrated with other surveillance systems to signal the advent of smart surveillance systems.

9.3.4 *Biometrics*

Biometrics refers to the use of measurements and analysis of human body characteristics to distinguish between individuals. There are two types of biometrics: those that use physical characteristics such as fingerprint, face or iris patterns and those that use behavioural characteristics such as voice, signature or gait patterns.[37] Biometrics can be used for identification, where an individual's pattern is matched to many records, or authentication, where an individual's pattern is checked against the one stored in their record. Fingerprinting systems are used to either identify or

---

[35] McElroy (2011).

[36] Wolfe (2010).

[37] Wei and Dongge (2006).

verify and are being used in an increasing number of applications, including national identity systems, criminal justice systems, immigration and border control, public transport, in schools and social assistance systems in some countries. These applications are becoming interlinked, and some countries are considering universal fingerprint databases linked with passports. Facial recognition technology works by matching an image of a person with an image stored on a database. Facial recognition technology involves capturing a still image of a person's face, or multiple still images of a person's face, and then using computer software to measure the distance between a number of nodal points on the individual's face.[38] Thus, like fingerprints, the individual's face is transformed into a mathematical template. Facial recognition technology is not particularly effective at identifying faces in a crowd, and works best when individuals voluntarily enrol and then co-operate with the identification system.[39] Iris recognition systems have consistently performed as the most reliable biometric identification technology. These systems work by converting an image of the iris into a sequence of 1s and 0s.[40] Once this sequence is collected, an IrisCode is used to represent the pattern, and an individual's identity is verified when two IrisCodes are compared.[41] Iris recognition has been used primarily for air travel, but could also be used as an access control system in other contexts. Finally, DNA profiling for matching has been implemented in the context of criminal justice throughout Europe, in the USA, Canada and Australia, as well as many other countries. In some jurisdictions, only violent criminals convicted of serious crimes can be included in a DNA database, whereas in other contexts, DNA samples can be taken, and stored, for anyone arrested.

Soft biometrics refers to biometric measurements that are behavioural and/or otherwise subject to change. Two often cited examples of soft biometrics include voice recognition systems and gait recognition systems. According to Wei and Lee, voice recognition systems work by capturing the voice of a person through a microphone and extracting certain features of their voice from the signals produced by their speech which can be compared to a database of known persons. [42] While this biometric is most commonly used for access control, it has also been used in the UK to check whether known offenders are complying with the terms of their curfew orders[43] or to check if football hooligans are at home during match times[44]. Yet, Wei and Li point out that problems such as background noise and people's sensitivity about having their speech recorded will likely prevent widespread roll out of this technology.[45]Gait recognition involves people being identified through a computer

[38] Stefani (2006).

[39] See Zureik and Hindle (2004); Introna (2009).

[40] Wei and Li, op. cit., 2006.

[41] Adkins (2007).

[42] Wei and Li, op. cit., 2006.

[43] *The Times*, "Joyrider, 14, is first tagging guinea pig", 17 July 2001.

[44] Fay (2005).

[45] Wei and Li, op. cit., 2006.

analysis of the way they walk via measurements of their step length, hip, knee and foot joint angles and speed. Although the accuracy of gait recognition has not yet been optimised, the technology has created quite a bit of interest because it has the potential to identify individuals at a distance, and to identify people without their co-operation. Therefore, like sensors, the range of different biometric technologies and applications available demonstrate the different types of data which can be collected and mined by surveillance systems, as well as the different types of usage to which they can be put. Furthermore, the different applications associated with biometric technology, from criminal justice to personal security and consumer applications, demonstrates the extent to which biometric identification or verification is becoming increasingly ubiquitous and unremarkable.

9.3.5 Location Determination Technologies

While a wide variety of location determination systems exists, all of them fall into two main classes of localisation techniques: (1) triangulation and (2) proximity sensing. One of the earliest location determination technologies was measuring the viewing angle of several known points (e.g., lighthouses, mountain peaks) and determining the intersection of the view lines on a map. Instead of measuring such distances directly, one typically measures signal propagation times t and then calculates the corresponding distance s through $s = v*t$ (given one knows the propagation speed v of the used signal). This is known as "Time of Arrival" (TOA). These techniques are used for the Global Positioning System (GPS) and for the triangulation of mobile phones or WiFi devices. "Sound ranging" systems also use the triangulation technique.[46] To determine a position on a surface, three points of reference are needed, and in order to determine a three-dimensional position, four such points are required. Since the mid-1990s, simpler civilian systems have emerged, with filters that can distinguish the sound of a firearm from all the other city sounds in neighbourhoods that are considered dangerous.[47] The sensors are small (can-sized) and placed on rooftops or light poles, and are virtually undetectable. They are typically linked directly to a police station, where they raise an instantaneous alarm as soon as a firearm has been fired, pinpointing the location with a precision of a few metres.

Proximity sensing systems work after a rather distinct principle: they do not aim at pinpointing objects or people in terms of co-ordinates, but at assessing their closeness to a known location. The location is thus a consequence of the neighbourhood relation with a known spot. Proximity sensing systems use physical phenomena with

---

[46] Strictly speaking, "triangulation" denotes the AOA technique, which measures the *angles* between the unknown location and several points of reference. Using *distances* would thus be called 'trilateration'. The term "triangulation", however, is commonly used to denote either of the two methods.

[47] ShotSpotter, "The ShotSpotter Gunshot Location System". http://www.shotspotter.com/technology.

limited ranges—when the corresponding phenomenon takes place, the neighbourhood is assessed. Examples include: the usage of magnetic induction in RFID (Radio Frequency Identification) systems to conclude upon the presence of an RFID tag in the vicinity of the antenna; the connection between a GSM base station and a cellular phone to assess the presence of the phone within the base station's cell; an existing connection between a laptop and a WiFi antenna to assess the laptop's presence within the range of the WiFi antenna; or—through low-power magnetic induction—the detection of an ID badge to assess its presence in the close "neighbourhood" (typically a few centimetres) of the access control antenna. The use of such location data in consumer applications, law enforcement, security and access control signals how pervasive this technology has become. Drivers and those navigating on their mobile phones use location determination technologies on a daily basis to find particular places as well as to suggest local services and amenities. RFID technology is used in a number of transportation, commercial and access control applications and in providing document authenticity and security.

9.3.6 Dataveillance

The term "dataveillance" denotes surveillance based on the electronic data traces typical for the modern world. Roger Clarke, who coined the term back in 1988, observed the increasing pervasiveness of such day-to-day data traces: "trends include the integration with EFTS [Electronic Funds Transfer System] of air-travel systems and telephone charging; road traffic monitoring, including vehicle identification, closely integrated with ownership and driver's-license records; computerization and integration of court records, criminal records, fingerprint records, and criminal-investigation systems;... and homes wired for reasons of employment, security, entertainment, and consumerism".[48] Electronic traces have since become ubiquitous: Employers can monitor employees' calls and e-mails; cellular phone companies have access not only to the calls but also the whereabouts of their customers; credit card companies know their clients' online and offline shopping habits; Internet service providers can inspect their subscribers' data traffic; operators of electronic highway tolls know when and where their subscribers drive. Clarke himself noted in 2003 the broadening and continuous sophistication of electronic data traces—and, thus, of potential dataveillance sources—in the 15 years that had passed since 1988 when he coined the term.[49] Clarke distinguishes between "personal dataveillance", the monitoring of the data of one specific person, and "mass dataveillance", the systematic investigation or monitoring of groups of people via their data traces.[50] *Personal dataveillance* represents the act of monitoring a specific targeted individual via his or her data. *Mass dataveillance* monitors the data traces of large groups of people in order to

[48] Clarke (1988).

[49] Clarke (2003).

[50] Clarke, op. cit., 1988.

identify individuals with a specific profile (e.g., individuals considered potentially dangerous): "mass dataveillance is concerned with groups of people and involves the generalized suspicion that some (as yet unidentified) members of the group might be of interest".[51]

The main method deployed for mass dataveillance is data mining. Definitions vary slightly, but *data mining* is usually understood as the "nontrivial extraction of implicit, previously unknown and potentially useful information from data"[52], or a "procedure by which large databases are mined by means of algorithms for patterns of correlations between data".[53] Such correlations indicate a relation between the data, without necessarily establishing causes or reasons—data mining is thus sometimes referred to as a discovery-driven approach as opposed to the more traditional assumption-driven approach.[54] Mass dataveillance is thus also closely related to *profiling*. Profiling is "a means of generating suspects or prospects from within a large population and involves inferring a set of characteristics of a particular class of person from past experience, then searching data-holdings for individuals with a close fit to that set of characteristics".[55] The main application domains of profiling are the targeted assessment of consumer behaviour, risk assessment for insurance and criminal profiling. Data mining is typically the first step in this process, as it defines the classes ("suspects or prospects") into which users can then be profiled. Profiling attempts to predict, or at least pre-empt, individual future behaviour by relying on the stereotypes learned during the data mining step, ultimately classifying individuals as potential risks or commercial windfalls.

While not exhaustive or completely mutually exclusive, this review of current surveillance technologies illustrates the different types of surveillance data that are available to different security, government, corporate and other stakeholders within the field of surveillance. Furthermore, because of the spread of different sectors of application (i.e., personal security, state security, public safety and criminal justice, consumer and entertainment sectors), surveillance technologies are becoming increasingly ubiquitous, particularly among European, North American and other industrialised populations. Increasingly, these different data sets and their use across different applications are being captured and integrated in order to produce more detailed and comprehensive profiles of individuals, which often enable the compilation of sophisticated behaviour analyses and the identification of individuals within complex smart surveillance systems. The following section examines some of the drivers for this integration of varied and detailed data.

---

[51] Ibid.

[52] Frawley et al. (1992).

[53] Hildebrandt (2008).

[54] Hildebrandt, op. cit., 2008.

[55] Clarke (1993). A closely related term is "social sorting". Lyon comments that surveillance is "a means of social sorting. It classifies and categorizes relentlessly, on the basis of various—clear or occluded criteria. It is often, but not always, accomplished by means of remote networked databases whose algorithms enable digital discrimination to take place". Lyon (2003).

9.4 Drivers of Surveillance

Surveillance technologies such as those described above do not simply come into existence—they are introduced into society by a range of stakeholders, who in turn are influenced in their actions by a diverse set of drivers. This section examines the technological, economic and policy drivers associated with the implementation of surveillance technology, and the main stakeholders influenced by these drivers. It also touches upon many of the key debates surrounding the implementation of surveillance systems, and how these have differed in particular contexts and applications. This information will be used to contextualise some of the emerging and developing smart surveillance technologies.

9.4.1 Surveillants, Surveilled and Other Stakeholders

We can identify five main stakeholders relevant to the surveillance society who have various reasons for implementing different smart surveillance technologies: authorities, industry, academia, policy-makers, the media and citizens:

1. *Governments and public authorities* are intimately involved in the introduction and procurement of smart surveillance systems in order to protect citizens and the state from illegal immigration, terrorism and crime, and as such they must often pass laws introducing or enabling new surveillance systems.
2. *Industry representatives* make up a large proportion of the stakeholders involved in the introduction of surveillance technology as they are interested in identifying a market for the products they develop and deriving economic gain from them. These typically come from a range of different links in the surveillance chain, such as developers, manufacturers and suppliers.
3. *Academics* are involved in the debates about surveillance technology. Academics develop new technologies or methods, explore applications for those technologies, develop standards and interoperability, explore the social implications of new technologies or encourage the take-up of new technologies.
4. *Policy-makers*, such as departments of defence or law enforcement often influence law and policy in relation to surveillance technologies, as do legislative committees. Other policy-makers, such as Data Protection Authorities or the European Article 29 Data Protection Working Party[56] (Article 29 WP) may comment on and produce guidelines for the use of surveillance technologies in society. The Article 29 WP has commented on the introduction of RFID technology, body scanners, electronic health records, passenger name records, video surveillance, smart phones, online social networking and electronic communications, as well as many other issues.

[56] The remit of the Article 29 WP is to provide expert advice to policy-makers in relation to data protection in Europe.

5. *The media* often distribute information and set the public agenda, particularly around generating demand for the implementation of surveillance systems. Journalists are often involved in implementing or operating surveillance technologies themselves in order to investigate or supplement a news story. However, the media also give voice to the potential negative privacy impacts of surveillance systems.
6. *Citizens and other groups of people* often demand the introduction of surveillance systems that target offenders, terrorists, illegal immigrants or other socially "undesirable" individuals. However, numerous dataveillance systems, cyber surveillance deployed by Internet service providers, or ubiquitous video and CCTV surveillance, are designed from the outset to target large parts of the society.

In addition to these categories of stakeholders, civil society organisations are also relevant to the introduction of smart surveillance systems, although they often focus on the potential negative aspects of these systems. While none of these stakeholder categories are exclusively supportive of or opposed to smart surveillance systems, civil society organisations, such as civil libertarians, human rights groups, privacy advocates and academic networks, are often the primary way in which details about how surveillance technologies may influence individual privacy are disseminated.[57] A range of civil rights organisations, e.g., the American Civil Liberties Union, Statewatch, European Digital Rights (EDRi-gram) and the Electronic Frontier Foundation (EFF), have generated information and undertaken legal action against governments, public authorities or other entities who implement surveillance technologies in ways which infringe upon privacy and other human rights.

A number of other drivers emanate from these different categories of stakeholders that also assist in contextualising the further development and deployment of smart surveillance technologies, including technological drivers, economic drivers and policy drivers, some of which were alluded to above.

9.4.2 *Technological Drivers*

Stakeholders are influenced by particular "drivers", some of which are technological. New discoveries, new standards and simply increases in available information can drive the introduction of particular technologies, where UK media and civil society organisation have described the introduction of CCTV and proposed introduction biometric identity cards "a solution looking for a problem".[58] Similarly, some have argued that the mass collection of information by governments in relation to air travel or consumer behaviour is driven by the simple fact that this information is available. Increasing interoperability is also a technological driver, in that the ability to link systems together increases the attractiveness of technologies for stakeholders

[57] Lyon, op. cit., 2007.

[58] See, for example, Liberty, *Liberty's Evidence to the Home Affairs Committee on the Government's Identity Card Proposals*, Dec 2003.

who are charged with procuring systems. For example, the US and Canadian governments encouraged the introduction of common standards to enable the mutually interoperable reading of biometric passports for their "smart border" programme.[59]

9.4.3 Economic Drivers

Another key driver of the introduction of surveillance technologies is economics. One such economic driver is stakeholder interest in maximising or maintaining profits. Zureik and Hindle note that "the economic payoff for the biometrics industry in the United States has been substantial" after September 2001.[60] This is particularly the case as the defence industry has sought new markets for technology it developed originally for military use. Unmanned aircraft systems, imaging scanners, biometrics and satellite surveillance provide some examples of such market augmentation. Rothstein and Talbott note that virtually all of the data surrounding the effectiveness of DNA databases in relation to criminal justice are compiled and released by "crime laboratories and other entities with an interest in promoting the maintenance or expansion of DNA databases".[61]

Researchers also note that decreases in the cost of surveillance technology are another driver of surveillance technology uptake. Because of a decrease in cost and an increase in convenience of use, fingerprinting is becoming increasingly popular for personal property protection, and in Asia and Europe, fingerprint readers are used to ensure that only legitimate owners are able to use their personal mobile phone.[62]

This also links to another driver of the introduction of surveillance technology—the protection of goods, services or property from theft, tampering or fraud. In addition to fingerprint readers, other technologies such as RFID and satellite tracking of vehicles represent further examples. Organisations installing surveillance technology also seek to use them for risk management or to avoid liability. McCahill finds that one of the uses of CCTV in shopping malls is to protect the management company from law suits as a result of trips, falls or other injuries as a result of spills or other obstacles.[63]

Government investment and other financial incentives are other drivers for the introduction of surveillance technology. Specifically, Zureik and Hindle note that the Department of Homeland Security in the US had a budget of $ 38 billion for investment in domestic security in 2004[64], while Webster notes that the UK government made approximately $\leq$ 200 million available for CCTV schemes between 1994 and 2003.[65]

---

[59] Zureik and Hindle, op. cit., 2004.

[60] Zureik and Hindle, op. cit., 2004, p. 123.

[61] Rothstein and Talbott (2006).

[62] Wei and Li, op. cit., 2006.

[63] McCahill (2002).

[64] Zureik and Hindle, op. cit., 2004, p. 121.

[65] Webster and William (2009).

9.4.4 Policy Drivers

Policy drivers also impel the introduction of surveillance technologies. These policy drivers include providing regional or national security, protecting citizens, reducing threats from crime and terrorism and co-operating with other governments or authorities.

Protecting citizens from crime and terrorism is discussed in the European Stockholm Programme[66], as well as the US and UK National Security Strategies. Zureik and Hindle note that, in their rhetoric, governments espouse the use of surveillance technology for citizen protection.[67] Furthermore, these policies need to respond to citizens' subjective feelings of safety and security. David Lyon discusses this driver in terms of perceived "risk" in society, where terrorism is a "dread risk" with a low probability of occurrence but high consequence. As a result, policy-makers and their constituents favour "zero risk" options such as hi-tech interventions to eliminate the threat.[68] The introduction of CCTV in many contexts is also a reaction to citizens' demands, especially if other, nearby areas already have CCTV systems, which is perceived to increase the threat from crime in the local area without CCTV.[69] Insurance companies may also demand the use of surveillance technologies as a condition of insurance coverage. These may include access control systems for dangerous goods, security systems to protect private property or dataveillance systems to detect unusual activity. Finally, and perhaps paradoxically, some surveillance systems are introduced to meet privacy demands. One example is the introduction of body scanners at airports, notably those with privacy enhancing software or other safeguards, which passengers seem to prefer to physical pat-down searches by security officials.[70] This squares with a key goal of the Stockholm Programme, which foregrounds the EU's commitment to provide both security and fundamental rights protection for citizens.

As alluded to above, co-operating with other governments or authorities is also a policy driver for the introduction of surveillance technology. Perhaps most famously, the introduction of RFID-enabled biometric passports in Europe was driven largely by a US declaration that this technology was necessary in order to enable visa-free entry to the USA for European citizens. The European Security Strategy, in particular, re-affirms a commitment to work with other partners, including the US and the North Atlantic Treaty Organisation (NATO), to devise interoperable solutions to meet common security needs.

Each of these drivers explains some of the reasons why stakeholders are supporting research into and the development of emerging and future smart surveillance assemblages. They are indicative of overall trends in the development of surveillance

---

[66] Council of the European Union (2010).

[67] Zureik and Hindle, op. cit., 2004.

[68] Lyon, op. cit., 2008, p. 503.

[69] McCahill, op. cit., 2002.

[70] Jones (2010).

technologies, including providing technological, economic and political reasons for supporting further research and development of surveillance systems.

9.5 Emergent Assemblages

The saying that "predictions are difficult, especially about the future" applies a fortiori to a domain as vast, heterogeneous and dynamic as surveillance. Nevertheless, by analysing current trends in both European and U.S. surveillance research, as well as current societal trends and the drivers of research such as those indicated above, we can attempt to extrapolate future smart surveillance both from a technological and a functional point of view. These emergent technologies and new types of assemblages reinforce and widen the trends presented in the previous two sections.

9.5.1 Major Smart Surveillance Research Initiatives

A good starting point for this future-oriented task is represented by the recent research initiatives funded by the European Commission within the Seventh Framework Programme (FP7, 2008–2013), and United States agencies such as the Defense Advanced Research Projects Agency (DARPA) and the National Science Foundation (NSF). Some of these research initiatives include foresight studies designed to identify technological trends in sectors such as surveillance and security.

As a first step, we collected relevant projects in a comprehensive list that identified 38 projects funded by the European Commission (FP7, mainly under the Security[71] and ICT Themes[72]) and 20 US projects, funded by DARPA[73] and the NSF. The absolute number of projects does not directly reflect the budget devoted to a specific research area though—DARPA projects usually have a wider scope (and a considerably larger budget) than individual EU projects. All US projects are technology-oriented, as are the majority of the EU ones. In the European FP7 program, however, and unlike the DARPA projects, the ethical analysis of surveillance is plainly represented. Of the total of 36 surveillance-relevant research projects identified, no less than seven investigate the ethics of state surveillance at different abstraction levels: from concrete technological proposals for a better privacy-compliance of video surveillance up to the effects of today's and tomorrow's surveillance on existing human rights' standards.

---

[71] The complete list of projects funded under the theme security is available at http://cordis.europa.eu/fp7/security/projects_en.html.

[72] See the FP7 projects dynamic database developed by the HIDE project and available on the HIDE website at http://www.hideproject.org/references/fp7_projects.html.

[73] Information taken from DARPA *Financial Year 2012 Budget Estimates*, available on the DARPA website.

As a second step, we then distilled this information in order to identify trends, core research areas and critical parts. We therefore extracted from each project—dependent both on its semantic breadth and its size—up to three main aims (e.g., "person identification" or "activity recognition"). We also identified the core technologies (for non-technological projects: the methods) used to achieve these aims. A subsequent mind mapping exercise revealed shared aims and common technologies across all research projects. Finally, we created groups of semantically related project aims (such as "person identification", "activity recognition", "person tracking" and "intrusion detection"). The more ubiquitous the presence of such a group within individual projects, the more likely it is to point towards a future trend. Figs. 9.4 and 9.5 summarise our findings, structuring both the aims of the individual projects, as well as the technologies used and/or envisioned to achieve these goals.

In FP7, the European Commission funds surveillance research mainly within the scope of two themes (SECURITY and ICTs), but surveillance-related projects can be found in other segments of the Cooperation programme, such as Transport and Space. The main research goals of the enlisted projects include the development of systems for the automated identification or tracking of individuals or of objects that can be related to individuals, activity recognition, software that identifies "suspicious" behaviours or intentions, and automated identification of illegal trespassing.

DARPA, on the other hand, groups its current research focus into nine so-called "strategic thrusts"[74]:

1. Robust, Secure, Self-forming Networks;
2. Detection, Precision ID, Tracking, and Destruction of Elusive Targets;
3. Urban Area Operations;
4. Advanced Manned and Unmanned Systems;
5. Detection, Characterization and Assessment of Underground Structures;
6. Space;
7. Increasing the Tooth to Tail Ratio;[75]
8. Bio-Revolution;
9. Core Technologies.

The main goals of DARPA research projects in the field of smart surveillance include the development of new theories on machine learning and reasoning that could enhance the capabilities of future surveillance systems, cognitive and ubiquitous powerful computing, video surveillance and threat detection systems, automatic information processing systems, new sensors, social networking monitoring, and communication surveillance. The list of DARPA projects shows a considerable difference with the EU research focus. The US Agency pays particular attention to the

[74] See Defense Advanced Research Projects Agency (DARPA) (2009).

[75] Wikipedia explains that the Tooth to Tail Ratio is a military term that refers to the amount of military personnel ("tail") it takes to supply and support each combat soldier ("tooth"). One of the stated goals of DARPA is increasing the tooth to tail ratio (reducing the amount of logistics and support personnel necessary in proportion to combat personnel without reducing combat effectiveness).

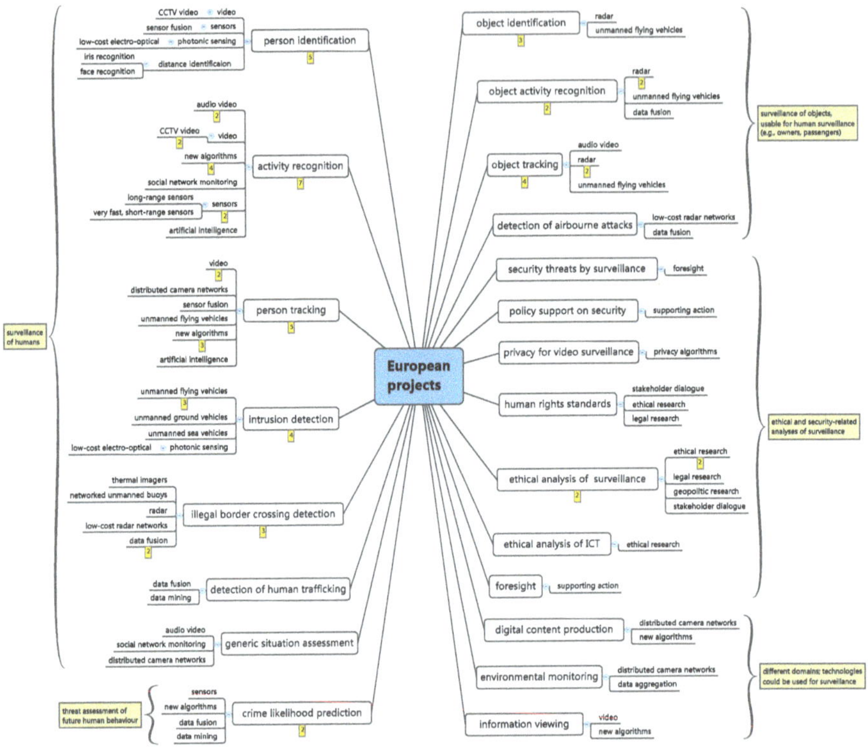

Fig. 9.4 European projects exploring different surveillance-related aspects. The first-level indicates the main aims of the projects, while the second layer indicates the technological means to achieve them. The small numbers indicate the frequency of occurrence across all surveyed projects

creation of advanced tools for what could be called "surveillance 2.0": apart from the development of traditional video surveillance technologies and threat detection systems, large research efforts of DARPA are devoted to the development of systems for the (entirely or partially automatic) monitoring of social networks.

While the majority of the enlisted EU projects, and the totality of the US ones, are technical, i.e., they focus on engineering issues and technological development and demonstrations, part of the European research effort is also devoted to the analysis of the broader ethical and legal implications of security technologies. The European Commission has funded research and supported activities on social implications of security technologies since its Fifth Framework Programme.[76] Starting with the current FP7, the Commission has included an "ethics, security and society" theme in the Security Programme under Activity 6 (Security and Society). The ESRIF report thus states that "ethical issues and full respect for privacy, liberty and civil rights are

[76] E.g., Changing landscape of European liberty and security (CHALLENGE), a project which took place from 2004 to 2008; European liberty and security (ELISE), 2004–2008; Bioethical Implications of Globalisation (BIG), 2002–2006.

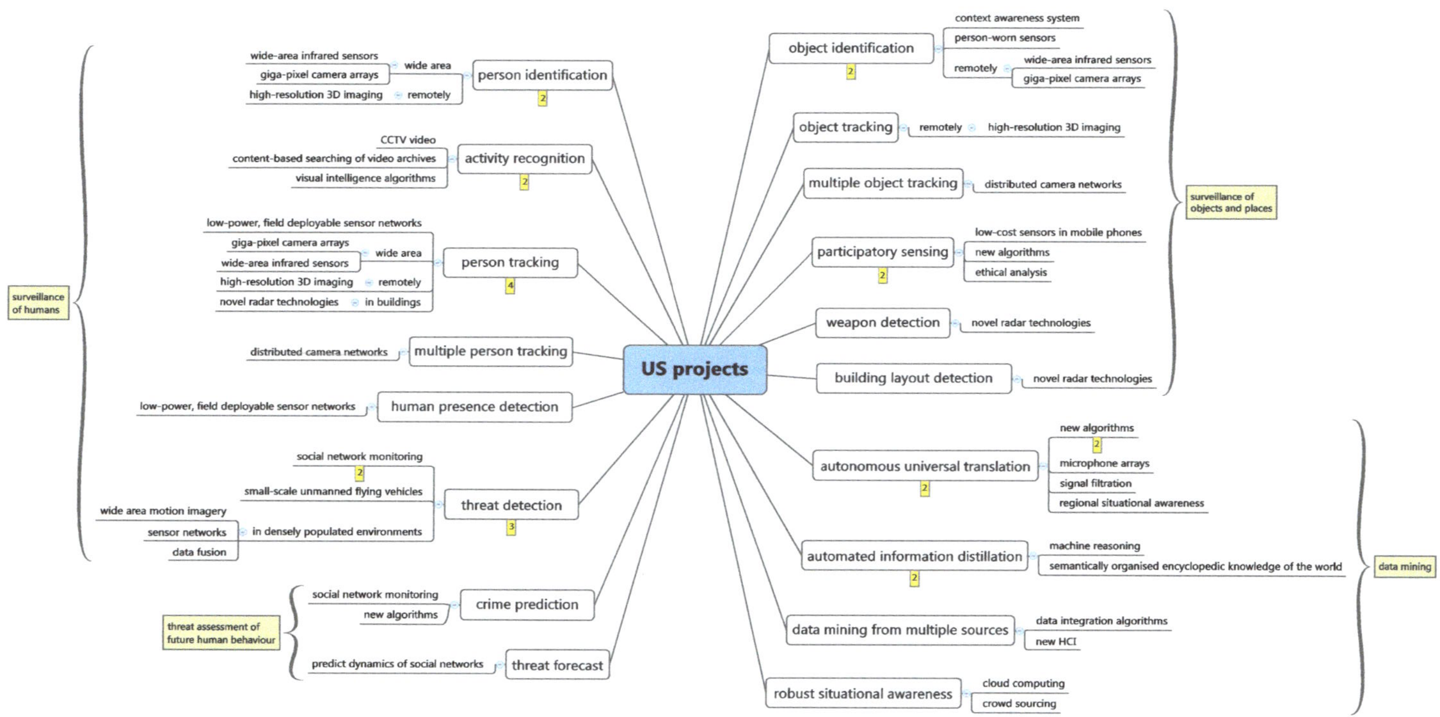

Fig. 9.5 Surveillance-related research in the US (DARPA and NSF). In contrast to EU-funded research, there is a notable absence of projects targeting ethical analysis

aspects that cannot be neglected in all present and future technological developments. A balance must be achieved between the privacy rights of citizens and the need to protect Europe and its citizens against threats."[77]

Despite its lack of projects focusing directly on ethical assessments, DARPA is committed to take into consideration all non-technical, sensitive aspects of its research:

> There is often a tension between novel concepts and an underdeveloped ethical, legal, and societal framework for addressing the full implications of such research. This is a problem not unique to DARPA. Other agencies have faced it, such as NIH, during the Human Genome Project. If we do our research well, we will necessarily bump up against these concerns.[78]

In order to address privacy implications, DARPA states that the agency will

> consistently examine the impact of its research and development on privacy, responsibly analyse the privacy dimension of its on-going research endeavours with respect to their ethical, legal and societal implications (ELSI), transparently respond to the findings of its assessments for unclassified work, and ensure independent review of its classified work, in accordance with a commitment to shared responsibility for addressing the privacy issue.[79]

To fulfil its responsibilities both to innovation and ethical assessment, DARPA has taken some initiatives, such as the creation of an internal independent privacy review panel that works in liaison with the Department of Defense Privacy Office, and the establishment of an ELSI working group together with the National Science Foundation.

Most projects financed by the European Commission and DARPA focus directly on different aspects of *human surveillance* (cf. Figs. 4 and 5). Five European and two US projects have *automatic identification* of persons amongst their main goals; nine projects (seven EU and two US) focus on the *automatic recognition of human activity*; a further 10 projects (five each) on the *tracking of persons*, including the simultaneous tracking of several persons; and another nine projects (eight EU and one US) aim at *intrusion detection*, either for a particular property or across the border.

Some projects aim at the *monitoring of objects*. While at first glance they might thus not seem relevant to an analysis of surveillance, they actually can be significant, either because the objects under scrutiny (such as luggage within an airport) can easily be matched to their owners, turning thus an object monitoring system into a human tracking and/or activity recognition one, or because the technology developed for such a project (such as unmanned flying drones fitted with a variety of sensors and cameras) can easily be used for human surveillance as well.

A second core area focuses on what is often called *big data*: combining the information from various sources (including the direct surveillance sources above), searching for patterns or stereotypes, and generating higher-level information. Dataveillance could be used to exploit big data, but the two terms are not synonymous. Dataveillance typically concerns the surveillance of individuals or groups via data trails, whereas the exploitation of big data may not necessarily concern the

[77] European Security Research and Innovation Forum (ESRIF) (2009).

[78] Defense Advanced Research Projects Agency, "DARPA's S&T Privacy Principles".

[79] Ibid.

surveillance of individuals or groups (although it could). While the European projects in this domain focus more on combining many individual information sources, the DARPA projects aim rather at generating new or better information from existing sources. There are two DARPA projects aiming at the machine-based distillation of information, and another two projects aiming at the autonomous translation of speech recorded in noisy natural environments (such as a crowd of people).

Note that in the European FP7 program, and unlike the DARPA projects, the *ethical analysis* of surveillance is plainly represented. From the total of 38 surveillance-relevant research projects identified, no fewer than seven investigate the ethics of state surveillance at different abstraction levels: from concrete technological proposals for a better privacy-compliance of video surveillance up to the effects of today's and tomorrow's surveillance on human rights.

9.5.2 *Emergent Technologies*

To achieve the aims of the projects listed above, numerous, heterogeneous technologies are under research within the different projects. While some projects have the exclusive development of a new technology at their core, most projects use them jointly with existing technologies to achieve a surveillance goal.

9.5.2.1 New Sensors

One of the most important technologies for surveillance is obviously video—a total of 14 projects use video data. While one project actually aims at developing new low-cost electro-optical components for the identification of persons and of possible intrusions, most projects rather use existing cameras and focus their technological efforts elsewhere, often in the development of smart surveillance algorithms evaluating the video input. Other types of sensors developed in European and US projects include:

- biometric sensors for remote identification and authentication,
- novel radar technologies for the identification of persons and objects (such as weapons) remotely, for example, inside buildings,
- sensor networks for the autonomous transmission of information between nodes in the area under scrutiny,
- new microphone arrays for voice recording in natural environments and subsequent automatic translation, and
- an architecture for "participatory sensing", which uses small custom sensor boxes (e.g., for sensing air quality) or even regular smart-phones (e.g., to measure noise levels) to "crowd-source" large-scale sensing tasks in a community/peer-to-peer/grassroots fashion.

9.5.2.2 Unmanned Vehicles as Mobile Sensor Platforms

Sensors need a physical platform from which to operate. While for numerous applications (for example, for biometric access control) a static platform is adequate, a mobile sensor platform opens new surveillance possibilities.

The recent surge of interest in unmanned aerial vehicles (drones) is well represented in European and US projects. Six projects in total make use of UAVs as mobile sensor platforms (cf. Figs. 9.4 and 9.5). As with video cameras, while most projects use existing drones in their technological mix aimed at surveillance, some try to advance the technology itself, such as developing very small-scale (a few centimetres) autonomous drones.

9.5.2.3 New Powerful Algorithms

The automatisation of surveillance (in other words, the emancipation from the processing limits imposed by human operators and from the risk of error due to their fatigue) lies at the core of all advanced surveillance projects. In all examined projects, software systems take care of identification, monitoring, tracking or activity recognition—human operators are sometimes consulted in a second step for the fine-tuning of already filtered events.

Due to these processing capabilities, but also to powerful, wide-area and detailed sensorial coverage, some of the projects display an impressive capacity for surveillance. DARPA's Wide Area Video Surveillance project, for example, can choose 130 independent targets within a Giga-pixel camera array (providing both video and infrared imagery) and automatically follow their movements. While the system is aimed at battlefield surveillance, such systems could be deployed for other surveillance tasks as well. DARPA also funds projects that explicitly develop algorithms for surveilling social networks, with the aim to infer current or future threats.

9.5.3 Future Smart Surveillance Assemblages

Combining the surveillance technologies most prevalent in current research projects with already existing technologies, and taking into account current societal, political and economic trends, we can arrive at a number of future smart surveillance systems and assemblages that are feasible, if not likely, to emerge over the next decade. Four such future smart surveillance scenarios are illustrated below.

9.5.3.1 Border and Crowd Control with Drone-Mounted Sensors

The detection of illegal trespassing ranks high among the aims of current European projects. No fewer than eight of them have "intrusion detection", "illegal border crossing detection" or "detection of human trafficking" as a core function related to the security of borders or infrastructures. To achieve such functions, a large number of the projects rely on sensors or networks of sensors mounted on unmanned aerial, sea or land vehicles.

Police forces already use drones for the video surveillance of rallies or sport events that might lead to riots, or for the security of sensitive meetings.[80] Video-equipped drones are also used for border control. It is likely that ever smaller and cheaper drones will be able to patrol increasingly long border stretches—not only North American ones, but the long maritime borders of Europe as well. The next logical step for UAVs used for crowd control is the development of smart CCTV algorithms for person tracking, facial and/or activity recognition—and indeed such development is the subject of on-going research.

Becoming more speculative, law enforcement might start using swarms of drones, the prospect for which is more related to societal acceptance than technological feasibility. Small crawling drones equipped with both optical and infrared cameras are already being produced,[81] and police forces in the US have been testing them for reconnaissance in dangerous environments. In a decade from now, such land-based drones might very well be equipped with iris scanners and be wirelessly connected to a biometric database, making them able to identify humans. Networked crawling, swimming and aerial drones might become a standard tool for law enforcement and counter-terrorist forces; the sensors with which they can be equipped, and the level of surveillance and intrusion they are allowed might become a matter of debate.

9.5.3.2 Lateral Surveillance with Drone-Mounted Sensors

Lateral surveillance has been defined as "peer monitoring" or "peer -to-peer surveillance of spouses, friends, and relatives".[82] Lateral surveillance could also be on the verge of a boom due to sensors mounted on unmanned vehicles, particularly UAVs. Hobby pilots of remote controlled (R/C) aircraft and helicopters routinely fly UAVs along the highest peaks of the Alps, along motorways, above private properties, and vertically along skyscrapers, as numerous clips on Internet film platforms show.[83] There is an obvious lateral surveillance potential to this development, for example, when filming unsuspecting targets from above their own property or through the windows of their apartment on the 45th floor.

In addition to cameras, UAVs can be equipped with other sensors used for different sorts of lateral surveillance. A slightly curious such example is represented by a rather large UAV carrying computation equipment strong enough to be used for communication surveillance. The drone can crack GSM encryption and carry

---

[80] For examples of how law enforcement authorities are using drones, see Finn and Wright (2012).

[81] See, for example, Recon Robotics.

[82] Andrejevic (2005).

[83] See, for example, the videos of the "Team Black Sheep". http://www.team-blacksheep.com/videos

out a man-in-the-middle attack for WiFi communication.[84] Other assemblages are possible: combining sensor-recorded data with public data from the yellow pages or the telephone book, it might be possible to match the data recorded by the UAV to known names and addresses. In a negative but possible scenario, such data could then be used for blackmailing these persons. By feeding the navigation system of flying drones with publicly available 3D models of a city, they could be sent autonomously for the targeted spying of a subject's known address.

9.5.3.3 Mandatory Crowd Sourcing

As discussed above, the ability to locate citizens (and/or some of the objects they own) in certain circumstances is, or will soon become, mandatory. Mobile telephony operators are required by law to be able to precisely locate their subscribers, and hand over this data to law enforcement authorities upon request. From 2015, all new vehicles sold in the EU will have to be equipped with the eCall system which will alert paramedics and the police in case of an accident and transmit the vehicle's whereabouts as well. Providers of mobile telephony are selling their customers' location data to producers of satellite navigation systems, who infer traffic jams from this information. Although the data is anonymised, the customer has nevertheless no possibility to opt out.[85]

Telephony providers might start handing over further data recorded by their customers, either because the law requires them or due to commercial interests. Such data could then be used for novel types of surveillance assemblages. Every telephone is, for example, inherently equipped with an audio sensor—its microphone. Law enforcement authorities in the US, and recently in Europe, have started to install sound ranging sensors in some cities for the automatic detection and localisation of gunshots. The costly part of such an operation is the deployment of the sensors throughout cities; the algorithms for filtering the sound of a firearm and for triangulating the position of the shooter are rather simple. If providers of mobile telephony were asked to run gunfire detection algorithms on the voice streams of their customers when they are making a call, the complex and costly part of gunfire detection would be easily tackled to almost the same results: most likely, there is always and anywhere someone talking on the phone. No software would need to be installed on the customer side; the algorithm would only analyse the voice streams within the premises of the providers. And the location, while not as precise as the one provided by dedicated audio sensors installed in known locations, would also be precise enough to constitute valuable information for automatically dispatched police forces. This assemblage represents a technologically feasible example for future mandatory crowd sourcing.

---

[84] Flacy (2011).

[85] TomTom, "Real-time traffic information". http://www.tomtom.com/landing_pages/traffic_solutions/web/

9.5.3.4 Smart "Blackbox" for Communication/SMS Surveillance

As discussed above, the surveillance of communication is becoming both easier in some aspects, and more difficult in others. However, on balance, it seems that the surveillance of communications is becoming rather more difficult than it was at the initiation of the Echelon programme of the US, UK, Canada, Australia and New Zealand, which has been credited with the ability to intercept a good deal of the worldwide radio and satellite communications.[86] Nowadays, however, fibre optic is the medium used by the vast majority of communications.[87] Such communications can only be intercepted by placing a wiretap at one of the points where such communication is being switched, making it harder for foreign intelligence agencies to access the inland communications of a third country; just pointing an antenna from their own territory towards the skies does not suffice any longer. At the same time, communication surveillance by law enforcement can also easily be avoided through the use of VoIP services such as Skype, which uses encryption algorithms that, as mentioned before, are currently considered impenetrable.[88] Finally, protesters or rioters are aware of the routine monitoring of social networks by law enforcement, and have started to warn of the usage of such media and spread the word of planned actions only by text messages sent to known friends.[89]

In this context, it is likely that law enforcement has to seek out new, smarter ways of communication surveillance. Given that the e-mail service Gmail scans its customers' e-mails for keywords and tries to find matching advertisements, it is conceivable that text messages could be scanned by the mobile telephony provider and potentially suspect messages presented to a human operator. The content of text messages could further be stored by the provider (either in plaintext or encrypted). When identical messages are being noticed in a short interval of time, indicating a possible call for public disobedience, this content would be forwarded to a human operator. As all messages have been stored, the senders and receivers would be easily found. Such "smart" communication surveillance system would not even have to be continuously turned on; it could be switched on before important political meetings, football matches or other sensitive events. Diverting all inland fibre-based communication from the telephony and Internet operators to secret services premises is another possibility; one that would allow these to eavesdrop on all the non-encrypted inland communications. Through keyword-searching algorithms and sensitivity levels set according to databases of suspected terrorists, such system could be an effective

---

[86] European Parliament, "Report on the existence of a global system for the interception of private and commercial communications (ECHELON interception system) (2001/2098(INI))", Rapporteur: Gerhard Schmid, A5–0264/2001, 11 July 2001. http://www.europarl.europa.eu/sides/getDoc.do?pubRef=-//EP//NONSGML+REPORT+A5-2001-264+0+DOC+PDF+V0//EN&-language=EN. See also Bamford (2009).

[87] European Parliament, op. cit.

[88] Even so, politicians and law enforcement authorities are pushing the new Internet companies to "co-operate" so that digital communications can be intercepted. See, for example, Savage (2010).

[89] As, for example, the organisers of an illegal party recently did in Zurich. See Schindler (2011).

counter-terrorist measure. Using automatic universal translators (as several are under research in DARPA projects) would make it language-independent.

9.6 Conclusions

This article has highlighted the ways in which both current and emerging technologies are increasingly being organised into assemblages or "smart surveillance" systems, where surveillance systems are becoming integrated, multi-modal, automated, ubiquitous and increasingly accepted by the public.[90] We have shown that contemporary surveillance involves different technologies and is used in different settings, for a range of purposes. In addition to more traditional criminal justice and national security applications, we find surveillance technologies, and often systems of surveillance technologies, in public spaces, mass transit, air travel, consumer space and combined with technologies or systems associated with communication. This means that as individuals travel back and forth to work or on errands, shop in-store or online, visit their town centre, communicate with friends and family, watch television, go on holiday, surf the Internet or even go for a hike near national borders, they are often subject to surveillance by a range of systems. As such, surveillance technologies have become part of our daily infrastructure and part of the activities that we undertake on a day-to-day basis. Such surveillance has "enter[ed] our daily life without notice, [and] become a common part of our socio-political and economic relations, so that we become acclimatised or accustomed to surveillance".[91] There is no doubt some surveillance yields social benefits, but equally there is no doubt that those controlling surveillance systems gain more power over those surveilled and targeted. Benjamin Goold speaks of the political dangers of surveillance and counsels that "We should resist the spread of surveillance not because we have something to hide, but because it is indicative of an expansion of state power. While individuals might not be concerned about the loss of autonomy that comes from being subjected to more and more state scrutiny, it is unlikely that many would be comfortable with the suggestion that more surveillance inevitably brings with it more bureaucracy and bigger, more intrusive government."[92] It is not just an expansion of state power; it is also an expansion of corporate power—especially of a few corporate behemoths such as Google and Facebook—as well as carrying an increased threat for malicious actions by individuals (e.g., stalkers) in the form of "lateral surveillance".

---

[90] Several Home Office studies have found evidence of strong public support for surveillance cameras. One found that "the level of support for CCTV remained high at over 70 % of the sample in all but one area" of the 13 schemes the study had assessed. Other research found that "levels of support for CCTV are high, although it was not clear that respondents were fully informed about how it functioned". House of Commons Home Affairs Committee (2007–2008). As evidence, the report cites Gill and Spriggs (2005) *Assessing the impact of CCTV* (London: Home Office Research, Developments and Statistics Directorate, 2005), p. ix; Spriggs, Argomaniz et al (2006).

[91] Wright et al., op. cit., 2010, p. 344, n. 3.

[92] Goold (2009).

From our review of surveillance technologies and applications, the ways in which they are being integrated into "assemblages" and, looking into the near future, developments on the horizon, one can only conclude that surveillance is indeed becoming ubiquitous in our societies, both online and in the physical world. Surveillance systems are capable of watching over us (the citizenry), tracking us and making assumptions about us in many different ways. Undoubtedly, as surveillance systems become more pervasive, they eat away at our privacy. And if one sees privacy as a cornerstone of democracy,[93] then surveillance systems are undermining the very political system some of them are supposed to protect. This is the great irony.

The ability of surveillance systems to monitor whatever we are doing and wherever we are going has greatly encroached upon privacy in the last 20 or 30 years. Privacy guru James Rule has commented that "The ridiculous ease of compiling data previously evanescent or unavailable challenges us to re-think our very idea of what is public and private."[94] Should citizens throw up their hands in despair? Have we lost our privacy to the rapacity of governments and multinationals? Is there nothing to be done to protect our privacy—and with it our democracy?

Like Benjamin Goold, James Rule argues that we can and should resist the depredations against our privacy. He makes some practical suggestions. Regarding surveillance by government entities, he says

> The new default condition for public policy should be: no government surveillance without meaningful individual consent or legislative authorization.... For government surveillance, elected officials should be required to authorize each appropriation of personal data in every government surveillance system.... The aim would be to *politicize* the working and extension of surveillance.[95]

Similarly, for the private sector

> a parallel precept should apply: no use of personal data for institutional surveillance without meaningful, informed consent from the individual.... Taking privacy seriously would entail that any commercialization of personal data, either from government files or private-sector records, would require active assent from the individual concerned. In the jargon of privacy-watchers, "opt in" would be the rule: no commerce in personal information from any source would be possible without adequate notice and explicit consent from the individual.[96]

Rule's propositions are congruent with what the European Commission has advocated in the proposed Data Protection Regulation, i.e., explicit user consent would be required for third-party use of personal data, a provision which is being opposed by the likes of Google, Yahoo, Amazon, Facebook and other large corporations whose growth has been fuelled by their use of our personal data. In some very real sense, the fate of the proposed Data Protection Regulation is intimately linked to the fate

---

[93] The Supreme Court of Canada has stated that "society has come to realize that privacy is at the heart of liberty in a modern state... Grounded in man's physical and moral autonomy, privacy is essential for the well-being of the individual". R. v. Dyment (188), 55 D.L.R. (4th) 503 at 513 (S.C.C.).

[94] Rule (2007).

[95] Ibid.

[96] Ibid., p. 196.

of our privacy and our sense of democracy. How pernicious the surveillance society becomes hangs in the balance.

Acknowledgements This paper has been prepared based in part on research undertaken in the context of the SAPIENT project (Project number: 261698), funded by the European Commission's Directorate General Enterprise. The views expressed in this paper are those of the authors alone and are in no way intended to reflect those of the European Commission.

References

Aaronovitch, David. 2009. The strange case of the surveillance cameras. The Times, 3 Mar 2009. http://www.timesonline.co.uk/tol/comment/columnists/david_aaronovitch/article5834725.ece.

Adkins, Lauren D. 2007. Biometrics: Weighing convenience and national security against your privacy. 2007. *Michigan Telecommunications Technology Law Review* 13:541–555. http://www.mttlr.org/volthirteen/adkins.pdf.

Andrejevic, Mark. 2005. The work of watching one another: Lateral surveillance, risk, and governance, surveillance & society 2 (4): 479–497 [481]. http://www.surveillance-and-society.org.

Bamford, James. 2009. *The shadow factory*. New York, Anchor Books, 161–163.

Bauman, Zygmunt, and David Lyon. 2013. *Liquid Surveillance*. Cambridge: Polity Press, 2–3.

Britannica. Electronic Eavesdropping, *Encyclopædia Britannica*, 2011. http://www.britannica.com/EBchecked/topic/183788/electronic-eavesdropping.

Clarke, Roger. 1988. Information technology and dataveillance. *Communications of the* ACM 31 (5): 498–512.

Clarke, Roger. 1993. Profiling: A hidden challenge to the regulation of data surveillance. *Journal of Law and Information* Science 4 (2): 403–419.

Clarke, Roger. 2003. *Dataveillance—15 years on*. Wellington: Privacy Issues Forum, 15–18.

Coroama, Vlad. 2006. The smart tachograph—Individual accounting of traffic costs and its Implications. *Proceedings of the 4th International Conference PERVASIVE 2006*, Dublin, Ireland, Springer, May 2006, 135–152.

Council of the European Union. 2010. The Stockholm programme—An open and secure Europe serving and protecting citizens, 5731/10, Brussels: 3 March 2010.

Defense Advanced Research Projects Agency (DARPA). 2009. Strategic Plan, May 2009.

Defense Advanced Research Projects Agency, "DARPA" sS & T Privacy Principles. http://www.darpa.mil/About/Initiative/DARPA%E2%80%99s_S_T_Privacy_Principles.aspx.

Diffie, Whitfield, and Susan Landau. 2009. Communications surveillance: Privacy and security at risk. *Communications of the ACM* 52 (11): 42–47.

Directorate of Airspace Policy. 2010. *CAP 722: Unmanned Aircraft System Operations in UK Airspace—Guidance*, Civil Aviation Authority, 6 Apr 2010.

European Security Research and Innovation Forum (ESRIF). 2009. Final Report, December 2009, 155. www.esrif.eu.

Fay, Joe. 2005. Dutch give football thugs a good talking to. The Register, 2 Sept 2005. http://www.theregister.co.uk/2005/09/02/dutch_hooligans/.

Finn, Rachel L., and David Wright. 2012. Unmanned aircraft systems: Surveillance, ethics and privacy in civil applications. *Computer Law & Security Review* 28 (2): 184–194. http://www.sciencedirect.com/science/journal/02673649.

Figueiras, J., and S. Frattasi. 2010. *Mobile positioning and tracking: From conventional to cooperative techniques*. Indianapolis: Wiley.

Frawley, W.J., G. Piatetsky-Shapiro, and C. J. Matheus. 1992. Knowledge discovery in databases: An overview. *AI Magazine* 13 (3): 57–70.

Flacy, Mike. 2011. Men build small flying spy drone that cracks Wi-Fi and cell data. *Digital trends*, 30 July 2011. http://www.digitaltrends.com/mobile/men-build-small-flying-spy-drone-that-cracks-wi-fi-and-cell-data/.

Gardner, Julian W., and Philip N. Bartlett. 1999. *Electronic noses: Principles and applications*. Oxford: Oxford University Press.

GeneWatch UK. The UK police national DNA database. http://www.genewatch.org/sub-539478.

Goold, Benjamin J. 2009. Surveillance and the political value of privacy. *Amsterdam Law Forum* 1 (4): 3–6, [5]. http://www.amsterdamlawforum.org/.

Haggerty, Kevin D., and Richard V. Ericson. 2000. The surveillant assemblage, *British Journal of Sociology* 51 (4): 605–622.

Hildebrandt, Mireille. 2008. Defining profiling: A new type of knowledge? In *Profiling the European Citizen: Cross-Disciplinary* Perspectives, eds. Mireille Hildebrandt and Serge Gutwirth, 17–46. Heidelberg: Springer.

House of Commons Home Affairs Committee. (2007–2008). *A Surveillance Society? Fifth Report of Session* 2007–2008, Vol. I. London: The Stationery Office Limited, 25.

Introna, Lucas D., and Helen Nissenbaum. 2009. *Facial recognition technology: A survey of policy and implementation issues*, The Center for Catastrophe Preparedness & Response, New York University. New York: 8 Apr 2009.

Jones, Jeffrey M. 2010. In U.S., air travelers take body scans in stride, *Gallup*, 11 Jan 2010. http://www.gallup.com/poll/125018/Air-Travelers-Body-Scans-Stride.aspx.

Langheinrich, Marc. 2009. Privacy in ubiquitous computing. In *Ubiquitous Computing Fundamentals*, ed. J. Krumm, 95–160. Chapman & Hall/CRC Press.

Lee, David. 2010. Using thermal cameras to secure the homeland, photonics.com, Feb 2010. http://www.photonics.com/Article.aspx?AID=40915.

Lyon, David. 2003. Introduction. In *Surveillance as social sorting: Privacy, risk, and digital discrimination*, ed. David Lyon, 8. London: Routledge.

Lyon, David. 2008. Biometrics, identification and surveillance. *Bioethics* 22 (9): 500.

Lyon, David. 2007. *Surveillance Studies: An Overview*, 15. Cambridge: Polity.

Marx, Gary T. 2002. What's new about the 'New Surveillance'?: Classifying for change and continuity. *Surveillance and Society* 1 (1): 9–29 [12].

Mathiesen, Thomas. 1997. The viewer society: Michel Foucault's 'Panoptique' Revisited. *Theoretical Criminology* 1 (2): 215–34.

McBride, Paul. 2009. Beyond orwell: The application of unmanned aircraft systems in domestic surveillance operations. *Journal of Air Law and Commerce* 74:628.

McCahill, Michael. 2002. The surveillance web: *The rise of visual surveillance in an English city*. Devon: Willan.

McElroy, Wendy. 2011.Commentary—'Pre-criminal' profiling may be coming soon to an airport near you. *TriValleyCentral.com*, 4 Aug 2011. http://trivalleycentral.com/articles/2011/08/04/florence_reminder_blade_tribune/top_stories/doc4e39b5da387f1946508835.txt.

Mireille Hildebrandt and Serge Gutwirth, eds. 2008. Profiling the European citizen: cross-disciplinary perspectives. Heidelberg: Springer, 17–46.

Naraine, Ryan. 2007. First look: Sentry remote and eBlaster 6.0, PC World. http://www.pcworld.com/article/139460/first_look_sentry_remote_and_eblaster_60.html.

Norris, Clive, and G. Armstrong. 1999. *The maximum surveillance society: The rise of CCTV*. Oxford: Berg.

Nouwt, Sjaak, Berend R. de Vries, and Dorus van der Burgt. 2005. Camera surveillance and privacy in the Netherlands. *Social Studies Research Network*, 2. http://ssrn.com/abstract=849205.

Petersen, J.K. 2007. *Understanding surveillance technologies: Spy devices, privacy, history & applications* (2nd ed.). Auerbach Publications: Boca Raton.

Prevelakis, Vassilis, and Diomidis Spinellis. 2007. The Athens affair. IEEE *Spectrum* 18–25.

Plungis, Jeff. Naked-image scanners to be removed from U.S. airports. Bloomberg News, 18 Jan 2013. http://www.bloomberg.com/news/2013-01-18/naked-image-scanners-to-be-removed-from-u-s-airports.html.

Polgreen, Lydia. 2011. Scanning 2.4 Billion eyes, India tries to connect poor to growth. The New York Times, 1 September 2011. http://www.nytimes.com/2011/09/02/world/asia/02india.html?_r=1&src=me&ref=general
Recon Robotics. Recon Scout XT. http://www.reconrobotics.com/products/recon-scout_XT.cfm.
Rothstein, Mark A., and Meghan K. Talbott. 2006. The expanding use of DNA in law enforcement: What role for privacy? *Journal of Law, Medicine & Ethics*, Summer. 155. http://ssrn.com/abstract=1512746.
Rule, James B. 2007. *Privacy in Peril*. Oxford University Press, 195.
Savage, Charlie. 2010. U.S. tries to make it easier to wiretap the internet. *The New York Times*, 27 Sept 2010. http://www.nytimes.com/2010/09/27/us/27wiretap.html?_r=1 & hp.
Schindler, F. 2011. Tumulte in der Zürcher Innenstadt – Polizei mit Grosseinsatz, *Tagesanzeiger*, 11 September 2011. http://www.tagesanzeiger.ch/panorama/vermischtes/Tumulte-in-der-Zuercher-Innenstadt–Polizei-mit-Grosseinsatz/story/22640435.
ShotSpotter. The ShotSpotter Gunshot Location System. http://www.shotspotter.com/technology.
Skyhook. How it Works. http://www.skyhookwireless.com/howitworks/.
Spriggs, Argomaniz, et al. 2006. *Public attitudes towards CCTV: results from the Pre-intervention public attitude survey carried out in areas implementing CCTV*, Home Office Online Report, October 2006, 49.
Stefani, John A. 2006. Finding Waldo. In Privacy and Security Technologies: An Interdisciplinary Conversation, eds. Katherine J. Strandberg and Danela Stan Raicu, 173–188. New York: Springer.
Want, R. 2009. An introduction to ubiquitous computing. In Ubiquitous Computing *Fundamentals*, ed. J. Krumm, 1–36. Chapman & Hall/CRC Press.
Webster, C. William R. 2009. CCTV policy in the UK: reconsidering the evidence base. *Surveillance & Society* 6 (1): 10–22. http://www.surveillance-and-society.org.
Wei, Gang, and Dongge Li. 2006. Biometrics: Applications, challenges and the future. In *Privacy and Security Technologies: An Interdisciplinary Conversation*, eds. Katherine J. Strandburg and Danela Stan Raicu, 135–150. New York: Springer.
Wolfe, Alexander. 2010. Wolf's Den: IBM patenting airport security profiling technology. *InformationWeek*, 19 Jan 2010. http://www.informationweek.com/news/government/security/222301388.
Wright, David, Michael Friedewald, Serge Gutwirth, Marc Langheinrich, et al. 2010. Sorting out smart surveillance. *Computer Law & Security Review* 26 (4): 343–354.
Zureik, Elia, and Karen Hindle. 2004. Governance, security and technology: The case of biometrics. *Studies in Political Economy, Vol. 73, Spring/Summer*, 113–137.

Chapter 10
Surveillance of Communications Data and Article 8 of the European Convention on Human Rights

Nora Ni Loideain

> The danger threatening democratic societies ... stems from the temptation facing public authorities to 'see into' the life of the citizens.[1]

10.1 Introduction

Article 8 of the European Convention on Human Rights ("ECHR" or the "Convention") guarantees respect for private life and correspondence, a legal instrument that has raised standards in the protection of individuals' freedoms through the reform of domestic laws and practices across Europe.[2] The following analysis of Article 8 comes at a pertinent time for the use of communications data by law enforcement authorities ("LEAs") within the EU due to technological advances in information processing, the advent of the Internet and regulatory developments.

The legislative framework, the EU Data Retention Directive[3], currently governing this area of surveillance is under review by the EU Commission.[4] This Directive requires the mandatory retention of every European citizen's communications data for up to two years for the investigation, detection and prosecution of serious crime, as defined by each Member State under their national law.[5] It is expressly provided under the Directive that compliance with the requirements of Article 8 of the ECHR is "a necessary measure" for this EU legal instrument.[6] Furthermore, challenges

[1] *Malone v. United Kingdom* (1985) 7 EHRR 14, concurring judgment of Judge Pettiti, 38.

[2] See Drzemczewski (1983); Gearty (1993); Keller and Stone-Sweet (2008).

[3] Council Directive (2006, OJ 2006 L 105 p. 54) (hereafter, the "Data Retention Directive"). Article 1(1) of the Directive specifically refers to "operators of publicly available electronic communications services or of public communications networks". These include fixed-line and mobile telephone companies as well as Internet service providers. They will be collectively referred to hereafter as "telecommunications operators".

[4] See EU Commission Directorate General for Home Affairs.

[5] Council Directive (2006, Art. 1 (1)).

[6] Council Directive (2006, Recital 9).

N. N. Loideain (✉)
Magdalene College, Cambridge, CB3 0AG, UK
e-mail: nl301@cam.ac.uk

DOI 10.1007/978-94-007-7540-4_10,

concerning the necessity and proportionality of the Data Retention Directive are pending before the Court of Justice of the European Union ("CJEU").[7] These pending assessments by the Commission and the CJEU will determine the future operation and scope of this legislation, matters that have already been the source of considerable unease amongst legislators, data protection authorities and telecommunications operators.[8]

Thus, Article 8 ECHR has an integral role to play in the review, development and enforcement of the law, policy and procedure that will protect the right to respect to private life of EU citizens in an environment where the technology to monitor their every communication and movement becomes more sophisticated.

This paper sets out and analyses the current scope of protection provided under Article 8 ECHR to this area of State surveillance and how the Strasbourg Court has applied the relevant jurisprudence in practice. The paper then proceeds to reflect on the need for the Court to exercise greater consistency and scrutiny in its application of Article 8 ECHR when reviewing the compliance of the use and retention of communications data by LEAs following its acquisition. As the Court has consistently emphasized, powers to instruct covert surveillance of citizens are only tolerated under Article 8 to the extent that they are strictly necessary for safeguarding democratic institutions.[9] In practice, this means that the Court must consistently review with rigour whether or not State have adequate and effective guarantees in place to prevent abuse of these powers.[10] The Court's role as "guardian of the Convention"[11] is particularly important in this respect in light of the fact that this State power is exercised in secret. Otherwise, a real risk exists that the standard of protection provided under the Convention in this area may become more illusory than real.

---

[7] Preliminary references concerning the Directive have been raised by the Constitutional Courts in Ireland and Austria. Questions concerning the role of Article 8 ECHR and the operation of the Directive have been raised in both of the preliminary rulings submitted to the CJEU. See Case C-293/12, Reference for a preliminary ruling from High Court of Ireland made on 11 June 2012—*Digital Rights Ireland Ltd v Minister for Communications, Marine and Natural Resources, Minister for Justice, Equality and Law Reform, The Commissioner of the Garda Síochána, Ireland and the Attorney General.* For details on the preliminary ruling sought by the Austrian Constitutional Court, see the press release by Verfassungsgerichtshofes, an Austrian privacy NGO, http://www.vfgh.gv.at/cms/vfghsite/attachments/2/7/9/CH0003/CMS1355817745350/press_release_data_retention.pdf.

[8] The review of the Directive is ongoing and follows an earlier assessment undertaken by the Commission of the legislation's implementation in Member States. Overall, the Commission found that the evaluation "demonstrated that data retention is a valuable tool for criminal justice systems and for law enforcement in the European Union". The Commission also found, however, that the Directive has thus far failed in its main objective as the "contribution" of the legislation to the harmonization of data retention across the EU has been "limited": Report from the Commission to the Council and the European Parliament, Evaluation report on the Data Retention Directive (Directive 2006/24/EC), COM(2011) 225 final Brussels (18.4.2011), 1. For further, see Ni Loideain (2012).

[9] *Klass v. Germany* (1978) 2 EHRR 214, para. 42.

[10] Kennedy v. United Kingdom *(2010) ECHR 682,* para. 153.

[11] *Malone v. United Kingdom* (1985) 7 EHRR 14, concurring judgment of Judge Pettiti, 43.

10.2 Approach of the Strasbourg Court

10.2.1 Application of General Principles

Article 8 of the Convention provides:

1. Everyone has the right to respect for his private and family life, his home and his correspondence.
2. There shall be no interference by a public authority with the exercise of this right except such as is in accordance with the law and is necessary in a democratic society in the interests of national security, public safety or the economic well-being of the country, for the prevention of disorder or crime, for the protection of health or morals, or for the protection of the rights and freedoms of others.[12]

The Strasbourg Court has decided a significant number of cases concerning Article 8 ECHR and the covert surveillance of communications by law enforcement and secret intelligence authorities.[13] The bulk of this case law has focused on the interception of content, particularly in relation to telephone conversations. In contrast, scrutiny of the acquisition and processing of "communications data" by the State has only been the focus of a few cases before the Court.[14]

The term communications data relates to information generated or processed as a consequence of a communication's transmission. The scope of this data does not include the content of the communication in question. There is no international consensus on a definition for communications data, or "traffic data" as it has been otherwise referred to, and these terms have been used interchangeably in the relevant literature. As a result, states and organisations, within and outside of the EU, have adopted different definitions of the concept for law enforcement purposes, varying in scope.[15] The obstacles posed by the legal and technical differences of these different frameworks in Member States across the EU for telecommunications operators was attributed as one of the main factors which led to the adoption of the Data Retention Directive.[16] Since the Directive came into effect, a trend has emerged among

---

[12] On Article 8 generally, see White and Ovey (2010, chs. 14–16); Harris et al. (2009, Harris *supra* n. 12, chs. 8–9); Feldman (2002, p. 523–542).

[13] The literature examining the area of covert surveillance and Article 8 is extensive. See, e.g., Harris et al. (2009, p. 400–404); Feldman (2002, p. 664–667). See also JUSTICE (2011, ch. 2).

[14] *Valenzuela Contreras v. Spain* (1999) 28 EHRR 483; *P.G. and J.H. v. United Kingdom* (2001) ECHR 546; *Copland v. United Kingdom* (2007) 45 EHRR 37.

[15] Young (2004–2005, pp. 346, 372–373).

[16] Council Directive (2006, Recitals 5–8).

policy-makers[17] and commentators[18] in the EU showing an increased use of the terms "communications data" and "telecommunications data".

There has also been a limited amount of case law in this area involving Internet-based communications, e.g. email and instant messaging *via* email and social networking sites, web browsing and 'VOIP' calls (Internet-based telephony, e.g. Skype).[19] As reflected in more recent case law, however, this is gradually changing as the use of these modern forms of communication surpasses the traditional landline telephone.

The use of covert surveillance by LEAs concerns the primary duty of the State under Article 8 not to interfere with the private life and correspondence of the applicant.[20] The application of Article 8 where a negative obligation arises involves a two-stage test.[21] The first stage involves the consideration of whether the complaint falls within the scope of Article 8(1). The second stage under Article 8(2) involves the requirement of safeguards to be put in place to prevent against the arbitrary use of this covert power by the State. Positive obligations have been found to arise in the context of the covert surveillance of communications where the intercepted conversation has been disclosed based on 'respect' for the protected interest of private life.[22] The Court then needs to determine whether the national authorities took the necessary steps to ensure effective protection of the applicant's right to respect for his private life and correspondence.[23]

As will be outlined below, the Court has progressively developed a general set of principles governing the covert surveillance of communications that constitute an interference under Article 8(1) and the requirements to be satisfied before a state can justify an interference by reference to the conditions of Article 8(2).[24]

---

[17] See the statutory legislation in the United Kingdom, Ireland and references by the European Commission in its evaluation of the Data Retention Directive: The Regulation of Investigatory Powers (Communications Data) Order 2003 (S.I. No.3127 of 2003) (United Kingdom); the Communication (Retention of Data) Act 2010 (Ireland); Report from the Commission to the Council and the European Parliament, Evaluation report on the Data Retention Directive (Directive 2006/24/EC), COM (2011) 225 final Brussels (18.4.2011), (hereafter, the "Evaluation of the Directive"), 11–12.

[18] See, e.g., Maras (2012, p. 447); Brown (2011, p. 95); Brown and Korff (2009, p. 119, 124).

[19] *Liberty v. United Kingdom* (2009) 48 EHRR 1 concerned legislation allowing for the interception of all public telecommunications, including email between the United Kingdom and any individual or transmitter outside of the UK; *Copland v. United Kingdom* (2007) 45 EHRR 37 concerned the surveillance of a state employee's Internet usage and emails.

[20] Starmer (1999, p. 129).

[21] For an analysis of the application of Article 8(2) to covert surveillance generally, see White and Ovey (2010, *supra* n. 12, pp. 365–371); Harris (2009, *supra* n. 12, pp. 397–404); Feldman (2002, *supra* n. 12, pp. 665–667).

[22] *K.U. v. Finland* (2009) 48 EHRR 1237, para. 42; *Craxi v. Italy (No. 2)* (2004) 38 EHRR 995.

[23] *Craxi v. Italy (No. 2)* (2004) 38 EHRR 995, para. 73; *Draksas v. Lithuania* (App. 36662/04) (31 July 2012), para. 62.

[24] Emmerson et al. (2007, pp. 281–284).

10.2.1.1 Engaging Article 8(1)

It has been well established by the Strasbourg Court that telephone conversations fall within the notions of 'private life' and 'correspondence'[25] and are protected under Article 8(1) whether they are made to, or received from, the home or a business premises.[26] An interception also interferes with the right to respect for private life and correspondence even if no use is subsequently made of the material.[27]

The Court has not considered it possible or necessary to attempt an exhaustive definition of the notion of 'private life'[28] but it has been held to cover the "physical and psychological integrity" of the individual.[29] As a result, this concept has been interpreted broadly. Its scope of protection applies to the interests within the inner circle of an individual's private life and extends to the zone of interaction of an individual with others.[30] The interception of communications constitutes a breach of physical and psychological integrity and amounts therefore to a breach of Article 8(1). Even if no physical intrusion into a private place occurs, surveillance can interfere with physical and psychological integrity and the right to respect for private life.[31] The mere fact that a secret regime of monitoring exists creates "a very real menace" for individuals that their exercise of the right to respect for their private and family life and their correspondence may be the subject of surveillance.[32]

The scope of the notion of correspondence is wide and was extended to apply to email and Internet usage in *Copland v. United Kingdom*.[33] This extension shows that these protected interests are keeping pace with technological developments in telecommunications[34] that have blurred the line between an individual's offline and online lives[35]. Mobile telephony with Internet access is a clear example of this development. The power to track an individual's offline and online communications and movements by monitoring every use of their mobile phone is a demonstration of this.

---

[25] *Klass v. Germany* (1978) 2 EHRR 214, para. 41; *Malone v. United Kingdom* (1985) 7 EHRR 14, para. 64; *Lambert v. France* (2000) 30 EHRR 346, para. 21; *Amann v. Switzerland* (2000) ECHR 87, para. 44; *Draksas v. Lithuania* (App. 36662/04), 31 July 2012, para. 52.

[26] *Niemietz v. Germany* (1992) 16 EHRR 97, para. 29; *Halford v. United Kingdom* (1997) 24 EHRR 523, para. 44; *Amann v. Switzerland* (2000) ECHR 87, para. 44.

[27] *Kopp v. Switzerland* (1999) 27 EHRR 91, para. 53.

[28] *Niemietz v. Germany* (1993) 16 EHRR 97; *Peck v. UK* (2003) 36 EHRR 719.

[29] *S and Marper v. United Kingdom* (2008) ECHR 1581, para. 66.

[30] *Niemietz v. Germany* (1993) 16 EHRR 97, para. 29; *P.G. and J.H. v. United Kingdom* (2001) ECHR 546, para. 56.

[31] Moreham (2008, pp. 44, 53).

[32] *Klass v. Germany* (1978) 2 EHRR 214, 29 (separate opinion of Judge Pinheiro Farinha who concurred with the Court's judgment but on different grounds).

[33] (2007) 45 EHRR 37, para. 41.

[34] Harris (2009, *supra* n. 12, p. 381).

[35] Gillespie (2009, pp. 552, 565).

10.2.2 *Conditions for Justifying an Interference with a Right: Article 8(2)*

10.2.2.1 'In Accordance with the Law'

The Court has been rigorous in its application of the 'accordance with the law' test in cases involving the covert surveillance of communications.[36] States seeking to justify challenges to such measures frequently fall at this hurdle of Article 8(2) before the necessity and proportionality tests are even considered.[37] The accordance with the law requirement concerns the principle of legality and has two main elements. Firstly, the interference must be in accordance with domestic law. Secondly, the 'quality' of the law, or the legality requirement, governing the exercise of this state power is requires that the measure must comply with the rule of law.[38] This means that the domestic law must provide protection against arbitrary interference with an individual's rights under Article 8.[39] This is an important requirement due to the lack of scrutiny and the risk of misuse of this secret power by LEAs.[40] In order for States to meet this aspect of Article 8(2) in relation to the interception of communications, the Court has developed a set of minimum safeguards that should be provided for under a statutory framework[41] and include:

> [T]he nature of the offences which may give rise to an interception order; a definition of the categories of people liable to have their telephones tapped; a limit on the duration of telephone tapping; the procedure to be followed for examining, using and storing the data obtained; the precautions to be taken when communicating the data to other parties; and the circumstances in which recordings may or must be erased or the tapes destroyed.[42]

The legality requirement consists of two key tests: accessibility and foreseeability. This means that the domestic law must be sufficiently accessible and clear in its terms

---

[36] Arai-Takahashi (2002, *supra* n. 61, p. 74).

[37] McHarg (1999, pp. 671, 685), observes, the requirements of Article 8(2) "are clearly designed to give some priority to rights by raising the hurdles which states must overcome in claiming public interest defences."

[38] *Khan v. United Kingdom* (2001) 31 EHRR 1016, para. 26; *P.G. and J.H. v. United Kingdom* (2001) ECHR 546, para. 44; *Copland v. United Kingdom* (2007) 45 EHRR 37, para. 46.

[39] *Khan v. United Kingdom* (2001) 31 EHRR 1016, para. 26.

[40] *Copland v. United Kingdom* (2007) 45 EHRR 37, para. 45; *Halford v. United Kingdom* (1997) 24 EHRR 523, para. 49.

[41] *Liberty v. United Kingdom* (2009) 48 EHRR 1; *Huvig v. France* (1990) 12 EHRR 547; *Amann v. Switzerland* (2000) ECHR 87; *Valenzuela Contreras v. Spain* (1999) 28 EHRR 483. The Court has accepted that common law, or case law, may also satisfy the legality test but only if it is sufficiently detailed, clear and precise.

[42] *Huvig v. France* (1990) 12 EHRR 547, para. 34; *Liberty v. United Kingdom* (2009) 48 EHRR 1, para. 62.

to give individuals an adequate indication to foresee the circumstances and the conditions in which public authorities are empowered to resort to measure in question.[43] In line with the principle of the rule of law, it is essential that covert surveillance measures must be based on law that is particularly precise with clear and detailed rules as the technology used is continually becoming more sophisticated.[44] The scope of the safeguards required in order to meet the accessibility and foreseeability tests will depend on the nature and extent of the interference with the private life and correspondence of the applicant.[45] The foreseeability requirement, in the "special context" of covert surveillance, does not mean, however, that an individual should be able to foresee when LEAs are likely to intercept his communications so that he can adapt his conduct accordingly.[46]

The foreseeability requirement is particularly relevant to the covert surveillance of communications by the State in two ways. Firstly, it acknowledges the inherent risks of arbitrariness involved in this area and secondly, it demands a level of transparency in the otherwise secret exercise of this power by public authorities. Another important effect of the foreseeability requirement is that the guarantees, which set out the scope of the manner of how this power will be exercised, must be set out in detail in law so they have a binding force which circumscribes the discretion of judges in the application of such measures.[47] The first hurdle of Article 8(2) has not therefore been overcome by States where there has been no legal framework governing the covert surveillance of telephone communications[48] or where the law did not indicate with sufficient clarity at the material time the extent of the discretion of the relevant public authorities or the way in which this discretion should have been exercised.[49]

The lack of an institution to effectively scrutinise any errors that could occur in the implementation of the surveillance measure will also not be in accordance with the law within the meaning of Article 8(2). The Court has expressed the view that this role

[43] *Malone v. United Kingdom* (1985) 7 EHRR 14; *Valenzuela Contreras v. Spain* (1999) 28 EHRR 483; *Liberty v. United Kingdom* (2009) 48 EHRR 1.

[44] *Huvig v. France* (1990) 12 EHRR 547, para. 32; *Weber and Saravia v. Germany* (2008) 46 EHRR SE5, para. 93.

[45] *P.G. and J.H. v. United Kingdom* (2001) ECHR 546, para. 46.

[46] *Leander v.* Sweden (1987) 9 EHRR 433, para. 51; *Liberty v. United Kingdom* (2009) 48 EHRR 1, para. 62.

[47] *Valenzuela Contreras v. Spain* (1999) 28 EHRR 483, para. 60.

[48] In *Halford v. United Kingdom* (1997) 24 EHRR 523, there was no domestic law in place regulating surveillance by the State of internal communications systems operated by public authorities. Interception of the applicant's telephone conversations in her place of work, police headquarters in this case, had no basis therefore in domestic law and could not be in accordance with the law within the meaning of Article 8(2) (see para. 51); see also *Copland v. United Kingdom* (2007) 45 EHRR 37.

[49] *Malone v. United Kingdom* (1985) 7 EHRR 14; *Valenzuela Contreras v. Spain* (1999) 28 EHRR 483, para. 60.

is best carried out by the judiciary.[50] It is not essential, however, under the requirements of the Convention that the judiciary carry out this role. In *Klass v. Germany*, the Court, having regard to the other safeguards provided for under legislation, accepted that the replacement of a judicial control with an initial control effected by an official qualified for judicial office and by the control provided by the Parliamentary Board and the G 10 Commission met the requirements of Article 8(2).[51]

10.2.2.2 A Legitimate Aim

Covert surveillance of communications can be justified under the Convention when undertaken in pursuit of one of the legitimate aims provided for under Article 8(2). Satisfying this condition of Article 8(2) is rarely a difficulty for States in light of the broad terms in which these legitimate purposes are framed[52] and due to the fact that applicants frequently concede the existence of a legitimate aim in this area.[53] The Court has accepted that the prevention of disorder, or crime, or the protection of the rights and freedoms of others would all be legitimate aims for telecommunications operators to provide LEAs with access to communications data without the consent of the subscriber.[54]

10.2.2.3 'Necessary in a Democratic Society'

An interference must be 'necessary in a democratic society' for the legitimate purposes as well as being lawful and serving a legitimate aim in order to justifiably restrict a right guaranteed under Article 8(1). States must establish this by showing that the impugned measure in question is a response to "a pressing social need" and that the interference with the protected rights is no greater than is necessary to address that pressing social need.[55] This requirement is otherwise referred to as the test of proportionality and consists of four principles.[56]

Firstly, the Strasbourg Court has held that the adjective 'necessary' is not synonymous with 'indispensable' or as flexible as 'reasonable' or 'desirable'.[57] The Court has accepted that powers allowing for the covert surveillance of individuals may be

---

[50] *Klass v. Germany* (1978) 2 EHRR 214, paras. 55–56; *Rotaru v. Romania* (2000) ECHR 192, para.59; *Kennedy v. United Kingdom* (2010) ECHR 682, para. 167.

[51] (1978) 2 EHRR 214, para. 56.

[52] McHarg (1999, *supra* n. 37).

[53] Beatson et al. (2008, p. 162).

[54] *K.U. v. Finland* (2009) 48 EHRR 1237, para. 49; *Copland v. United Kingdom* (2007) 45 EHRR 37, para. 48.

[55] *Leander v. Sweden* (1987) 9 EHRR 433, para. 58.

[56] *Silver v. United Kingdom* (1983) 5 EHRR 347, para. 97.

[57] (1983) 5 EHRR 347, para. 97.

permitted under Article 8 to the extent that they are strictly necessary for safeguarding democratic institutions.[58] In practice, this means that powers allowing for the covert surveillance of communications must have adequate and effective guarantees against abuse. The Court makes a case-by-case assessment of the guarantees that Contracting States have put in place governing the relevant surveillance measure. This review will take into account a number of factors including the circumstances of the case, such as the nature, scope and duration of the possible measures, the grounds required for ordering them, the authorities competent to authorise, carry out and supervise them, and the kind of remedy provided by the domestic law.[59]

Secondly, a "margin of appreciation" will apply in assessing the existence and extent of such necessity.[60] The margin of appreciation is a tool developed from the Court's interpretation of the Convention and its use has provoked controversy amongst commentators.[61] The concept raises the question regarding the extent to which the Strasbourg Court should defer to a State's interpretation of the situations it faces in permitting a limitation on the rights guaranteed by the Convention.[62] The scope of this margin is not unlimited as it is subject, to the final ruling of the Court in deciding whether these restrictions are compatible with the Convention. This assessment embraces both the legislation and the decisions applying it, including those given by the independent domestic courts.[63] In cases concerning covert surveillance by public authorities, the Court has tended to grant a wider margin to Contracting States in matters of national security in light of the sensitive and confidential nature of the information involved.[64] Thirdly, the phrase 'necessary in a democratic society' implies that there is 'a pressing social need' which must be 'proportionate to the legitimate aim pursued' if the interference is compatible with the Convention. Fourthly, Articles under the Convention that provide for an exception to a right guaranteed are to narrowly interpreted.[65] Thus Article 8(2) is subject to a narrow interpretation on the basis that it provides for an exception to rights guaranteed under Article 8(1).[66]

Powers of secret surveillance, characterising as they do the police state, are tolerable under the Convention only in so far as strictly necessary for safeguarding the

---

[58] *Klass v. Germany* (1978) 2 EHRR 214, paras. 42, 48.

[59] (1978) 2 EHRR 214, para. 50; *Weber and Saravia v. Germany* (2008) 46 EHRR SE5, para. 106; *Kennedy v. United Kingdom* (2010) ECHR 682, para. 153.

[60] *Silver v. United Kingdom* (1983) 5 EHRR 347, para. 97.

[61] See generally Arai-Takahashi (2002); Yourow (1996).

[62] White and Ovey (2010, *supra* n. 12, p. 325).

[63] *Lambert v. France* (2000) 30 EHRR 346, para. 30.

[64] *Klass v. Germany* (1978) 2 EHRR 214, paras. 48–49; *Leander v. Sweden* (1987) 9 EHRR 433, para. 59; *Segerstedt-Wiberg v. Sweden* (2007) 44 EHRR 14, para. 104; Feldman (2002, *supra* n. 12, p. 666).

[65] *Klass v. Germany* (1978) 2 EHRR 214, para. 42; *Silver v. United Kingdom* (1983) 5 EHRR 347, para. 97.

[66] *Klass v. Germany* (1978) 2 EHRR 214; *Rotaru v. Romania* (2000) ECHR 192; *Draksas v. Lithuania* (App. 36662/04) (31 July 2012).

democratic institutions.[67] Hence, States must ensure that adequate safeguards govern the use of these surveillance powers to prevent against their abuse or misuse.[68] The assessment of these safeguards is relative and may take into account the nature, scope and duration of the monitoring involved, the grounds required for its use, the authorities competent to permit, carry out and supervise the relevant measure and the kind of remedy provided by the national law.[69] This last condition of Article 8(2) has been met where sufficiently tight controls of this state power have included the imposition of strict conditions on the use of this power by the public authorities, both in its authorization and implementation.[70]

10.2.3 Collection, Storage and Use of Personal Data

The protection of personal data is of fundamental importance to a person's enjoyment of their right to respect for private and family life as guaranteed by Article 8 of the Convention.[71] Although informational aspects to privacy were not considered to fall within the ambit of Article 8 in its early jurisprudence, the Court has still adopted a broad interpretation of the scope of the right to respect in relation to personal information.[72] This approach reflects the Court's view of the Convention as "a living instrument" to be interpreted in light of present day conditions and applied in a manner that renders its guarantees practical and effective and not theoretical and illusory.[73]

In developing this scope of protection to personal data under Article 8(1), the Court has held that the object of the data protections standards under the Council of Europe Convention 1981[74] and the principal EU Data Protection Directive 1995[75] correspond with the broad interpretation of the right to respect for private life as protected under Article 8. Effectively, the Court has approached the protection of personal data as an extension of the right to privacy in its application of Article 8.[76]

---

[67] *Klass v. Germany* (1978) 2 EHRR 214, para. 42.

[68] *Segerstedt-Wiberg v. Sweden* (2007) 44 EHRR 14, para. 88.

[69] *Klass v. Germany* (1978) 2 EHRR 214, paras. 50.

[70] *Klass v. Germany* (1978) 2 EHRR 214; *Weber and Saravia v. Germany* (2008) 46 EHRR SE5; *Kennedy v. United Kingdom* (2010) ECHR 682.

[71] *Z v. Finland* (1997) 25 EHRR 371; *S and Marper v. United Kingdom* (2008) ECHR 1581. See generally, White and Ovey (2010, *supra* n. 12, pp. 374–377); Feldman (2002, *supra* n. 12, pp. 308–316).

[72] Feldman (2002, *supra* n. 12, pp. 306–308).

[73] *Tyrer v. United Kingdom* (1979–1980) 2 EHRR 1, para. 31; *Hirsi Jamaa v. Italy* (App.25579/05) (16 December 2010) (Grand Chamber), para. 175.

[74] *Rotaru v. Romania* (2000) ECHR 192, para. 43.

[75] *S and Marper v. United Kingdom* (2008) ECHR 1581, para. 50.

[76] European Union Agency for Fundamental Rights (FRA) (2010, p. 6).

The mere collection of personal information can amount to an interference with Article 8. It is irrelevant if this information has not been subsequently used or disclosed.[77] Private-life issues may even arise from the collection of information existing in the public domain if any systematic or permanent record comes into existence based on this material. For example, files collected by intelligence authorities on a particular individual will fall within the scope of Article 8, even where the information has not been gathered through intrusive or covert surveillance.[78] An interference with this right will also arise even if this information has been collected from information in the public domain, such as information relating to political opinion, affiliations or activities[79], once it has been systematically collected and stored by public authorities.[80]

The level of interference with an individual's private life that may be permitted under Article 8 may vary according to the type of information concerned[81] and the purposes for which it is being used, collected, stored or disclosed. A distinction will also arise between the collection of personal information, its storage and future use.[82] Thus, while the collection of personal information may amount to a justified interference with Article 8, the scope of its retention may not meet the conditions of Article 8(2).[83] It is important to bear this in mind in relation to the storage and disclosure of personal information, "where a long series of acts or failures to act may impact in different ways at different times on interests protected by Article 8(1)".[84]

The storage of information concerning the private life of an individual by the police or security intelligence authorities will amount to an interference with private life.[85] The subsequent use of this stored information has no impact on this finding.[86] In determining whether the personal information retained by the authorities involves any of the private-life aspects mentioned above, the Court will have due regard to the following factors: the specific context in which the information at issue has been recorded and retained; the nature of the records; the way in which these records are used and processed (this includes the possible use of these records in future[87]); and the results that may be obtained.[88]

---

[77] *Amann v. Switzerland* (2000) ECHR 87; *Copland v. United Kingdom* (2007) 45 EHRR 37.

[78] *Rotaru v. Romania* (2000) ECHR 192, paras. 43–44.

[79] (2007) 44 EHRR 2, para. 107.

[80] (2007) 44 EHRR 2, para. 72.

[81] *S and Marper v. United Kingdom* (2008) ECHR 1581, para. 120.

[82] Emmerson (2007,*supra* n. 24, p. 301).

[83] *S and Marper v. United Kingdom* (2008) ECHR 1581.

[84] Feldman (2002, *supra* n. 12, p. 313).

[85] *Amann v. Switzerland* (2000) ECHR 87; *Rotaru v. Romania* (2000) ECHR 192; *Segerstedt-Wiberg v. Sweden* (2007) 44 EHRR 2.

[86] *Amann v. Switzerland* (2000) ECHR 87, para. 69; *S and Marper v. United Kingdom* (2008) ECHR 1581, para. 67.

[87] *Van der Velden v. Netherlands* (App.29514/05) (7 December 2006); *S and Marper v. United Kingdom* (2008) ECHR 1581.

[88] *Friedl v. Austria* (1995) 21 EHRR 83, paras. 49–51; *Peck v. UK* (2003) 36 EHRR 719, para. 59; *S and Marper v. United Kingdom* (2008) ECHR 1581, para. 67.

States have a positive obligation to ensure secure storage of personal information about an individual. Disclosing information about an individual will also interfere with the informational aspects of private life and would amount therefore to a breach of Article 8(1).[89] A margin of appreciation will apply to national authorities in striking a fair balance between the relevant public and private interests. The scope of this margin will depend on the nature and seriousness of the interests at stake and the gravity of the interference. For example, disclosure of an applicant's telephone conversations monitored for the acceptable aims of safeguarding national security and the prevention of crime by a public authority to the media have resulted in a breach of this positive obligation.[90] Subsequent destruction of data used by the State, in addition to a refusal to inform the subject of the covert monitoring in question, will also raise issues under the Convention as such conditions may serve to conceal monitoring measures interfering with the applicants' rights under Article 8.[91]

As the Court has highlighted, "the need for these safeguards is all the greater where the protection of personal data is concerned, not least where such data are used for police purposes".[92]

10.3 Application of Principles to Communications Data

Before addressing the approach of the Strasbourg Court to Article 8 ECHR and the surveillance of communications, it is essential to provide some context highlighting the manner in which this area of surveillance has changed since it was first brought to the attention of the Court in 1984.[93]

10.3.1 Communications Data and State Surveillance

10.3.1.1 Evolving Nature of Communications Data

Improvements in information processing and the creation of the Internet have made the collection, access, storage and processing of communications data more efficient than ever before. The impact of 'digitization', which changed the format of information processing from analogue to digital, has played a particularly significant role in this development.[94] The analogue format of information processing meant that methods of access and storage varied among different types of information. This lack of

[89] *Z v. Finland* (1998) 25 EHRR 371, para. 99; *Peck v. UK* (2003) 36 EHRR 719, para. 77.

[90] *Craxi v. Italy* (2004) 38 EHRR 47, paras. 74–75; *Draksas v. Lithuania* (App.36662/04), 31 July 2012, paras. 60, 62.

[91] *Weber and Saravia v. Germany* (2008) 46 EHRR SE5, para. 79.

[92] *S and Marper v. United Kingdom* (2008) ECHR 1581, para. 103.

[93] *Malone v. United Kingdom* (1985) 7 EHRR 14.

[94] See, e.g., Diffie and Landau (2010); Solove (2004); Agre and Rotenberg (1997).

compartmentalization was considered to be the best protection against privacy invasion for a long time because every organization had its own system of information processing.[95] Digitization changed this by storing all information using the same format—a system of binary signals. Any type of information such as sound, video or text, once it was put into this format, could then be stored on any digital storage device. This standardization of information processing led to much greater efficiency and soon displaced analogue processing once technology was available to transfer analogue into digital information.[96]

Digitization precipitated the development of the Internet resulting in a global digital network of electronic communications providing instant access to people and information regardless of location. Similar to the impact of digitization for information processing generally, the impact of the Internet to modern telecommunications has irrevocably changed how we communicate with one another and the surveillance of these communications. The Internet links together a vast number of computers enabling the communications of hundreds of millions of people around the world and represents "a fundamental shift in our communications environment".[97] The automatic 'packet-switch system' involved in Internet communications, e.g. email, enables the tracking of each step in the process of an online communication as a result of advances in digital processing.[98] The very nature of how this network operates inherently facilitates the surveillance of communications data on a scale that is unprecedented.[99] In addition, Internet communications automatically generate abundant volumes of communications data. This data abundance has resulted in search engines having the capability to retain "a perfect memory"[100] of how each individual has used that facility resulting in a society where "Google knows more about us than we do ourselves."[101]

The development of social networking has also served to significantly extend the scope of communications data now being generated through the use of the Internet in the twenty-first century.[102] In 2001, the EU had 10 million Internet users. This number had increased to more than 350 million users by 2011.[103] A distinct change in how the Internet was used for communication began to take place in 2001 when users realized that the Internet "wasn't just a network to receive information, but one where you could produce and share information with your peers (often termed Web 2.0)".[104] Of course, recording aspects of one's life, or "life-logging", is by no means

[95] Mayer-Schönberger (2010a).

[96] Mayer-Schönberger (2010b, pp. 57–58).

[97] Naughton (1999, p. 40).

[98] For a clear outline of how this process works in practice, see Kerr (2005, pp. 211, 216).

[99] Diffie and Landau (2010, *supra* n. 94, p. 314).

[100] Mayer-Schönberger (2010b, *supra* n. 96, pp. 11–12).

[101] *Id.*

[102] Marsden (2011, p. 6).

[103] See Internet World Statistics: www.internetworldstats.com.

[104] Mayer-Schönberger (2010b, *supra* n. 96, p. 3).

a new concept.[105] The impact of digitization has since, however, transformed the depth, volume and types of data involved in this process on an unprecedented scale:

> Before the 20th century, life-logging was restricted to recordings on paper media and involved written accounts, such as books, diaries, or collections of letters between people as well as person-constructed images such as drawings or paintings . . . By the end of the 20th century, most of these life-log data were *digitally recorded with both the resolution and frequency of recording dramatically increasing year on year*. Paper diaries and letters gave way to blogs, e-mail, and social networking status updates with the significant difference that the latter were potentially recorded forever and with a vastly more complete history than the episodic fragments of days gone by.[106]

The trend of open disclosure by individuals of their information online has since become synonymous with the rise of the online environment of social networking sites ("SNS"), e.g. Facebook. SNS were a response from the private sector to the increasing use of the Internet in everyday communications in the late 1990s.[107] Worldwide, Facebook reports that it has more than 1 billion users with more than 550 million of these users accessing this website every day.[108] More than 230 million Internet users in the EU subscribe to Facebook.[109] 10 million users submit requests to access their Facebook accounts every second.[110] Consequently, every communication made on a social networking site provides for the accumulation of a considerable amount of communications data due to its ubiquitous use, particularly sensitive information concerning personal and professional relationships and other information relating to an individual's private life.

10.3.1.2 Surveillance of Communications Data

Policy-makers have attempted to emphasize the absence of content when arguing that the acquisition of communications data by public authorities represents a less serious intrusion for the privacy of communications. In other words, there has been a focus on developing the perception that the interception of communications still poses a more serious threat to privacy than the use and analysis of communications data. For example, the Data Retention Directive refers to the fact that its scope does not apply to the content of communications in three different provisions of the legislation.[111] This argument is based on the understanding that access to specific items of communications data is considered to be less sensitive as they provide less personal information than the content of an individual's communications. Surveillance

---

[105] European Network and Information Security Agency (ENISA) (2011, p. 5).

[106] *Id*(emphasis added). For further analysis on the value of communications data from SNS for state surveillance, see Omand and Miller (2012).

[107] For further, see Marsden (2011, *supra* n. 104, ch. 3); Howard (2008, p. 14).

[108] See "Key facts" on the Facebook website: http://newsroom.fb.com/content/default.aspx?NewsAreaId=22.

[109] Internet World Statistics: www.internetworldstats.com/stats4.htm.

[110] Howard (2008, *supra* n. 109, p. 16).

[111] Data Retention Directive, Recital 13; Article 1(2); Article 5(2).

of this less sensitive information should, therefore, be afforded a lesser standard of protection under the law.[112]

This narrow view of communications data reflects an outdated understanding of the current technological environment in which the covert surveillance of telecommunications now operates. Instead, the nature of communications data still seems to be considered in the context of when the traditional landline telephone was the prevailing method of communication.[113] The capacity of communications data to reveal sensitive personal information was significantly more limited at this time, particularly before the advent of the Internet and mobile telephony.

In a society now dominated by these forms of digital communication, patterns of contacts are far less visible to the individuals and far more susceptible to monitoring by public and private organisations. This produces a fundamental shift of power away from individuals and into the hands of the State.[114] This is particularly the case with the access to and processing of communications data. Information that could have been considered by an individual as not particularly sensitive or revealing, e.g. the time, device, or location of an email, may be much more sensitive when placed in a narrative of their overall communications data over an extended period of time. In a communication with another individual, we say what we choose to share but the processing of the communications data from the sending of that message can also disclose actions, movements and intentions.[115] Consequently, the justification for arguing for a lower level of legal protection for the access and analysis of modern communications data by law enforcement carries little weight:

> Generally speaking, the law treats the interception of communications as a much more serious interference with privacy than access to communications data. However, in many cases, the information about a phone call, e.g. the time the call was made, who it was made to, how long the call lasted and so forth, can be far more useful to investigators than what was actually said.[116]

There is a consensus among technologists and legal scholars that major changes in telecommunications technology have dramatically altered the capacity of communications data to reveal sensitive information, even more than content.[117] For example, the analysis of communications data as a result of these technological developments can reveal more details about a communication. Young compares communications

---

[112] An example of the lesser standard of protection afforded to access to communications data is the fact that such information, unlike access to content data, can be obtained without a warrant. See, e.g., the relevant statutory provisions governing this area in the UK: Regulation of Investigatory Powers Act 2000, Pt. 1.

[113] Escudero-Pascual and Hosein (2004, pp. 77–78).

[114] Diffie and Landau (2010, *supra* n. 94, p. 331).

[115] Escudero-Pascual and Hosein (2004, *supra* n. 117, p. 82).

[116] JUSTICE (2011, *supra* n. 12, p. 18, para. 20); see also Young (2004–2005, *supra* n. 15, p. 378).

[117] Diffie and Landau (2010, *supra* n. 94, p. 314); Breyer (2005, *supra* n. 139, pp. 370–371).

20021021070824178 165 0187611205 6139574222 - ---------001------003sth 46 5145281768-----0013 1410260

(Caller at (613) 957-4222 makes a phone call at 7:08:24 AM on October 21, 2002 to recipient at (514) 528 1768 for 3 minutes and 20 seconds.) …

Fig. 10.1 Traffic Data on a Plain Old Telephone System (POTS)

time GMT=20010810010852 Cell ID=115 MAC ID=00:02:2D:20:47:24 (A)
time GMT=20010810010852 Cell ID=115 MAC ID=00:02:2D:04:29:30 (B)
time GMT=20010810010852 Cell ID=115 MAC ID=00:60:1D:21:C3:9C
time GMT=20010810010853 Cell ID=129 MAC ID=00:02:2D:04:29:30
time GMT=20010810010854 Cell ID=129 MAC ID=00:02:2D:1F:53:C0
time GMT=20010810010854 Cell ID=129 MAC ID=00:02:2D:04:29:30 (B)
time GMT=20010810010854 Cell ID=129 MAC ID=00:02:2D:20:47:24 (A)
time GMT=20010810010856 Cell ID=41 MAC ID=00:02:2D:0A:5C:D0
time GMT=20010810010856 Cell ID=41 MAC ID=00:02:2D:1F:78:00
time GMT=20010810010900 Cell ID=154 MAC ID=00:02:2D:0D:27:D3

(On August 10, 2001 at 1:08:52 AM, cellphone user A was in radio cell 115 (Dorval Airport) with cellphone user B and both traveled together at 01:08:54 am to cell 129 (Hilton Hotel).)[118]

Fig. 10.2 Traffic Data From Two Callers on a Wireless Network (~GSM)

data from a traditional landline phone (Fig. 10.1) with digital communications data from a mobile phone (Fig. 10.2):

As Young observes, the privacy implications of the data in Fig. 10.1 compared to Fig. 10.2 are considerably less serious. There is less information available to collect, use and disclose for the purposes of surveillance. The information, however, from both examples falls under the definition of communications data despite the fact that the resulting information is contextually very different.[119]

The consequences of digitization for telecommunications, particularly the development of smart devices, have significantly changed the nature of personal information that the surveillance of communications data can now provide in two main ways.

First, the scale and detail of personal information that can be acquired from the long-term retention of traffic and location data, makes the nature of communications data immeasurably more intrusive than access to the content of a single letter, e-mail or a telephone call. Whilst monitoring by the State of citizens' communications is

[118] Young (2004–2005, *supra* n. 15, pp. 379–380), using a sample from a presentation by A. Pascual, "Access to 'Traffic' Data: When Reality is Far More Complicated Than a Legal Definition" (11 October 2002).

[119] Id.

by no means novel, the difference in the threat posed by modern telecommunications technologies is their efficiency and power.[120] Before, technology was mainly administered by humans and, in terms of monitoring, only what was different was noticed. Today, digitized data processing is ubiquitous and automatic and collects everything: "Then the default was that searchable records were not collected; now the default is that all monitoring produces searchable records".[121]

The power to reveal sensitive personal information from the communications data of landline telephones is significantly more limited compared to what is now possible following the advent of the Internet and mobile telephony.[122] Additionally, the rate at which landline telephones are now being in modern society has declined to such an extent that monitoring their use is increasingly of little benefit for law enforcement purposes. The Data Retention Directive regulates the retention of communications data from what are now commonly referred to as "smart" telecommunications devices: phones, and now tablets, that are both mobile and capable of accessing the Internet. Use of these smart devices has also become ubiquitous in modern society. This has resulted in a sea change in the prevailing use of fixed devices for communications, from the stationary telephone and personal computer, to smartphones and tablets.

These changes have significant implications for enhancing the value of communications data for public authorities, particularly in the area of law enforcement. The "electronic exhaust" left behind has led to the development of what has been described as "a rich tracking ecosystem".[123] These developments also present unique privacy challenges: "more than other types of technology, mobile devices are typically personal to an individual, almost always on, and with the user. This can facilitate unprecedented amounts of data collection."[124] For example, Malte Spitz, a German Green Party representative, demonstrated the scope of this surveillance in a request made to his mobile phone provider. In this request he sought a record of the communications data collected and retained from the use of his mobile phone. Over the course of six months, this communications data tracked his geographical location and the use of his phone more than 35,000 times building a detailed narrative of his movements and his communications.[125]

Secondly, an individual item of communications data, which would otherwise be peripheral, could reveal the underlying content in the body of that communication, particularly the combination of such data, e.g., the duration and time of a phone call or the size and subject line and time of an e-mail.[126] It is now possible to construct a

---

[120] Lessig (1999, p. 151).

[121] Id.

[122] Gillespie (2009, pp. 560–561).

[123] Bray (2013, pp. 68–69).

[124] See the U.S. Federal Trade Commission ("FTC") (2013, pp. 2–3).

[125] Spitz (2012).

[126] See, e.g., McPhie (2005, p. 33).

profile of an individual that may be considered to be invasive of privacy from isolated items of information that would *not* be considered private or personal.[127]

We now live in a time of "Big Data", "digital dossiers"[128] and "dataveillance"[129] where information concerning every individual is systematically stored and monitored in massive computer databases by a host of public authorities and private sector organisations. These dossiers are the result of "aggregation", one of the privacy harms that have emerged from advances in information processing. Solove describes this development as part of his taxonomy of privacy as follows[130]:

> Aggregation is the gathering of information about a person. A piece of information here or there is not very telling, but when combined, bits and pieces of data begin to form a portrait of a person. The whole becomes greater than the parts. This occurs because combining information can reveal new facts about a person that she did not expect would be known about her when the original, isolated data was collected.[131]

This data could be used to identify an individual's personal and professional relationships, their racial or ethnic origin, political opinions, religious or philosophical beliefs, trade-union membership or data concerning health or sex life.[132] If accessed illegally by third parties, or 'hacked', this communications data could be used to cause other harms, including fraud through identity theft.[133] The much more significant issue, however, relates to the possibility of this information being used for profiling and how this may lead to the creation or reinforcement of unequal treatment in society. It is these insidious threats of discrimination and manipulation, Lessig warns, that ought to be of concern.[134]

Two factors suggest that more challenges concerning the surveillance of communications data by law enforcement are likely to come before the Strasbourg Court in future. Firstly, the impact of technological developments in telecommunications is likely to continue to increase the capacity of communications data to reveal more personal information. Consequently, this will increase its value and the frequency of its use by law enforcement agencies. Further, there is a high probability that this may become the prevailing surveillance technique for law enforcement in light of continuing developments in telecommunications and the increasing abundance of data generated in daily communications.

Secondly, the blanket mandatory retention of communications data by telecommunications operators established under the Data Retention Directive enables and

---

[127] Michael (1994, p. 10).

[128] Solove (2008, p. 13).

[129] Clarke (1988, pp. 498, 499) defined as "dataveillance" as "the systematic use of personal data systems in the investigation or monitoring of the actions or communications of one or more persons".

[130] Solove's taxonomy of privacy focuses on activities that can and do cause privacy problems; Solove (2004, *supra* n. 130, ch. 5).

[131] Solove (2008, *supra* n. 130, p. 118).

[132] See Article 29 Data Protection Working Party, 5/2009 Opinion on online social networking, (12.06.2009).

[133] *FTC Report* (2013), n. 126.

[134] Lessig (1999, n. 122, pp. 153–154).

thereby entrenches the current use of communications data by law enforcement in Europe. The increased role of the telecommunications sector in facilitating access to communications data means that the State now plays a limited role in its acquisition. LEAs "can obtain most of the data they need from the comfort and safety of their own desks, with a few clicks of a mouse, a fax, or a phone call to a telecommunications or Internet service provider."[135] This ease of access arguably encourages the frequency of its use making the covert surveillance of communications data facilitated by telecommunications operators an increasingly entrenched practice. For example, the number of requests by LEAs across EU Member States amounts to tens of thousands every year. The total number of requests recorded in 2008 to telecommunications operators from twelve Member States, less than half of the EU membership, came to 1, 392, 281.[136] As Breyer observes, the mandatory retention of communications data has added a new dimension to traditional law enforcement powers:

> The analysis of traffic data may reveal details of a person's political, financial, sexual, religious stance, or other interests. Therefore, it is fully justified to describe blanket traffic data retention as a new dimension in surveillance, as compared to traditional police powers. Data retention does not only apply in specific cases. Instead, society is being pre-emptively engineered to enable blanket recording of the population's behaviour, when using telecommunications networks.[137]

10.3.2 A Less Serious Infringement?

Contracting states have argued before the Strasbourg Court that the surveillance of communications data is a less serious interference with the right to respect to private life and correspondence when compared to telephone tapping.[138] One policy-maker has gone so far as to claim that the Court itself considers the acquisition of communications data by LEAs to be a lesser infringement of Article 8 than the interception of communications based on its decision in *Malone v. United Kingdom*.[139] Although no explanation has been provided for this argument, similar concerns have been highlighted elsewhere in the literature based on the interpretation that the Court left open the issue in *Malone* as to whether or not the practice of metering itself amounted to an interference with Article 8(1).[140]

---

[135] Slobogin (2011, p. 2).

[136] Evaluation of the Directive *supra* n. 3, p. 35.

[137] Breyer (2005, pp. 365, 365).

[138] See *Malone v. United Kingdom* (1985) 7 EHRR 14; *Valenzuela Contreras v. Spain* (1999) 28 EHRR 483; *P.G. and J.H. v. United Kingdom* (2001) ECHR 546.

[139] See the Secretary of State for the Home Department (2012, p. 100): "[I]t is clear from *Malone* that the Court considers the acquisition of communications data to be a less serious infringement of privacy rights than the interception of communications".

[140] Bygrave (1998, p. 247, 263) highlights the contrasting concurring opinion of Judge Pettiti who was unequivocal in his finding that the use of metering for any aim other than "its sole accounting

It was established in *Malone v. United Kingdom* that the use of communications data by LEAs falls under the same scope of protection provided under Article 8(1) to telephone conversations on the basis that it is "an integral element" of the communications made by telephone.[141] The case concerned a challenge to the regime governing the surveillance of communications data, in the form of traffic data derived from telephone usage, otherwise referred to as "metering" by the police in England and Wales.

The practice of metering involved the use of a device called a meter check printer that registered telephone numbers dialled and the time and duration of each call. The main aim of the process was for the telecommunications provider to ensure that the subscriber was correctly charged. The applicant was an antique dealer who was charged and convicted of handling stolen goods. He alleged that his telephone had been metered on behalf of the police. This belief was based on the evidence that when the applicant was charged the police searched the premises of about twenty people whom he had recently telephoned. The Government, however, denied that the police had either caused the applicant's telephone calls to be metered or had undertaken any search operations based on the use of this information.

No statutory requirement was in place imposing the mandatory retention of these records for law enforcement purposes. A practice did exist, however, whereby the Post Office, the relevant telecommunications' provider at the time, would make and provide such records to the police at their request.[142] The Government argued that metering did not amount to an interference with any right guaranteed by Article 8 on the grounds that the Post Office collected communications data from metering for billing purposes and no interception of telephone conversations was part of this process.

The Court rejected this argument. While it accepted that metering could be used for the legitimate purpose of ensuring accurate billing to its subscribers, this did not mean that any use of data obtained from metering could not give rise to an issue under Article 8. The threat posed by advances in information processing, long before 'Big Data' or the ability to predict an individual's future conduct based on the monitoring of their current communications through the analysis of narrative data became part of the modern data surveillance environment, were astutely observed by Judge Pettiti as far back as 1984 in his concurring ruling[143]:

> The comprehensive metering of telephone communications (origin, destination, duration), when effected for a purpose other than its sole accounting purpose, albeit in the absence of any interception as such, constitutes an interference in private life. On the basis of the data thereby obtained, the authorities are enabled to deduce information that is not properly

purpose" constitutes an interference with private life; see *Malone v. United Kingdom* (1985) 7 EHRR 14, 43.

[141] (1985) 7 EHRR 14, para. 84.

[142] (1985) 7 EHRR 14, paras. 17, 56.

[143] Judge Pettiti's foresight of the impact of technological developments in the area of State surveillance of communications is highlighted by Murphy and O'Cuinn (2010, pp. 601, 619).

> meant to be within their knowledge. It is known that, as far as data banks are concerned, the processing of 'neutral' data may be as revealing as the processing of sensitive data.[144]

The Court then interpreted the information obtained by metering, particularly the numbers dialled, to be an integral element in the communications made by telephone. This formed the basis for the finding that the release of such information to the police without the consent of the subscriber amounts to an interference with a right guaranteed by Article 8.[145] The Court has consistently held since that an interference with the right to respect for private life and correspondence has arisen in cases where the police have accessed communications data without the consent of the subscriber.[146] In *Valenzuela Contreras v. Spain*[147] the Court, citing *Malone* as authority, rejected a suggestion by the Spanish Government that a lesser standard applies under Article 8 to the practice of metering.[148]

In *Copland v. United Kingdom*, this scope of protection was also extended to the use by public authorities of communications data from Internet and email usage.[149] The focus of the Court's assessment in this case was the negative obligation on the UK Government, taking responsibility for a State-administered college as a public body for the purposes of the Convention, not to interfere with the interests of private life and correspondence protected under Article 8(1).

The applicant was an employee of the college whose telephone, email and Internet usage at work were monitored in order to allegedly ascertain whether she was making excessive use of work facilities for personal purposes. There was no policy in force at the college at the material time regarding the monitoring of telephone, email or Internet use by employees.[150] Data relating to the applicant's Internet usage included an analysis of the websites visited by the applicant, times and dates of website visits and their duration. The Government claimed that this surveillance had not taken place for longer than a month. The applicant contested the short length of this period but did not suggest what amount of time she suspect may have been involved.[151] The analysis

---

[144] *Malone v. United Kingdom* (1985) 7 EHRR 14, 43.

[145] (1985) 7 EHRR 14, paras. 83–84, 89.

[146] *Valenzuela Contreras v. Spain* (1999) 28 EHRR 483; *P.G. and J.H. v. United Kingdom* (2001) ECHR 546.

[147] (1999) 28 EHRR 483.

[148] As held by the Court, (1999) 28 EHRR 483, para. 47: "The tapping of Mr Valenzuela Contreras's telephone line ... constitutes an 'interference by a public authority' within the meaning of Article 8 § 2 in the applicant's exercise of his right to respect for his private life and correspondence. Indeed, that point was not disputed. Nor is it decisive in that regard that, as the Government intimated, only a 'metering' system was used (see the *Malone* judgment cited above, p. 38, para. § 87)." Bygrave's concern regarding the ambiguity of the Court's stance on metering and Article 8(1) in *Malone* is understandable given that his observation was made prior to the ruling of *Valenzuela Contreras*. The argument made recently in 2012 by the UK Secretary of State carries much less weight in light of the Court's subsequent case law and the absence of any supporting authority for this interpretation of *Malone*.

[149] (2007) 45 EHRR 858, para 41.

[150] (2007) 45 EHRR 858, para 15.

[151] (2007) 45 EHRR 858, para 11.

of the applicant's emails amounted to communications data as it did not include content. Specifically, the government claimed that the monitoring concerned email addresses, including the dates and times of those emails, and that the surveillance had taken place for a few months.[152]

The Court found that the collection and storage of the applicant's personal information relating to the applicant's telephone, email and Internet usage, without her knowledge, amounted to an interference with her right to respect for her private life and correspondence within the meaning of Article 8.[153] As there was no domestic law regulating the surveillance at the relevant time, the interference was not in accordance with the law and could not be justified under Article 8(2). Accordingly, the Court found that a violation of Article 8 had taken place.

The general case law of the Court shows that an approach of strict scrutiny is applied to the rights that most closely reflect the Convention's fundamental values.[154] In applying the 'in accordance with the law' test to covert surveillance by law enforcement authorities, the values of the rule of law and democracy have been invoked in judgments concerning both the interception of content and/or communications data.[155] Regardless of the type of information involved, the Court has found that applicants have been denied "the minimum degree of legal protection to which citizens are entitled under the rule of law in a democratic society" where the law has not contained adequate and effective safeguards.[156]

Despite being in line with domestic law, Article 8 was violated in *Malone* due to an absence of legal rules governing the scope and manner of the discretion enjoyed by the police and the practice of metering. The foreseeability requirement was not met and the domestic law was found therefore not to be in accordance with the law within the meaning of Article 8(2).[157] Similarly, the foreseeability requirement was not satisfied in *Valenzuela Contreras*.[158] The Court found that the national law, written and unwritten, governing the practice of metering did not indicate with sufficient clarity the extent of the discretion of the police or the way in which it should have been exercised at the material time.[159]

In *Copland v. United Kingdom*, the Court found that the same scope of protection that applies to the processing of personal data by LEAs also applies to personal information relating to the telephone, e-mail and Internet "usage".[160] The same

[152] (2007) 45 EHRR 858, para 13.

[153] (2007) 45 EHRR 858, para 44.

[154] Beatson et al. (2008, *supra* n. 53, p. 146).

[155] *Malone v. United Kingdom* (1985) 7 EHRR 14; *Valenzuela Contreras v. Spain* (1999) 28 EHRR 483.

[156] *Valenzuela Contreras v. Spain* (1999) 28 EHRR 483, para. 61.

[157] (1985) 7 EHRR 14, paras. 87–88.

[158] The surveillance undertaken by the police in this case involved the metering of several phones. The metered information consisted of lists and times of numbers dialled and received in order to establish whether the applicant was harassing his former fiancée.

[159] (1999) 28 EHRR 483, paras. 60–61.

[160] (2007) 45 EHRR 37, para. 44.

protection, therefore, applies to the storage and collection of communications data by LEAs. This extends the scope of protection under Article 8(1) to communications data 'metered' from telephones established by the Court in *Malone* to communications data derived from email and Internet use.

This means that the principles adopted by the Court in respect of the case law concerning the processing of personal data by LEAs apply equally to the processing of communications data. Thus, the right to respect for private life and correspondence is continuously engaged when LEAs seek, collect, store, process, compare, or disseminate personal information obtained through the covert surveillance of communications data.[161]

10.4 Reflections

It is welcome that the Court has acknowledged that the acquisition and processing of communications data warrants the same protection as personal data under Article 8(1). In order for this protection to be effective and meaningful in practice, however, the use and subsequent processing involved in the monitoring of communications data needs to be subject to greater consistency and scrutiny in line with the general principles and safeguards of Article 8(2).

The Court did not find a violation of Article 8 in relation to the internal police guidelines governing the storage and destruction of the communications data that was used by LEAs in *P.G. and J.H. v. United Kingdom*.[162] This finding is curious in light of the absence of statutory provisions outlining the procedures to be followed for examining, using and storing the communications data obtained in the contested domestic legislation.[163] The need for such provisions to be legally binding and accessible and foreseeable to the public forms part of the minimum requirements that contracting states are required to satisfy in order to prevent abuses of this covert power of surveillance by public authorities.

The ruling in *P.G. and J.H.*, however, deviated from the established principle of Convention jurisprudence in this area and found that these internal policy guidelines were sufficient in place of specific statutory provisions. This does not reflect the previous or subsequent approach of the Court in its application of the legality test under Article 8(2) in this area.[164] This meant that the applicants in *P.G. and J.H.* at the material time did not enjoy the minimum degree of protection to which citizens are entitled to under the rule of law in a democratic society.[165] It is submitted that this finding should have been otherwise. Contrary to the finding of the Court, the

[161] *Leander v. Sweden* (1987) 9 EHRR 433, para. 48; *Kopp v. Switzerland* (1999) 27 EHRR 91, para. 58; *Amann v. Switzerland* (2000) ECHR 87, para. 69.

[162] (2001) ECHR 546.

[163] (2001) ECHR 546, para. 47.

[164] In contrast, see the rulings of *Khan v. United Kingdom* (2001) 31 EHRR 1016 and *Liberty v. United Kingdom* (2009) 48 EHRR 1.

[165] *Malone v. United Kingdom* (1985) 7 EHRR 14, para. 79.

measures governing the storage and subsequent processing of this information should in fact have amounted to a violation of Article 8.

The Court has already acknowledged the "the rapid development of telecommunications technology" in the context of how it has brought about the emergence of new types of crime as a factor in its ruling that the privacy of telecommunications and Internet users guaranteed under Article 8 should be scaled back as it is not an absolute right.[166] As a result of the impact of this factor, the Court has found that the scope of Article 8 must give way on occasion to the legitimate aim of the prevention of crime or the protection of the rights and freedoms of others.[167] Equally, the increased level of intrusiveness posed by the rapid development of telecommunications technology needs to be taken into account by the Court when assessing the proportionality of the use, storage and subsequent processing of communications data by LEAs.

The nature and extent of the communications data used and stored was raised in *Copland v. United Kingdom* but the Court avoided dealing with the necessity and proportionality elements of Article 8(2) as the interference did not overcome the quality of law hurdle.[168] We now live, however, in an era of digital dossiers where information derived from communications data can no longer be considered as isolated pieces of neutral data and where the use of such information is likely to become more entrenched with the establishment of the Data Retention Directive. The Court needs to ensure that it takes these developments into account in future.

The possible future use of communications data obtained and stored by law enforcement authorities should also be considered in the Court's test of proportionality under Article 8(2) and not just in establishing whether there has been an interference with Article 8(1).[169] This was one of the concerns that arose for the Court in the case of the long-term retention of DNA profiles and other materials by the LEAs in the Grand Chamber decision of *S and Marper v. United Kingdom.* As Ian Brown observes, the Court's concerns "over the use of DNA profiles have clear applications to the detailed information revealed about individual's private lives by communications data".[170]

In line with the principle of the Convention as a living instrument, the Court needs to consider the cumulative effect of the technological and legal developments that have taken place since its first ruling on the subject in 1984.[171] As communications technology has advanced, so too has the amount of communications data available about an individual's private life and correspondence. Communications data obtained, therefore, from metering the antiquated landline telephone dealt with

---

[166] *K.U. v. Finland* (2009) 48 EHRR 1237, para. 22.

[167] (2009) 48 EHRR 1237, para. 49.

[168] (2007) 45 EHRR 37.

[169] *S and Marper v. United Kingdom* (2008) ECHR 1581, paras. 71–72; *Van der Velden v. Netherlands* (App. 29514/05) (7 December 2006).

[170] Brown (2011, *supra* n. 18, p. 102).

[171] *Malone v. United Kingdom* (1985) 7 EHRR 14.

by the Court decades ago[172] is "nothing ... when compared to what is today recorded digitally in respect of every mobile phone call, text message or Internet session".[173]

References

Agre, Phillip, and Marc Rotenberg, eds. 1997. *Technology and privacy: The new landscape*. Cambridge: MIT Press.

Article 29 Data Protection Working Party. 2009. 5/2009 Opinion on online social networking. (12.06.2009).

Arai-Takahashi, Yutaka. 2002. *The margin of appreciation doctrine and the principle of proportionality in the jurisprudence of the ECHR*. Antwerp: Intersentia.

Beatson, Jack, Stephen Grosz, Tom Hickman, Rabinder Singh, and Stephanie Palmer. 2008. *Human rights: Judicial protection in the United Kingdom*. London: Sweet & Maxwell.

Bray, O. 2013. The app effect: How apps are changing the legal landscape. *Computer and Telecommunications Law Review* 66:68–69.

Breyer, Patrick. 2005. Telecommunications data retention and human rights: The compatibility of blanket data retention with the ECHR. *European Law Journal* 11 (3): 365–375.

Brown, Ian. 2011. Communications data retention in an evolving internet. *International Journal of Law and Information Technology* 19 (2): 95–109.

Brown, Ian, and Douwe Korff. 2009. Terrorism and the proportionality of internet surveillance. *European Journal of Criminology* 6 (2): 119–134.

Bygrave, Lee A. 1998. Data protection pursuant to the right to privacy in human rights treaties. *International Journal of Law and Information Technology* 6:247–284.

Clarke, Roger A. 1988. Information and technology and dataveillance. *Communications of the ACM* 31 (5): 498–512.

Council Directive (EC) No 24/2006. (OJ 2006 L 105 p. 54).

Diffie, Whitfield, and Susan Landau. 2010. *Privacy on the line: The politics of wiretapping legislation*. Cambridge: MIT Press.

Drzemczewski, Andrew. 1983. *European human rights convention in domestic law: A comparative study*. New York: Oxford University Press.

Emmerson, Ben, Andrew Ashworth, and Alison MacDonald. 2007. *Human rights and criminal justice*. London: Sweet & Maxwell.

Escudero-Pascual, Alberto, and Gus Hosein. 2004. Questioning lawful access to traffic data. *Communications of the ACM* 47 (3): 77–82.

European Network and Information Security Agency (ENISA). 2011. *Risks and benefits of emerging life-logging applications: Final Report*. Heraklion: ENISA.

European Union Agency for Fundamental Rights (FRA). 2010. *Data Protection in the European Union: The Role of National Data Protection Authorities*. Belgium: FRA.

EU Commission Directorate General for Home Affairs website: http://ec.europa.eu/dgs/home-affairs/what-we-do/policies/police-cooperation/data-retention/review-of-data-retention-directive/index_en.htm.

Feldman, David. 2002. *Civil liberties and human rights*. Oxford: Oxford University Press.

Gearty, Conor. 1993. The European court of human rights and the protection of civil liberties: An overview. *Cambridge Law Journal* 52:89–127.

Gillespie, Alisdair A. 2009. Regulation of internet surveillance. *European Human Rights Law Review* 4:552–565.

---

[172] (1985) 7 EHRR 14; *Valenzuela Contreras v. Spain* (1999) 28 EHRR 483; *P.G. and J.H. v. United Kingdom* (2001) ECHR 546.

[173] JUSTICE (2011, *supra* n. 12, p. 71).

Harris, David J., Michael O'Boyle, Edward P. Bates, and Carla M. Buckley. 2009. *Law of the European Convention on Human Rights*. Oxford: Oxford University Press.
Howard, Bill. 2008. Analyzing online social networks. *Communications of the ACM* 51 (11): 14–16.
JUSTICE. 2011. *Freedom from suspicion: Surveillance reform for a digital age*. London: JUSTICE.
Keller, Helen, and Alec Stone-Sweet. 2008. *A Europe of rights: The impact of the ECHR on national legal systems*. Oxford: Oxford University Press.
Kerr, Orin. 2005. A thinly veiled request for congressional action on e-mail privacy: United States v. Councilman. *Harvard Journal of Law and Technology* 19 (1): 211–230.
Lessig, Lawrence. 1999. *Code and other law of cyberspace*. New York: Basic Books.
Maras, Marie-Helen. 2012. The economic costs and consequences of mass communications data retention: is the data retention directive a proportionate measure? *European Journal of Law and Economics* 33 (2): 447–472.
Marsden, C.T. 2011. *Internet co-Regulation: European law, regulatory governance and legitimacy in cyberspace*. Cambridge: Cambridge University Press.
Mayer-Schönberger, Viktor. 2010a. Delete! Oxford Internet Institute, University of Oxford. http://podcasts.ox.ac.United Kingdom/delete-audio. Accessed 10 May 2010.
Mayer-Schönberger, Viktor. 2010b. *Delete: The virtue of forgetting in the digital age*. Princeton: Princeton University Press.
McHarg, Aileen. 1999. Reconciling human rights and the public interest: Conceptual problems and doctrinal uncertainty in the jurisprudence of the European court of human rights. *Modern Law Review* 62 (5): 671–96.
McPhie, David. 2005. Almost private: Pen registers, packet sniffers, and privacy at the margin. *Stanford Technology Law Review* 1:1–20.
Michael, James. 1994. *Privacy and human rights: An international and comparative study, with special reference to developments in information technology*. Paris: UNESCO.
Moreham, Nicole A. 2008. The right to respect for private life in the European convention on human rights: A re-examination. *European Human Rights Law Review* 1:44–79.
Murphy, Therese, and Gearoid O'Cuinn. 2010. Works in progress: New technologies and the European Court of Human Rights. *Human Rights Law Review* 10 (4): 601–638.
Naughton, John. 1999. *A brief history of the future: The origins of the Internet*. London: Phoenix.
Ni Loideain, Nora. 2012. Assessing the evaluation of the EC Data Retention Directive. In *Human rights and risks in the digital era: Globalization and the effects of information technologies*, ed. Christina M. Akrivopoulou and Nicolaos Garipidis, 67–79. Hershey: Information Science Reference.
Omand, David. 2012. Intelligence and security in the digital age. Conference Paper, University of Cambridge, 1 Mar 2012.
Omand, David, Jamie Bartlett, and Carl Miller. 2012. *Intelligence: DEMOS Report*. London: DEMOS.
Report from the Commission to the Council and the European Parliament. 2011. Evaluation report on the Data Retention Directive (Directive 2006/24/EC), COM(2011) 225 final Brussels. (18.4.2011).
Secretary of State for the Home Department. 2012. Draft Communications Data Bill, CM 8359. London: Home Office. (June 2012).
Slobogin, C. 2011. The Law Enforcement Surveillance Reporting Gap. Seminar Paper, Indiana University Bloomington—Center for Applied Cybersecurity Research, 10 April 2011, 2.
Solove, Daniel J. 2004. *The digital person: Technology and privacy in the information age*. New York: New York University Press.
Solove, Daniel J. 2008. *Understanding privacy*. London: Harvard University Press.
Spitz, Malte. 2012. Your phone company is watching you. TEDGlobal Conference Paper, Edinburgh. http://www.ted.com/talks/malte_spitz_your_phone_company_is_watching.html. Accessed June 2012.
Starmer, Keir. 1999. *European Human Rights Law: The Human Rights Act and the ECHR*. London: Legal Action Group.

Tene, Omer. 2008. What Google Knows: Privacy and internet search engines. *Utah Law Review* 4:1433–1492.

U.S. Federal Trade Commission ("FTC") Staff Report. 2013. *Mobile privacy disclosures: Building trust through transparency*. Washington D.C.: FTC. (Feb 2013).

White, Robin C.A., and Claire, Ovey. 2010. *The European Convention on Human Rights*. Oxford: Oxford University Press.

Young, Jason M. 2004–2005. Surfing while Muslim: Privacy, freedom of expression and the unintended consequences of cybercrime legislation. *Yale Journal of Law and Technology* 7 (2): 346–421.

Yourow, Howard C. 1996. *The margin of Appreciation Doctrine in the dynamics of European human rights jurisprudence*. The Hague: Kluwer.

Part IV
Understanding Data Protection and Privacy

Chapter 11
Realizing the Complexity of Data Protection

Marion Albers

11.1 Introduction

Realizing the Complexity of Data Protection sounds a bit off-putting. Would it not be a better approach to lay down a few simple principles that would provide legal guidance for processing personal data? In contrast to such thinking, the present contribution advances the thesis that data protection is by its nature an extraordinarily complex field and therefore requires multi-level and complex regulation. Yet at its core, the legal framework is still characterized by out-dated concepts going back to when data protection first emerged. This applies both to the understanding of fundamental rights relevant to data protection and to the basic approaches to regulation. Modern data protection calls for new legal approaches.

The article provides a legal analysis both of the influential legal intellectual approaches and of the central legal provisions and also identifies areas where reconceptualizations are needed. Other analyses of legal rules would be equally interesting: for example, from a political-science perspective concerning the impact of lobbyism or from an engineering perspective regarding the transformation of data protection into technological concepts. However, the legal perspective, which addresses the understanding of legal rules in terms of legal theory, doctrinal constructions and methodological approaches, is just as important for data protection. After all, the law substantially shapes data protection by means of patterns that can be explained in a manner intrinsic to the legal system. Yet every sophisticated legal approach is also characterized by the fact that it is able to incorporate insights from other disciplines, that is, to guarantee that the law is appropriately receptive and that it is compatible with concepts across various disciplines. Precisely this is what data protection law must be able to achieve, as data protection lies at the intersection of numerous disciplines. This is another reason why sufficiently complex regulatory concepts are necessary.

M. Albers (✉)
Fakultät für Rechtswissenschaft, Universität Hamburg, Rothenbaumchaussee 33,
20146 Hamburg, Germany
e-mail: marion.albers@uni.hamburg.de

S. Gutwirth et al. (eds.), *Reloading Data Protection*,
DOI 10.1007/978-94-007-7540-4_11,

Legally speaking, data protection is characterized both by fundamental rights and by legal principles and provisions. The norms cannot be understood by examining only their wording. More important are the concepts underpinning them that guide the understanding of the norms. The present article analyzes them less from the perspective of legal method, which involves the interpretation of certain rules in a way that is consistent with the method, but more from a legal-theory and doctrinal perspective. The deliberations center on German and European law. They put German law into focus as a continental European legal system oriented toward codification in which data protection law was developed fairly early and has in the meantime been elaborated to become an extensive complex of norms addressing fundamental rights as well as legal rules. An impact on European law arises from reciprocal influences in law-making and from the network of jurisdiction among the German Federal Constitutional Court, the European Court for Human Rights, and the European Court of Justice.

The analysis starts by presenting the conception of fundamental rights (Sect. 2.1) as well as of protected interests (Sect. 2.2). Beyond the right to privacy, the right to informational self-determination has become a guiding principle, especially in the German legal system. Sect. 3 demonstrates to what a large extent the concepts of fundamental rights influence the approaches, principles and legal constructs of data protection law. My hypothesis is that the elementary patterns of thinking must be constructed in a different way in order to achieve appropriate data protection law. Data protection will then prove to be a highly complex and novel field involving particular challenges for law. This hypothesis shall be explained in Sect. 4 with three aspects in mind: Firstly, the object of data protection is complex, namely not only personal data, but a network consisting of several basic elements: data and information, knowledge and the flow of data and information, decisions and the consequences of decisions (Sect. 4.1). Secondly, data protection cannot be reduced to a uniform legally protected good. It encompasses a complex bundle of interests and legal positions aiming at protecting the individual in his or her sociality (Sect. 4.2). Thirdly, data protection requires complex concepts of regulation that must not only coordinate data protection law with the issue-related substantive legal norms appropriately, but must also take up basic elements of risk regulation or technology law (Sect. 4.3). After all, data protection law is anything but bureaucratic. It is modern and exciting, and at the same time requires additional elaboration in many respects.

11.2 Guiding Paradigms of Data Protection Based in Fundamental Rights

The concept of data protection emerged in the 1970s against the background of central mainframe computing systems. At the European level as well as in Germany the first sets of legal rules were developed.[1] With huge amounts of data being processed in these systems in a predefined sequence, the idea was that the individual steps of

---

[1] See the Convention No. 108 for the Protection of Individuals with regard to Automatic Processing of Personal Data, 28 January 1981; Mayer-Schönberger 1997, 219, 220 ff.; Bygrave 2002, 94 ff. As

data processing, including data collection, data storage, and data use, would have to be controlled. Attention was focused on regulating the individual steps of data processing.[2]

This background and the patterns of thinking associated with it not only formed the basis for early data protection rules, but also for the substance of fundamental rights which were concretized or developed in response to the challenges processing personal data electronically. Starting in the 1970s, the right to privacy began to be interpreted in a new way.[3] In Germany, the Federal Constitutional Court derived the right to informational self-determination in its 1983 decision concerning the census ("Volkszählungsurteil").[4] Later, fundamental rights quickly became the guiding principles for the general understanding of data protection by law.

For their part, fundamental rights are linked to certain patterns of observation and thinking. The traditional understanding of fundamental rights is connected to liberal paradigms. According to this notion, fundamental rights are about protection against encroachments by the state. Although extensions of the functions of the fundamental rights, e.g., rights to protection by the state or institutional guarantees, are recognized in principle by now protection against encroachments is often considered the primary dimension of protection in fundamental rights; it still is the leading approach. However, this approach has prerequisites and limitations influencing the substance and the functions fundamental rights can have. The newly derived right to informational self-determination can illustrate this very clearly. In the following section, the traditional concept of fundamental rights and its limitations will be elucidated as well as the characteristics of the right to informational self-determination.

11.2.1 The Traditional Concept of Fundamental Rights

11.2.1.1 Protection Against Encroachments as a Central Pattern of Fundamental Rights

According to the "classical" view based on liberalism, fundamental rights serve primarily as protective rights of the individual against interventions by the state.[5] The persons protected enjoy certain freedoms or legal positions. State measures interfering in these freedoms can be fended off by means of legal remedies, provided they are not covered by constitutional law.

---

to the modernization see www.coe.int/t/dghl/standardsetting/dataprotection/modernisation_en.asp. See also Nouwt 2009, 275, 286 ff. For data protection history in Germany see Abel 2003, Chap. 2.7 Rz. 1 ff.; Simitis 2011, Rn. 1 ff.

[2] Influential in Germany: Wilhelm Steinmüller/Bernd Lutterbeck/Christoph Mallmann/Uwe Harbort/Gerhard Kolb/Jochen Schneider, *Grundfragen des Datenschutzes*: Gutachten im Auftrag des Bundesministeriums des Innern, 1971, BTDrucks. VI/3826, Anl. 1.

[3] See, among others, Westin 1970, p. 42.

[4] BVerfGE 65, 1, 42 ff.; Dec 15, 1983, Census Judgment.

[5] Negative liberty, see i. e. Berlin 1969, 118 ff. With regard to the jurisdiction of the FCC see BVerfGE 7, 198, 204 f.—Lüth—68, 193, 205.

The traditional view of protection against encroachments as a central pattern of fundamental rights is reflected more or less distinctly in their codification, i.e. in the European Convention on Human Rights (ECHR), in the Charter of Fundamental Rights of the European Union (Charter) or in the German Basic Law (Grundgesetz; GG). The jurisdiction of the European Court of Human Rights, of the European Court of Justice and of the German Federal Constitutional Court has elaborated the function of the fundamental rights to protect against encroachment in numerous decisions.

In terms of their structure, fundamental rights involve on the one hand the scope of protection and—on the other—the reservation allowing legal regulation provided that such regulation meets all constitutional requirements. For example, their scope of protection safeguards the right to respect for private life or the free development of one's personality[6], freedom of expression[7], and the inviolability of the secrecy of telecommunications.[8] The crucial point is that the classical concept takes these freedoms as a given. The role of the state is reduced to the function of limiting freedom with regard to public good or the rights of others. The reservations included in fundamental rights allocate this task primarily to the legislature and enables it to limit the guarantees of freedoms by means of constitutional statutory regulations.[9] All interventions by the state require a statutory basis. This basis must take the relevant constitutional requirements, especially the principle of the clarity and certainty of provisions and the principle of proportionality, into account as must the executive branch in any decision founded upon that statutory basis.

11.2.1.2 Limitations of the Concept

The understanding of fundamental rights as protection against encroachments on rights seems to be a far-reaching, optimal protection of freedom. But in fact, it has

---

[6] Article 8 (1) ECHR: "Everyone has the right to respect for his private and family life, his home and his correspondence."; Article 7 EU Charter: "Everyone has the right to respect for his or her private and family life, home and communications."; Article 2 (1) GG: "Everybody has the right to the free development of his or her personality [. . .]".

[7] Article 5 (1) GG: "Everyone has the right freely to express and disseminate his or her opinions in speech, writing and pictures [. . .]".

[8] Art. 10 (1) GG: "The secrecy of communication by letters and of telecommunication is inviolable."

[9] See Article 8 (2) ECHR: "There shall be no interference by a public authority with the exercise of this right except such as is in accordance with the law and is necessary in a democratic society in the interests of national security, public safety or the economic well-being of the country, for the prevention of disorder or crime, for the protection of health or morals, or for the protection of the rights and freedoms of others."; Article 52 (1) EU Charter: "Any limitation on the exercise of the rights and freedoms recognised by this Charter must be provided for by law and respect the essence of those rights and freedoms. Subject to the principle of proportionality, limitations may be made only if they are necessary and genuinely meet objectives of general interest recognised by the Union or the need to protect the rights and freedoms of others."; Article 2 (1) GG: "[. . .] provided that they do not interfere with the rights of others or violate the constitutional order or moral law." or Article 5 (2) GG: "These rights shall be subject to the limitations laid down by the provisions of the general laws and to statutory provisions for the protection of young people and to the obligation to respect personal honour."

many prerequisites and is accompanied by limitations due to its structure. As the traditional concept of protection against encroachments builds on liberal conceptions of fundamental rights, "freedom" is understood as a pre- or non-state sphere, i.e. as a sphere antecedent to the state or external to it, and as "natural freedom."[10] Consequently, fundamental rights are directed solely toward the protection of freedoms that already exist. The social preconditions of individual freedom or, expressed more radically and more precisely, the social foundation and embeddedness of individual freedom are not given consideration. The approach also results in specific subject matters or interests which are to be protected by fundamental rights. They are conceived of as individualistic, i.e. from a perspective focusing on the individual and in the form of an individual good that is not already structurally limited.[11] Protection is granted, in particular, to individual self-determination, to individual decisions and behavior at will, to one's own body, or to property.

In liberal thought on fundamental rights, the state appears exclusively as an institution that limits individual possibilities to decide or to act freely. Such limitations must be justified, namely by the parliament passing a law that pursues a legitimate goal, that is as precise as possible both in its legal requirements and its legal consequences, and that is commensurate with the principle of proportionality. Such laws guide and limit the decisions of the executive. Just as the concept of freedom and the scope of protection of individual rights are shaped in specific ways, the role of laws and the requirements of the design of laws are tailored exclusively toward justifying limitations on freedom. The multi-dimensional role of the legislation as well as of laws is disregarded.

11.2.2 Informational Self-Determination as a Protected Interest

The classical-liberal concept of fundamental rights also characterizes the form in which the goods to be protected by data protection are described. This applies especially clearly to the right to informational self-determination. This right is the decisive fundamental right in the realm of data protection in Germany. However, it is also being mentioned with greater frequency in the transnational and European debate as a central right worthy of protection.[12] Several scholarly debates are about how to understand or how to concretize this right. This section analyzes the right to informational self-determination with a view to the influential jurisdiction of the German Federal Constitutional Court, which has developed and established it. At least in this respect, the construction of this right and the description of its scope of protection are based upon traditional doctrinal concepts and are therefore insufficient.

[10] Böckenförde 1974, 1529, 1532; Lübbe-Wolff 1988, 75 ff.

[11] Albers 2005, 30 ff.

[12] See, i.e., Schwartz 1989, 675 (677 ff., 701). See also Raab and Goold 2011, 17. With distinguishing considerations Rouvroy and Poullet. (Fn. 1), 45, 52 ff. For an overview of the constitutional rights in European countries see Leenes et al. 2008.

The Federal Constitutional Court derived the right to informational self-determination from the general right of personality guaranteed by Art. 2 in conjunction with Art. 1 GG[13] in its 1983 decision concerning the census.[14] The right to informational self-determination confers on the individual the power to, in principle, determine for himself or herself the disclosure and use of his or her personal data.[15] Individuals have the right to decide themselves whether and how their personal data is to be divulged and used, in other words: a right to self-determination about processing of data relating to them.

How did the Federal Constitutional Court arrive at this subject matter to be protected called "informational self-determination"? Its precursor is the right to privacy, which is also anchored in Art. 2 in conjunction with Art. 1 GG and was recognised in the case-law of the Federal Constitutional Court since the 1970s. The Federal Constitutional Court originally conceived this right employing the spatial imagery of areas of retreat walled off from the outside world, or similarly isolated situations for interaction and communication, and as the right to be let alone, or as the right to keep events in this isolated sphere confidential.[16] The right to privacy centered on a spatially as well as thematically specified area which is to remain, in principle, free of undesired inspection. This was the traditional, narrow concept of privacy. This concept drew the same broad criticism as it did in the American privacy debate. The first point of criticism emphasized the relativity of the sphere of personal privacy: it could be described only in terms "relative" to those receiving information.[17] Therefore, what was to be protected was not a predetermined sphere, but the capacity of the individual to decide to whom to disclose which information. Alan Westin couched this idea in these terms as early as 1972.[18] The second point of criticism highlighted the fact that the need for protection was less about the private sphere as the context in which certain data emerge but rather about which information could be derived from data obtained and how that information could be used.[19] In other words, what is decisive is not the context data originate from but rather the context in which the information is used. The Federal Constitutional Court responded to these central points of criticism of the rather narrow concept of the right to privacy understood as a protected sphere by developing the idea of a right to informational self-determination

---

[13] Article 2 GG: "Everybody shall have the right to the free development of his or her personality [. . .]"; Article 1 GG: "Human dignity shall be inviolable. To respect and to protect it shall be the duty of all state authority."

[14] BVerfGE 65, 1, 42 ff.; Dec 15, 1983, Census Judgment. Subsequent decisions are, amongst others, BVerfGE 78, 77, 84 ff.; 84, 192, 194 ff.; 113, 29, 46 ff.; 115, 166, 188 ff.

[15] BVerfGE 65, 1, 43. Analyzing the decision and its background: Albers (Fn. 11), 149 ff.; see also Rouvroy and Poullet (Fn. 12), 52 ff.

[16] BVerfGE 27, 1, 6 ff; 27, 344, 350 ff.; 32, 373, 378 ff.; 33, p. 367 376 ff.; 44, 353, 372 ff. See also Warren and Brandeis 1890, 193–220.

[17] See Schlink 1986, 233, 242; Solove 2004, 212 f.

[18] Westin 1970, 42.

[19] See Simitis 1971, 673, 680.

which centers on individual decision capacities as well as on the context of use.[20] The Court also took up the acknowledged constitutionally protected goods of autonomy and freedom of decision and action, arguing as follows: free decision and action are possible only under certain circumstances. If a person is unsure whether deviating behaviors may be stored as information and used to his/her disadvantage, he/she will try not to attract attention by such behavior and is no longer free to act at will.[21] That is why the protection of fundamental rights must cover the protection against information and data processing by the state. The Federal Constitutional Court then shaped this extent of protection with reference to freedom of decision and action. Just as people can decide about their actions, they also have the right to determine how "their" personal data will be processed.

What characterizes this right to informational self-determination? It reaches beyond the classical understanding of the right to privacy. Its core element is a relatively abstract and therefore far-reaching individual right to make decisions ranging from disclosure of data to their processing and to their use. Even if the right to informational self-determination is derived from the right to the free development of his or her personality and from human dignity[22], its scope of protection is shaped likewise a property right.[23] Similar to some American conceptions of privacy—"Privacy," *Charles Fried* writes, "is the *control* we have over information about ourselves [...], is control over knowledge about oneself."[24]—informational self-determination is thought of as a right of control over personal data. The holders of fundamental rights also have the right to know by whom and for what purposes personal data referring to them are processed[25], but that right is accessory in the context of the concept.

The fundamental right protects this right to decide over the disclosure, processing and use of personal data as an individual protection against any encroachment. It follows from such a scope of protection that, as a matter of principle, every step in processing personal data is to be considered as an encroachment on the right to informational self-determination. Therefore, every step in processing personal data must be based either on consent or—more important[26]—on a constitutional legal basis which has to meet the requirements of the principles of clarity and

---

[20] For literary sources of the Court's decision see Hermann Heußner (former judge at the FCC preparing the Census Decision), 1984, 279 (280 f.). Amongst others, the ideas of Westin have been received by the members of the Court, see Ernst Benda (former President of the FCC participating at the Census Decision) 1974, 23 (32).

[21] BVerfGE 65, 1, 43.

[22] See the considerations of Rouvroy/Poullet (Fn. 12), 52 ff.

[23] It is true that the FCC also stated: "The individual does not have a right in the sense of an absolute, unlimitable mastery over 'his' or 'her' data; he/she is rather a personality that develops within a social community and is dependent upon communication.", BVerfGE 65, 1, 46. However, these grounds refer to the reservation allowing to limit the scope of protection by means of statutory rules; they do not alter the shaping of the scope of protection.

[24] Fried 1968, 475, 482.

[25] BVerfGE 65, 1, 46.

[26] The core of the right to informational self-determination is not that consent has to play a key role. Theoretically and practically more important is that a constitutional legal basis is necessary to justify data processing.

determinedness and of proportionality.[27] Additionally, the Federal Constitutional Court emphasized the principle of specifying the purposes of data processing in advance and the principle that further data processing is bound to the original purpose.[28] These consequences already show the far-reaching influence such a concept of informational self-determination has on data protection laws.

11.3 Influence on Data Protection Approaches and Principles

In Germany, the right to informational self-determination is very firmly entrenched and has many ramifications and marks the approaches, principles, legal constructs and laws pertaining to data protection to this day. The respective patterns of thinking have also influenced the Data Protection Directive of the European Union and the fundamental right expressed in Art. 8 of the EU Charter of Fundamental Rights via reciprocal influences in law-making. Similarly, they affect court rulings via the network of jurisdiction among the German Federal Constitutional Court, the European Court for Human Rights, and the European Court of Justice.[29] In this section, important implications these patterns of thinking have on the approaches, principles, and legal constructs of data protection are highlighted.

Informational self-determination, shaped as the individual right to decide over the disclosure, processing and use of personal data, centers on data, specifically the individual piece of personal data, and in the broader sense its processing in a sequence of pre-defined steps—collection, storage, alteration, use, transfer. Additionally, "data" and "information" are treated as though they were synonyms. This reflects an ontic concept of information, namely the idea that information is a kind of depiction of reality and that data could be treated as if they were objects. Views of this kind occur in the basic approaches and in the legal definitions of data protection law. For example, the German Federal Data Protection Act (Bundesdatenschutzgesetz; BDSG) does not distinguish between data and information (§ 3 I BDSG) and focuses on the lawfulness of the collection, storage, use or transfer of personal data (§§ 4, 13 ff., 27 ff. BDSG). Similarly, both the Directive 95/46/EC[30] and the Proposal of the Commission for a General Data Protection Regulation[31] define personal data as any information relating to an identified or identifiabe natural person or data subject

---

[27] BVerfGE 65, 1, 44 ff.

[28] BVerfGE 65, 1, 46.

[29] For Data Protection in the Case Law of the EctHR and the ECJ see de Hert and Gutwirth, (Fn. 1), 3, 14 ff.; Siemen 2006, p. 51; Schweizer 2009, 462, 464 ff.

[30] Directive 95/46/EC of the European Parliament and of the Council of 24 October 1995 on the protection of individuals with regard to the processing of personal data and on the free movement of such data, Official Journal L 281/31.

[31] Proposal of the European Commission of 25 January 2012 for a Regulation of the European Parliament and of the Council on the protection of individuals with regard to the processing of personal data and on the free movement of such data (General Data Protection Regulation), COM (2012) 11 final.

(Art. 2 of Directive 95/46/EC; Art. 4 GDPR-Proposal) and lay down conditions under which the processing of data, defined as any operation or set of operations which is performed upon personal data, is lawful (Art. 2, 5 ff. of Directive 95/46/EC; Art. 2, 5 ff. GDPR-Proposal).

It is only owing to this ontic concept of information and data that the protected interest can be formulated analogously to the concept of property, namely as the right of disposal over processing of personal data carried out by others. The idea of informational self-determination as the exercise of individual control over data or information can be found throughout data protection law.

As the individual's authority to decide about any processing of personal data is protected, every step in processing personal data requires either consent or a legal basis. Both German and European data protection law include the principle that, apart from their legitimate use with a person's consent, personal data must not be processed in the absence of a legal basis (§ 4 I BDSG, Art. 8 II 1 EU Charter, Art. 7 of Directive 95/46/EC). This basis has to permit, in sufficiently precise form, such processing for legitimate purposes. Explicit purposes are to be specified for data processing in advance (§§ 4 III, 13, 14 I, 15 I, 16 I BDSG; Art. 8 II 1 EU Charter, Art. 6 (b) of Directive 95/46/EC, Art. 5 (b) GDPR-Proposal). Further data processing is principally bound to these purposes or at least may not be incompatible with these purposes (§§ 14 I, II, 15 I, III, 16 I, IV BDSG; Art. 6 (b) of Directive 95/46/EC, Art. 5 (b) GDPR-Proposal).[32] The entire approach is guided by the idea that courses of action and decision-making processes could be almost completely foreseen, planned and steered by legal means. In Germany, this has resulted in a far-reaching juridification and in a multitude of data protection laws, which, however, often simply map the data processing steps.

11.4 The Complexity of Data Protection: Analyses and Consequences

Data protection law has been in flux for some time now. Changes in basic societal and technical conditions have often been pointed out. But the issue is by no means simply one of adaptation to changes in external conditions. At a fundamental level, the patterns of thought and description used in data protection law must be reflected upon critically and reconceptualized.

This shall be explained for three points in particular: firstly, for the subject matter at hand; secondly, for the description of the protected interests; and thirdly, for the

[32] The requirement that personal data must not be further processed in a way incompatible with the specified purposes sets lower standards than the requirement that further data processing is principally bound to the purposes specified in advance. Additionally, the meaning of "incompatible" requires interpretation. See for the functions of the principle of specifying purposes and of binding data processing to the purposes specified in advance Albers (Fn. 11), 168 f., 498 ff.

concepts for regulation. As a result, it will emerge that in all respects data protection requires an innovative approach, is highly complex, and poses unprecedented challenges for law.

11.4.1 The Complexity of the Subject Matter: Data and Information, Knowledge and Flow of Data and Information, Decisions and Consequences of Decisions

The goal of data protection is not the protection of data but of the individuals to whom the data refer. The object of protection, then, is not the personal data per se. We must expand this isolated view by including several elements: at a basic level the element of information; in the structural dimension knowledge; in the temporal dimension the flow of data and information; and in the broader context decisions and consequences of decisions.

Concepts of "data" and "information" are described in multifarious and discipline-dependent ways.[33] In the (social) context of data protection it is at least important to realize that data and information are not synonymous. On the contrary, they must be strictly differentiated. Data might be described as characters recorded on a data carrier, including written documents or videos as well as data digitally stored on hard drives or mobile data storage devices. Data, forms of storage, and processing operations are characterized by the various media, technologies, and networks.[34] Due to their objectification, data can be conceived of distinctly and provide a starting point for legal regulation. Nonetheless, data are not meaningful per se, but rather as "potential information". Their information content is not an intrinsic attribute of the data themselves.[35] It is created only by means of interpretation in the particular context of interpretation.

Information involves meaning, and pieces of information are elements of meaning. Units of information may base on data (or on observations or communications) but data only attain meaning by being explained and interpreted by the recipient or data user who uses data to obtain information. Devising meaning depends on the individual situational conditions for interpretation as well as on the context of the knowledge and interpretation.[36] Information is context-dependent in an elementary way. Although this insight may be well-established today, people hardly face up to the difficulties this entails for legal regulation and for a description of the object to be regulated.

---

[33] See, for example, Floridi 2010, 19 ff.

[34] See, among others, Waldo et al. 2007, 88 ff.

[35] See with regard to communication Ashby 1963, 124: "The information conveyed is not an intrinsic property of the individual message."

[36] Albers 2002, 61, 67 ff. See also Bateson 1972, 315 ff.

Due to the fact that information requires interpretation, which takes place in a particular context of knowledge and interpretation and is dependent on the individual, situational conditions of interpretation, information refers to the structures and processes within which it can be created in the first place. In the structural dimension, knowledge is involved in generating information. Knowledge is founded upon texts, files, archives, registers, databases, expert systems, but also upon institutional, organizational or procedural arrangements. It makes interpretation possible, and limits the possibilities of interpretation. Knowledge is a factor and a product of the context in which handling of information and data occurs and it influences this handling inherently.[37] Whether or not data processing poses risks to the person the data refer to also depends on the knowledge that exists or can be developed in a particular context or in a particular case. That is why data protection must also take the knowledge level into account.[38] In the temporal dimension, the procedural character of data processing comes into play as well. Data and information are constantly generated anew and altered during processing operations. In addition, a collection of personal data reveals its social and legal meaning only when one views it along with its linkages to other data, its use, or its transfer to other agencies. For example, one can understand what it means if personal telecommunications data is stored longer than necessary for billing (in the context of data retention)[39] only with the duties of telecommunications companies in mind to transmit personal data to the security authorities which then use the data for further investigations against the respective person.[40]

The ways in which data and information are handled, the knowledge and the processing operations are impacted by the media, technologies, and networks employed. Whether data are stored in paper files, automated electronic files, or in network systems has an influence on, for instance, the quantity and the form of data that can be stored and easily accessed, the potentials for interlinking them, or the possibilities for transmitting data. Media, technologies, and networks can increase the dangers individuals are subject to, but can certainly also limit such vulnerability by putting technical barriers and safeguards related to data processing into place.

What matters not least is the connections between information or knowledge on the one hand and the decisions made by the public or private bodies processing the data on the other. In the end, information and knowledge serve as bases for certain decisions and actions. Such decisions have consequences. They may have an adverse effect on the person to whom the data and information refer in the form of a limitation of his/her freedom. And protection from unjustified disadvantages is one of the reasons for data protection.

---

[37] Albers 2012, § 22 Rn. 14 ff.; Trute 2010, 11 ff.

[38] See also Mireille Hildebrandt, *Who is Profiling Who? Invisible Visibility*, in: Gutwirth et. al. (Fn. 1), 239, 240 ff.

[39] Article 3 of Directive 2006/24/EC of the European Parliament and of the Council of 15 March 2006 on the retention of data generated or processed in connection with the provision of publicly available electronic communications services or of public communications networks [. . .], Official Journal L 105/54.

[40] See BVerfGE 125, 260 (318 ff.). For a critical review of this Decision see de Vries et al. 2011, 3 ff.

As a result, data protection deals with highly complex subject matter: It is necessary to operate with the differentiation between data and information. The dimension of knowledge and the temporal dimension of data and information flow must be regarded as well as the decisions and consequences of decisions. In other words, any new concept would be misguided if it just focused on information rather than on data, and simply substituted one term for the other. On the contrary, data remains an important reference point for legal regulation. But data must be conceived of within a network of several fundamental elements and is not the only reference point. Data protection aims at regulating data processing, but precisely also at regulating the generation of information and knowledge, at influencing the decisions based on such generation, and at preventing adverse consequences for the individuals affected.

11.4.2 The Complexity of the Protected Interests of Affected Individuals

This brings us to the second point: How can we describe the protected interests of affected individuals? At the center of the legal discussion are a few very abstractly stated descriptions of legally protected goods which are related to fundamental rights: Private life or privacy[41], protection of personal data, informational self-determination. Art. 8 ECHR, the right to respect for private life[42], has been concretized to various claims against collection and storage of personal data or claims to be informed about data that refer to oneself. However, legal rulings of the European Court of Human Rights (ECtHR) unfold from case to case; the contents of what constitutes the right to respect for "private life" as a legally protected good is compiled merely casuistically.[43] Looking at Art. 7 of the EU Charter[44], Art. 16 (1) of the TFEU and Art. 8 (1) of the EU Charter[45] the right to respect for "private life" and the right to the "protection of personal data"—each one a very abstractly formulated legally protected good—stand side by side. To date, the European Court of Justice avoids a clear cut differentiation[46] and only specifically describes objectives of protection and legally

---

[41] For an analysis of the concept of „information privacy" in the UK see Raab and Goold (Fn. 12).

[42] See Fn. 6.

[43] See the references in Fn. 29.

[44] See Fn. 6.

[45] Art. 8 (1) of the EU Charter: "(1) Everyone has the right to the protection of personal data concerning him or her. (2) Such data must be processed fairly for specified purposes and on the basis of the consent of the person concerned or some other legitimate basis laid down by law. Everyone has the right of access to data which has been collected concerning him or her, and the right to have it rectified. (3) Compliance with these rules shall be subject to control by an independent authority."

[46] See ECJ, Rs. C-92/09 u. C-93/09, *Schecke and Eifert vs. Land Hessen*, http://curia.europa.eu, §§ 45 ff. The differentiation is necessary but not easy due to the interplay between Art. 7 EU Charter in conjunction with Art. 52 (3) EU Charter, Art. 8 ECHR on the one hand and Art. 8 EU Charter on the other.

protected goods to a very limited extent. The Federal Constitutional Court focuses on the "informational self-determination" derived from Art. 2 in conjunction with Art. 1 GG as a legally protected good. Just as well, German academic approaches have long been centered on patterns of thought such as informational self-determination, authority to decide about processing of personal data, and individual control. In recent years, there has been some movement, and a new discussion regarding the rights which data protection should safeguard has commenced. One widespread criticism argues that control is simply not possible because of the factual circumstances and the conditions of the internet. But the approach taken by this criticism is not sufficiently profound. The idea of control over one's own data fails not only because it would no longer be practicable. It fails because it does not fit the subject matter to be protected. A reconceptualization is needed which leaves the classic concept of basic rights behind. The interests which data protection is to safeguard cannot be grasped using an individualistic perspective; a multidimensional understanding of fundamental rights is required; and as a result, data protection includes a bundle of rights which must be described in a new way.

11.4.2.1 From Individualistic Patterns to the Protection of the Individual in Sociality

Protection of fundamental rights in terms of the way government agencies or other private parties handle personal information and data is different from the legally protected good in the traditional understanding of fundamental rights. It is true that a holder of fundamental rights exists. But the object of protection is not the holder's freedom of decision or of action, which would be impaired by state intervention. Instead—as the analysis of the subject matter has just demonstrated—the holder is to be protected in terms of personal information and data, which are generated and processed by others in particular contexts. Government agencies or other private bodies are structurally involved in this, due to the mere fact that data and information must be interpreted. Personal information or data cannot be assigned to the person in question like an object belonging to him or her.[47] Individualistic patterns of assignment fall short.

Reasoning why and to what extent the person in question is to be protected must rather stem from a supraindividual perspective, namely by taking a categorizing view of the context and of adverse consequences that are to be expected with regard to the person to whom data, information and knowledge refer. The fact alone that a piece of data refers to a person does not yet predicate a person's need for protection. The need for protection arises in particular in relation to negative effects of handling the personal data and the information gained from it. Legally protected goods and encroaching mechanisms require their own separate patterns of description. In addition, protection directed solely at defending against and refraining from processing personal data is insufficient. The person protected may also be interested in personal

---

[47] More thoroughly Allen 2000, 861, 865 ff.

data being made available so that an agency has the information at its disposal which it needs for a correct decision. And it is just as important that the person affected is informed about processing of personal data and information and can influence it. Hence, individuals need not only defensive rights, but also rights to know, to obtain information, to participate, and to exert influence. The subject matter to be protected by data protection based on fundamental rights must therefore be designed differently and be more diverse than the legally protected goods in terms of the "classical" concept of fundamental rights and the "classical" concept of protection against encroachments. Appropriate data protection requires a more sophisticated conception of fundamental rights.

11.4.2.2 The Necessity of Building Upon a Multidimensional Understanding of Fundamental Rights

Extensions of the functions of the fundamental rights and of the scope of their protection which go beyond the traditional understanding of fundamental rights are recognized in principle by now. Modern codifications, for example the EU Charter of Fundamental Rights, reflect the diversity of dimensions of protection in their catalogs of fundamental rights.[48] The German Federal Constitutional Court has derived positive obligations of the State, for example obligations to provide for the minimum income needed to exist and especially the state's duty to protect (Schutzpflicht) as well as the so-called "Drittwirkung" by which fundamental rights indirectly influence the legal relationships between private persons. Nevertheless, the court rulings of the Federal Constitutional Court, the European Court of Human Rights, and the European Court of Justice are tentative in this regard. Protection against encroachments is still considered the primary dimension of protection in fundamental rights. That is one of the reasons why, in the case of data protection, the protected interests are shaped likewise a property right. Doctrinal reasons are also evident with regard to the rather hesitant acknowledgement of fundamental rights of access to personal data[49] or of institutional guarantees. In scholarly debates, the foundations, the extent and the details of the further dimensions of fundamental rights' protection beyond the traditional understanding are the subject of heated controversy.

---

[48] See, for example, Art. 14, Art. 27 ff. EU Charter.

[49] In Germany, the first Senate Decision of the FCC which fundamentally derived rights to know not only from the guarantee to access to the courts, Art. 19 (4) GG, but from Art. 2 (1) in conjunction with Art. 1 (1) GG was not earlier than in 2008, see BVerfGE 120, 351 (362 f.); prior to that see BVerfG (Chamber Decision), NJW 2006, 1116 (1117 ff.). The ECtHR has recognised rights to access to personal files and to obtain information earlier, however, mostly in special cases, see for the rights of persons to receive the information necessary to understand their childhood and development *Gaskin vs. United Kingdom*, Judgment of 7 July 1989, Application No. 10454/83, for the right of access to health-related (not necessarily personal) data ECtHR, *McGinley and Evan vs. UK*, Judgment of 9 June 1998, Application Nos. 21825/93, 23414/94 —, Rn. 98 ff; see also ECtHR, *Segerstedt-Wiberg*, Judgment of 6 July 2006, Application No. 62332/00—, Rn. 99 ff. The Court argues cautiously: "Although the object of Article 8 is essentially that of protecting the individual against arbitrary interference by the public authorities, it does not merely compel the State to abstain from such interference: in addition to this primarily negative undertaking, there may be positive obligations inherent in effective respect for private or family life. In determining whether or not

However, the abstract guarantees of fundamental rights are open to interpretation and permit to elaborate diverse dimensions of protection. The classical understanding is a concept which is too narrow and has dysfunctional prerequisites and limitations. Fundamental rights are not only about protection against encroachments, but also about rights to know and to obtain information, about rights to participate and to influence decisions, about rights to be protected by the state, or about institutional guarantees. As individualistic patterns of assignment and the idea of control over one's own data fall short and as the subject matter to be protected is multifarious, data protection has to base upon the further development of the functions and the contents of fundamental rights.

11.4.2.3 The Bundle of Protected Interests

"Data protection" is a rather vague concept. Some scholars emphasize that data protection simply describes the tool for safeguarding legally protected freedoms like autonomy or freedom of decision. Others assume that it points to the good or goods to be protected. It could also be understood as covering both: the means of protection and, as an umbrella term, the legally protected interests. Anyway, when it comes to the goods to be protected, data protection should not be understood as a merely instrumental concept which protects other freedoms known from the traditional concept.[50] Instead, it is necessary to leave behind the descriptions using an individualistic approach, to wit: self-determination, freedom of decision, property. The interests to be protected should be designed so that they gain their meaning when the sociality of the individual in question is taken into account. This is responsive to the subject matter elucidated above: data, information, knowledge. Hence, data protection is about protection from the creation of personality profiles, protection of a person's reputation, protection from stigmatization and discrimination, protection of normative justified expectations of privacy, protection against identity theft, protection against surveillance and protection of contextual integrity.[51] These examples illustrate that data protection does not encompass a uniform legally protected good. On the contrary, there are complex and manifold interests that are to be protected. Their wide range and contextual dependencies have already been worked out in the context of the "privacy" debates in the US, for example by Daniel Solove and Helen Nissenbaum, among others.[52]

---

such a positive obligation exists, the Court will have regard to the fair balance that has to be struck between the general interest of the community and the competing interests of the individual, or individuals, concerned [...]" (*McGinley and Evan vs. UK*, Judgment of 9 June 1998, Application Nos. 21825/93, 23414/94—, Rn. 98).

[50] Of another opinion: Britz 2010, 569 ff.; Poscher 2012, 178 ff.

[51] The fundamental right to the guarantee of the confidentiality and integrity of information technology systems which has been derived from Art. 2 in conjunction with Art. 1 GG by the Federal Constitutional Court in 2008—BVerfGE 120, 274—points in the right direction, but it should be understood merely as a part of data protection.

[52] Solove 2008; Nissenbaum 2008, 119 ff.; Nissenbaum 2010. See also Rössler 2001.

A closer analysis reveals that the dangers posed by processing of personal data and information and the needs for protection that data protection responds to have be identified at different levels.[53] At a basic level, the crucial problem centers on information and data processing that is all-encompassing, unlimited, and not transparent. As long as one is confronted with a situation of this kind, then no suitable estimate can be made in what contexts what information is being generated and how such information is being used or what negative consequences individuals will have to face in specific constellations. This problem of unlimited and intransparent data processing must be countered by legal regulation providing basal limits and transparency. Only on this basis is it possible to work out interests to be protected which exist in quite specific contexts due to quite specific disadvantages.

At the basic level, Orwell's "Big Brother,"[54] Bentham's "Panopticon,"[55] and Kafka's "The Trial"[56] might be illustrative as widely known, culturally anchored metaphors that—although these narratives are of course rooted in quite different contexts—take up different facets of the dangers just mentioned above. Daniel Solove has pointed out that the "Big Brother metaphor is definitely effective at capturing certain privacy problems"[57] but that it is the Kafka metaphor which captures those elements of threats to privacy which deal with certain data collection and circulation by others or other entities "without having any say in the process, without knowing who has what information, what purposes or motives those entities have or what will be done with that information in the future."[58] This illustrates that, at the basic level, there are already multifarious problems data protection shall countervail. Speaking legally, they are not solved by merely assigning an individual right to control personal data to the data subject. In keeping with the dangers to liberty, duties of the legislative branch and requirements of legal regulation are necessary. The legislation must regulate data processing in an appropriate way and safeguard that handling personal information and data does not take place in an unrestricted, unlimited, and intransparent way as well as it has to ensure that the individuals affected have the possibility to obtain sufficient knowledge about and sufficient influence on processing of personal data and information. At this level the state is anything but kept out.

At a second concrete level, it is about individual and specific interests to be protected, which arise for the affected person in concrete contexts in terms of adverse consequences. The capability to describe the dangers as well as the specific interests to be protected at this second level requires that basic regulation occur at the first level. An example is the problem of the domestic intelligence service monitoring a public meeting, with negative consequences for the freedom of assembly. Another example is the protection of individuals from media intrusion by publishing personal data or pictures. At this level, rights as protection against encroachments are applicable.

[53] See Albers (Fn. 11), 353 ff.

[54] Orwell 2008.

[55] Bentham 1995, 29 ff.

[56] Kafka 2002.

[57] Solove 2001, 1393, 1399.

[58] Solove (Fn. 57), 1426.

Nevertheless, duties to protect have to be derived, too, as well as an overall concept beyond traditional approaches is necessary.

The result shows that data protection outlines a complex bundle of interests worthy of protection. Data protection bases upon a multi-dimensional understanding of fundamental rights and requires entirely new descriptions of the protected interests: in place of legally protected goods conceived of in an individualistic way, it is about individual legal positions *in* sociality, or, in other words: the individual's social positions to be protected by fundamental rights. The bundle of protected interests and positions must still be worked out in greater detail and will also have to be dynamically adapted time and again to new dangers.

11.4.3 The Complexity of Appropriate Concepts for Regulation

The third point section of this paper shall demonstrate how complex appropriate concepts for regulation must be. To date, concepts are still characterized by the image of central mainframe computers that process data using programs in a predefined sequence. Legally, informational self-determination as the good to be protected and the reservation allowing legal regulation lead to the idea that every step in processing personal data must be justified by consent or legally regulated by means of a basis in law. But meanwhile, the pitfalls of consent are recognized as well as the multitude of laws is more and more criticized as a flood of legislation. More problematic than the quantity of laws is that the regulations often simply map the data processing steps and that the approach is characterized by the belief in planning prevalent in the last century when people were convinced that it was possible to regulate things precisely using legal means.[59]

However, fundamental rights as basis of data protection do not result in being forced to understand laws against the backdrop of their traditional role. As well as allowing the development of new legally protected goods, fundamental rights permit a multidimensional understanding of the reservations and of regulations. Legal norms do not only limit freedoms. They can also create freedoms in the first place, make them concrete, and influence their social conditions and prerequisites. Data protection law must be founded on the diverse functions and diverse forms of law. Regulation concepts must include a wide range of constituent elements, which utilize the entire spectrum of legal forms and instruments. They are therefore complex on their own terms and in addition, they have to be interwoven. Further factors make clear how challenging appropriate data protection laws are.

---

[59] For new challenges with regard to ubiquitous computing which affects the current principles of data protection see Čas (Fn. 1), 139, 141 ff.

11.4.3.1 A Wide Range of Regulation Elements

Rather than merely steering the steps of processing data, appropriate regulation concepts require many different elements. Regulation of data processing stages will still play an important part in the future. This form of regulation is, however, supplemented and augmented by other constituent elements: data protection through system design, data protection through the development and use of technology, organizational and procedural precautions, expanded functions of data protection officers, or quality assurance mechanisms such as data protection audits. In addition, there is a variety of affected individuals' rights to know, to obtain information, to participate and exert influence. The fact that data protection law includes a large number of constituent elements is generally recognized by now. But up to the present, elements of different origins have tended to exist side by side. In the future, they must dovetail and be interwoven appropriately. This is an ambitious task. Moreover, the constituent elements are rather complex themselves and call for highly varied instruments. This can be exemplified by data protection through system design, by data protection through technology and by individual rights to information.

Data protection through system design refers to a level preceding regulation of the steps of data processing. In summarizing broad discussions, it can be described as "data protection functionality incorporated into systems and procedures".[60] The leading idea is that regulating the steps of data processing is not sufficient because data processing takes place within certain social systems, within organizational structures and procedures and under specific technical conditions.[61] This predetermined context influences which and how many personal data is needed, how long data has to be stored, how many people have access to them and how transparent data processing is. Therefore, the legal regulation and shaping of this context prior to the subsequent processing of data and information is not less important than the regulation of the data processing operations. That makes also clear that "system design" does not refer solely to technical systems or procedures; organizational structures or decision procedures have to be taken into consideration as well.[62] Hence, data protection through system design aims at the legal shaping of the social, organizational, procedural and technical contexts in which personal data and information are handled. It has a broad scope: from the shaping of administrative competences to which data processing operations are oriented, to organizational and procedural approaches, to the technical setup of data processing equipment. Understood in this way, data protection through system design is an evidently ambitious task to fulfill. The German Federal Data Protection Act, for example, attempts to realize it by the general principle of data avoidance and data minimization (§ 3a BDSG): Systems shall be designed in a way that as few personal data are needed as possible. Whether these principles really make sense as overall principles is contested.[63] This points to

[60] Köhntopp 2001, 55, 56.

[61] See also Point 4.1 of this chapter.

[62] The scholarly elaborations are heterogeneous in this respect.

[63] More closely Albers (Fn. 37), Rn. 106 ff.

the difficulty that the realization of data protection through system design depends on a—not yet achieved[64]—clear and convincing elaboration of protection objectives and protected interests. All in all, system design as regulation element takes data protection law beyond the traditional patterns of regulatory law.

Whilst the social risks of mainframe computing systems and data processing technologies once were the reason for developing data protection concepts, technologies in the meantime are considered to be also a tool for realizing data protection. Privacy friendly or privacy enhancing technologies play an important role both in European and in national law.[65] But data protection through technology places high demands on law. The first problem is that it has to be ensured that technology with which the normative standards for the handling of personal information and data can be fulfilled is available at all. Technological developments cannot be commanded. Indirect incentives and mechanisms for exerting influence must be drawn upon, e.g. giving financial support, institutionalizing bodies or procedures for developing privacy friendly technologies or issuing quality seals and product certificates. These "soft law"-instruments might influence technology development but their influence is limited. Assumed that applicable technologies are available data protection through technology shaping defines requirements for the selection, use, and configuration of data processing networks, systems, programs, or storage media. In advance of concrete processing operations, these requirements are to ensure that normative rules are already technically established or can at least be fulfilled. Data protection through technology shaping overlaps with data protection through system design. It includes, for example, requiring data protection-friendly default settings. Data protection through the use of technology encompasses requirements of the forms of technology that accompany and secure the regulation of the steps of data processing, for instance the obligation to use encryption procedures when transmitting data. Just as data protection through system design, data protection through technology development, shaping and use is an ambitious task. And just as well, it depends on clearness about protection objectives and protected interests and, including forms of "soft law" and diverse instruments, it takes data protection law beyond the traditional patterns of regulatory law.

The rights of affected persons to information about the collection and use of personal data seem to be—although they are directed towards positive actions of the state or of private persons processing personal data—rather uncomplicated. However, they fulfill different functions: They are intended to convey to the data subjects the information they need regarding what others know about them so they can orient themselves in their social environment. They open up opportunities to participate and to influence the data and the knowledge. They safeguard the possibility of legal remedies. Due to these different functions they must be guaranteed and carried out on several levels and in a variety of forms: as general information about tasks and organizational structures of authorities or bodies processing data, as duties to

[64] See Point 4.2 of this chapter.

[65] See, i.e., Report from the Commission, First report on the implementation of the Data Protection Directive (95/46 EC), COM (2003) 265 final, 15 f.

inform or duties of notification, or as rights to access to information or to documents. Additionally, the exercise of rights to information in practice depends on social and individual prerequisites, which can be influenced only indirectly by means of law.

To conclude with another regulation element, which has to be refined: Data protection cannot be guaranteed solely by mechanisms that accord the persons affected individual protection and individual redress mechanisms. Appropriate institutional guarantee mechanisms have to be established as well[66] so that it must be decided, e. g., under which conditions they make sense and how they should be combined with individual rights and legal remedies of the data subject.

11.4.3.2 Further Characteristics of Data Protection Law

Concepts for regulation of data protection become complex not least due to the fact that data protection law must be coordinated with already existing issue-related legal norms containing, for example, tasks and competences in a particular field. A thoroughly coordination is necessary because of the close linkages between data, information, knowledge, and decisions.[67] Data protection is not a special field of law that could stand in isolation beside the substantive fields of law. Rather, data protection law pertains to a fundamental cross-cutting dimension. The need to coordinate with the substantive provisions also points to the need to differentiate within data protection law itself. For example, one must consider the questions of when sector-specific regulations are necessary, when general regulations fit best or to what extent uniform data protection law for the public and private realms makes sense.

A number of additional factors make appropriate concepts for regulation even more challenging. In contrast to the original concepts of data protection, it is in fact not possible to readily predict the handling of personal data and information, the knowledge generated from them, and the ensuing decisions. The idea that these processes could be almost completely foreseen, planned and steered by legal means[68] has turned out to be too simple. Processing of data and information, generating information and knowledge, coming to decisions on the basis of information and knowledge include dynamics and uncertainty at many points. This is all the more the case with a view to the use of technologies. Consequently, it is less the steering idea which characterizes or should characterize data protection law than, similar to environmental law, the idea of risk regulation.

As an innovative and highly dynamic field, data protection law needs to be, in terms of legal theory, “reflexive law” and, from a doctrinal point of view, a mixture of stability and dynamics. This is reflected, for instance, in the delegation of legislation competences, in the use of legal terms which are vague and need to be concretized, in normative references to dynamically adapted technical standards, in rules allowing for experimentation, in evaluation procedures or in other tools to ensure the capacity to learn and develop.

[66] More profoundly Mayer-Schönberger 2010, 1853, 1873 ff.

[67] Point 4.1 of this chapter.

[68] See Point 3. of this chapter.

Last but not least, data protection law cannot be understood against the background of the traditional ideas of hierarchical law implementation or enforcement. There are a number of general theoretical approaches aiming at superseding concepts of central steering by more flexible concepts of law. From a political-science point of view has been analyzed, how the substance of data protection law is made concrete by the interactions among different actors—the legislative, executive and judicial branches, data protection agencies, data users, data subjects.[69] An appropriate normative conception has to be responsive to the interplay of actors generating and concretizing law whilst, at the same time, keeping the normative perspective. All in all, data protection law proves to be a field of law in which new approaches are required.

11.5 Outlook: Data Protection Law as a Central New Field of Law

In sum, data protection law is a new, highly complex field of law in which a considerable amount of elaboration must still be carried out regarding its subject matter, the interests protected and appropriate concepts for regulation. Elaborating the law also depends on insights from other disciplines, for example the social sciences, the technological sciences, or information science. All this makes studying data protection law so exciting.

References

Abel, Ralf-Bernd. 2003. Geschichte des Datenschutzrechts. In *Handbuch Datenschutzrecht*, ed. Alexander Roßnagel. München: Beck (Chapter 2.7).

Albers, Marion. 2002. Information als neue Dimension im Recht. *Rechtstheorie* 33:61–89.

Albers, Marion. 2005. *Informationelle Selbstbestimmung*. Baden-Baden: Nomos.

Albers, Marion. 2012. Umgang mit personenbezogenen Informationen und Daten. In *Grundlagen des Verwaltungsrechts Vol.II, 2nd ed.*, eds. Wolfgang Hoffmann-Riem, Eberhard Schmidt-Aßmann, Andreas Voßkuhle. München: Beck, § 22.

Allen, Anita L. 2000. Privacy-as-data control: Conceptual, practical, and moral limits of the paradigm, *32 Connecticut Law Review* 861–875.

Ashby, William. 1963. *An introduction to cybernetics*, 5th ed. London: Chapman & Hall.

Bateson, Gregory. 1972. *Steps to an ecology of mind. Collected essays in Anthropology, Psychiatry, Evolution, and Epistemology*. Chicago: University of Chicago Press.

Benda, Ernst. 1974. Privatsphäre und "Persönlichkeitsprofil". Ein Beitrag zur Datenschutzdiskussion. In *Menschenwürde und freiheitliche Rechtsordnung*, eds. Leibholz, Faller, Mikat, Reis. Tübingen: Mohr. 23–44.

Bennett, Colin, J., Charles, D. Raab. 2003. *The Governance of Privacy*. Aldershot: Ashgate.

Bentham, Jeremy. 1995. *The Panopticon Writings* (Edition Miran Božovič). London.

Berlin, Isaiah. 1969. Two concepts of liberty. In *Four essays on liberty*, ed. Isaiah Berlin. Oxford: Oxford University Press.

[69] See Bennett and Raab 2003; Raab 1993, 89 ff.

Böckenförde, Ernst-Wolfgang. 1974. Grundrechtstheorie und Grundrechtsinterpretation. *Neue Juristische Wochenschrift* 1529–1538.

Britz, Gabriele. 2010. Informationelle Selbstbestimmung zwischen rechtswissenschaftlicher Grundsatzkritik und Beharren des Bundesverfassungsgerichts. In *Offene Rechtswissenschaft*, ed. Wolfgang. Hoffmann-Riem, 562–596. Tübingen: Mohr Siebeck.

Bygrave, Lee A. 2002. *Data protection law*, The Hague: Kluwer.

Čas, Johann. 2011. Ubiquitous computing, privacy and data protection: Options and limitations to reconcile the unprecedented contradictions. In *Computers, privacy and data protection: An element of choice*, ed. Serge Gutwirth, Yves Poullet, Paul de Hert and Ronald Leenes, 139–169. Dordrecht: Springer.

de Hert, Paul and Serge, Gutwirth. 2009. Data protection in the case law of Strasbourg and Luxemburg: Constitutionalisation in action. In *Reinventing data protection?* ed. Serge Gutwirth, Yves Poullet, Paul de Hert, Cécile de Terwagne, Sjaak Nouwt, 3–44. Dordrecht: Springer.

de Vries, Katja, Rocco Bellanova, Paul de Hert, and Serge Gutwirth. 2011. The German constitutional court judgment on data retention: Proportionality overrides unlimited surveillance (Doesn't It?). In *Computers, privacy and data protection: an element of choice*, ed. Serge Gutwirth, Yves Poullet, Paul de Hert, and Ronald Leenes, 3–23. Dordrecht: Springer.

Floridi, Luciano. 2010. *Information. A very short introduction*, New York: Oxford University Press.

Fried, Charles. 1968. Privacy. *Yale Law Journal* 77:475–493.

Heußner, Hermann. 1984. Das informationelle Selbstbestimmungsrecht in der Rechtsprechung des Bundesverfassungsgerichts, *Die Sozialgerichtsbarkeit* (SGb). 279–285.

Kafka, Franz. 2002. *Der Proceß*. Frankfurt a. M.: S. Fischer.

Köhntopp, Marit. 2001. Datenschutz technisch sichern. In *Allianz von Medienrecht und Informationstechnik?* ed. Alexander Roßnagel, 55–66. Baden-Baden: Nomos.

Leenes, Ronald E., Bert-Jaap Koops and Paul de Hert, eds. 2008. *Constitutional rights and new technologies. A comparative study*. The Hague: Asser Press.

Lübbe-Wolff, Gertrude. 1988. *Die Grundrechte als Eingriffsabwehrechte*. Baden-Baden: Nomos.

Mayer-Schönberger, Viktor. 1997. Generational development of data protection in Europe. In *Technology and privacy: The new landscape*, ed. Philip E. Agre, Marc Rotenberg. Cambridge: MIT Press, 219 ff.

Mayer-Schönberger, Viktor. 2010. Beyond privacy, beyond rights—toward a systems theory of information governance. *98 California Law Review* 1853–1885.

Nissenbaum, Helen. 2008. Privacy as contextual integrity. *79 Washington Law Review* 119–157.

Nissenbaum, Helen. 2010. *Privacy in context. Technology, policy, and the integrity of social life*. Stanford: Stanford University Press.

Nouwt, Sjaak. 2009. Towards a common European approach to data protection: A critical analysis of data protection perspectives of the Council of Europe and the European Union. In *Reinventing data protection?* eds. Serge Gutwirth, Yves Poullet, Paul de Hert, Cécile de Terwagne and Sjaak Nouwt, 275–292. Dordrecht: Springer.

Orwell, George. 2008. *Nineteen Eighty-Four*. London: Penguin.

Poscher, Ralf. 2012. Die Zukunft der informationellen Selbstbestimmung als Recht auf Abwehr von Grundrechtsgefährdungen. In *Resilienz in der offenen Gesellschaft*, eds. Hans-Helmuth Gander et al., 167–190. Baden-Baden: Nomos.

Raab, Charles. 1993. The governance of data protection. In *Modern governance: New government—society interactions*, ed. Jan Kooiman, 89–103. London: Sage.

Raab, Charles and Benjamin Goold. 2011. Protecting information privacy. *Equality and human rights commission research report series*. research report 69.

Rössler, Beate. 2001. *Der Wert des Privaten*. Frankfurt am Main: Suhrkamp.

Rouvroy, Antoinette, and Yves Poullet. 2009. The right to informational self-determination and the value of self-development: Reassessing the importance of privacy for democracy. In *Reinventing data protection?* ed. Serge Gutwirth, Yves Poullet, Paul de Hert, Cécile de Terwagne, Sjaak Nouwt, 45–76. Dordrecht: Springer.

Schlink, Bernhard. 1986. Das Recht der informationellen Selbstbestimmung. *Der Staat* 25:233–250.
Schwartz, Paul. 1989. The computer in German and American constitutional law: Towards an American right of informational self-determination. *American Journal of Comparative Law* 37:675–701.
Schweizer, Rainer. 2009. Die Rechtsprechung des Europäischen Gerichtshofes für Menschenrechte zum Persönlichkeits- und Datenschutz. *Datenschutz und Datensicherheit (DuD)* 462–468.
Siemen, Birte. 2006. *Datenschutz als europäisches Grundrecht*. Berlin: Duncker & Humblot.
Simitis, Spiros. 1971. Chancen und Gefahren der elektronischen Datenverarbeitung. *Neue Juristische Wochenschrift* 673–682.
Simitis, Spiros. 2011. Einleitung: Geschichte—Ziele—Prinzipien. In *Kommentar zum Bundesdatenschutzgesetz, 7th ed*, ed. Simitis, Baden-Baden: Nomos 2011.
Solove, Daniel. 2001. Privacy and power: Computer databases and metaphors for information privacy. *Stanford Law Review* 53:1393–1462.
Solove, Daniel. 2004. *The digital person*. New York: NYU Press.
Solove, Daniel. 2008. *Understanding Privacy*. Cambridge: Harvard University Press.
Trute, Hans-Heinrich. 2010. Wissen—Einleitende Bemerkungen. In *Wissen—Zur kognitiven Dimension des Rechts, Die Verwaltung, Beiheft 9*, ed. Hans C. Röhl, 11–38.
Waldo, James, Herbert S. Lin, and Lynette I. Millett, eds. 2007. *Engaging privay and information technology in a digital age*. Washington: The National Academies Press.
Warren, Samuel D., Louis D. Brandeis. 1890. The right to privacy. *4/5 Harvard Law Review* 193–220.
Westin, Alan F. 1970. *Privacy and Freedom*. 6th. ed. New York: Atheneum.

Chapter 12
Forgetting About Consent. Why The Focus Should Be On "Suitable Safeguards" in Data Protection Law

Gabriela Zanfir

12.1 Introduction

When one reads the proposal for a data protection regulation (DPR) released by the European Commission in 2012[1], one finds 56 references to the notion of "consent" (including the Preamble). By comparison, Directive 95/46[2] (DPD—Data Protection Directive) contains 12 such references. One explanation for the exponential growth of the regulation of consent is the energy put in the last decade into analyzing if and why consent is pivotal in data protection law in general[3], what does freely given, informed and unambiguous consent mean[4] or whether consent is revocable[5], just to give a few examples. Despite of all the attention consent enjoyed from academia and advisory bodies, the truth is that it represents just one of the six legal grounds to process personal data (one of five for sensitive data)[6]. Moreover, as Kightlinger showed, consent plays a limited role in the DPD's treatment of the requirements imposed on data controllers for data quality, fairness of processing or data security[7]. For instance, the controllers have to comply with obligations such as the one to inform the data subject pursuant to Article 10 and Article 11 of the DPD, regardless of the legal basis for the data processing.

---

[1] European Commission (2012b).

[2] Directive 95/46 of the European Parliament and of the Council of 24 October 1995 on the protection of individuals with regard to the processing of personal data and on the free movement of such data, OJ L 281, (23 November 1995), 31-50.

[3] See Article 29 Working Party (2011); Brownsword (2009); Bygrave and Schartum (2009); Feretti (2012); Le Métayer and Monteleone (2009).

[4] See Manson and O'Neill (2007); Whitely and Kanellopoulou (2010).

[5] See Curren and Kaye (2010).

[6] Article 29 Working Party (2011), *supra* in note 3, p. 34.

[7] Kightlinger (2007–2008).

G. Zanfir (✉)
Faculty of Law and Administrative Sciences, University of Craiova, Craiova, Romania
e-mail: gabriela.zanfir@gmail.com

S. Gutwirth et al. (eds.), *Reloading Data Protection*,
DOI 10.1007/978-94-007-7540-4_12, © Springer Science+Business Media Dordrecht 2014

Even in data protection's most legitimizing provision as a fundamental right, Article 8 of the Charter of Fundamental Rights of the European Union (the Charter), consent is enshrined as an alternative for the bases of fair processing. Article 8(2) of the Charter states that data must be processed "on the basis of the consent of the person concerned or some other legitimate basis laid down by law".

In addition, a significant part of the future data protection law in the European Union makes no reference whatsoever to consent: the proposal for a Directive regarding data protection in criminal matters[8] (the draft directive), also contained in the data protection reform package issued by the European Commission.

While this paper does not aim to minimize the role of consent in the legal philosophy of the informational self-determination, it proposes a more practical approach to what efficient protection of personal data means. In the end, informational self-determination can be considered as rooting in free will, which can be expressed by consent, withdrawal of consent, action or inaction with regard to the processing of personal data.

The first section of the article analyzes the *status quo* of consent in the Data Protection Directive (12.2), with references to the improvements brought by the DPR proposal, emphasizing the background value of consent as a legal basis for processing data in the European Union. After embracing the fact that there is more likely for data processing to happen under consent-free conditions than subject to consent, the second section looks at the aims of data protection and explores the ways to accomplish those aims (12.3). The final section will structure a possible set of "suitable safeguards" to keep the data processing fair, based on the current European data protection general legal framework, but also on the recent proposals for future data protection legislation: rights of the data subject, purpose requirements and accountability mechanisms (12.4). The conclusion (12.5) will show that the focus in giving effect to data protection law should be on stronger rights for the data subject, on clear purpose and time limitation related to it for data processing and on several rights of the data subject and correlative obligations of the controllers and processors, which are applicable regardless of the legal basis for the data processing.

12.2 The *Status Quo* of Consent in the Data Protection Directive

Pursuant to Article 7 DPD, personal data may be processed only if the data subject has unambiguously given his consent, or processing is necessary for the performance of a contract to which the data subject is party, or processing is necessary for compliance with a legal obligation to which the controller is subject, or processing is necessary for the performance of a task carried out in the public interest or in the exercise of official authority vested in the controller, or processing is necessary for the purposes of the legitimate interests pursued by the controller or by the third party or parties to whom the data are disclosed, except where such interests are overridden by the

[8] European Commission, COM(2012) (2012a)

interests for fundamental rights and freedoms of the data subject. This enumeration means, in fact, that "most instances of processing will be able to be justified under the criteria in paras b-f of the provision",[9] which do not include consent.

The number of "or"-s offered as an alternative for data processing based on consent must be disappointing, *prima facie*, for all the data protection enthusiasts who link informational self-determination primarily to the consent of the individual concerned. They usually stumble upon the first enumerated criteria for lawful processing, a fact that was translated in the doctrine by considering consent "a cornerstone"[10] or "pivotal"[11] for data protection law.

When read carefully, Article 7 DPD reveals itself as allowing the processing of personal data on almost any ground, a door opened gradually from exceptions provided by law, to the "legitimate interests pursued by the controller". The only criterion offered for assessing the legitimacy of the interests is a balance between them and the "interests for fundamental rights and freedoms" of the data subject, which is quite an evasive criterion. The alternative prerequisites are formulated broadly, thereby reducing significantly the extent to which data controllers are hostage to the consent requirement in practice.[12]

12.2.1 The Unsettled Position of Consent

The attributes envisaged for consent in the Data Protection Directive—"freely given", "specific", "informed and unambiguous" were subject to doctrinal debates[13] and to the intervention of the Article 29 Working Party.[14]

Even the authors who consider data processing consent a crucial component of data protection law which gives effect to the goal it purports, admit that the way in which it is currently devised in the law and its application provide an insufficient protection for individuals and an inadequate safeguard for the values it aims to protect *vis-à-vis* the realities of marketplace practices and economic interests.[15] Moreover, as Bygrave and Schartum explain, a large range of extra-legal factors undermines the privacy interests that consent mechanisms are supposed to promote or embody, as the degree of choice presupposed by these mechanisms will not often be present for

---

[9] Bygrave (2002).

[10] See Feretti (2012) (n 3) at 484; See Métayer and Monteleone (2009) (n 3) at 136.

[11] See Manson and O'Neill (2007) (n 5) at 112; They are referring to the UK Data Protection Act, which transposes the provisions of the Data Protection Directive, stating that the Act "assigns individual consent a large, indeed pivotal role in controlling the lawful acquisition, possession and use of personal information"; See also Brownsword (2009) (n 4) at 109.

[12] See Bygrave (2002) (n 9) at 66.

[13] See Le Métayer and Monteleone (2009) (n 3) at 139.

[14] See Article 29 Working Party (2011) (n 3).

[15] See Feretti (2012) (n 3) at 505.

certain services or products, particularly offered by data controllers in a monopoly or near-monopoly position.[16]

Taking into account consent is considered to "remain key to inform a properly functioning policy for the enhancement of individual autonomy"[17] and that its concrete mechanisms are, nevertheless, unclear, academics sought solutions to make consent rules work properly. They proposed the insertion of "collective consent"[18] in data protection law, or even "privacy agents"[19] who are to handle other people's consent, besides solutions like removing the psychological barriers to provide consent by providing comprehensive normative disclosure limits, making it explicit that data subjects may always be allowed to refuse consent or withdraw it at a later stage without negative consequences or strings attached.[20]

In the DPR proposal, the European Commission clarifies most of the concerns regarding the conditions for valid consent, while distributing it, in a form or another, throughout the whole act as a sign of strengthening the position of the data subject with regard to data processing, even if, *de facto*, its role is still an alternative to other forms of lawful processing.

12.2.2 The Reply of the DPR Proposal

The proposal for a Data Protection Regulation has been received extremely different by privacy specialists. While some see it as failing to provide either significant legal certainty or simplification, adding administrative burden and leaving a substantial risk of fragmentation,[21] others see it as a "cause for celebration for human rights",[22] considering that "once finalized the new instrument is expected to affect the way Europeans work and live together".[23] Surprisingly, though, both extreme approaches agree on one point: the provisions for consent have been significantly improved.

The skeptics underline that the draft regulation "helpfully removes the unnecessary and confusing distinction between *explicit* consent and other consent (see Articles 8 and 7 of the DPD, respectively)",[24] while the others also consider that the

[16] See Bygrave and Schartum (2009) (n 4) at 160. In line with their idea, Feretti (2012) (n 4) at 488, also makes a point from underlying that "the inclusion of data processing consent in the general terms and conditions of sale or services can be a common, yet subtle or elusive, method of obtaining consumer consent notwithstanding whether a transaction occurs online and irrespective of the opt-in/opt-out dichotomy".

[17] See Feretti (2012) (n 3) at 500.

[18] See Bygrave and Schartum (2009) (n 3) at p. 170.

[19] See Le Métayer and Monteleone (2009) (n 3) at pp. 140–142.

[20] See Feretti (2012) (n 3) at p. 501.

[21] Traung (2012).

[22] de Hert and Papakonstantinou (2012).

[23] de Hert and Papakonstantinou (2012), p. 131.

[24] See Traung (2012) (n 21) at p. 38.

Commission substantially reinforced the individual consent requirement, enhancing its definition by means of requiring explicit consent.[25]

Thus, the new definition of consent is considered clarifying, especially if read in conjunction with Recital 25 of the draft regulation.[26] According to Article 4(8) of the DPR proposal, "the data subject's consent means any freely given specific, informed and explicit indication of his or her wishes by which the data subject, either by a statement or by a clear affirmative action, signifies agreement to personal data relating to them being processed".

Consent is again enumerated as one of the six bases for lawful data processing in Article 6(1), point a), of the draft regulation, which proposes an interesting addition by declaring that consent is such a lawful basis if it is given "for one or more specific purposes".

One of the most important innovations of the draft regulation are the clear conditions for consent in Article 7, as it introduces procedural provisions regarding the proof of the data subject's consent—the burden of proof rests on the controller, the explicit option of the data subject to withdraw consent and rules intended to counterbalance the power positions held by some controllers, such as employers. Hence, consent shall not provide a legal basis for the processing, where there is a significant imbalance between the position of the data subject and the controller. All of these improvements regarding the conditions for consent are responses to the critiques of the provisions in the DPD.[27] However, there are already concerns regarding the entering into force of Article 7 as it is currently drafted, exactly because the requirements towards data processors appear to be quite demanding.[28]

But what is indeed remarkable regarding consent in the DPR proposal is its widespread echo throughout the whole draft. While the DPD only specifically refers to consent in Article 2—its definition, Article 7—lawful processing basis, Article 8—sensitive data and Article 26—derogation rules for data transfers to third countries without an adequate level of protection, the draft regulation introduces a panoply of functions for consent individually or for processing pursuant to consent, with regard

[25] See de Hert and Papakonstantinou (2012) (n 22) at p. 135.

[26] Recital 25 specifically states that silence or inactivity should not constitute consent and that consent is considered as being explicitly given either by a statement or by a clear affirmative action by the data subject, ensuring that individuals are aware that they give their consent to the processing of personal data, including by ticking a box when visiting an Internet website or by any other statement or conduct which clearly indicates in this context the data subject's acceptance of the proposed processing of their personal data.

[27] Even the European Commission criticized the effects in practice produced by the wording of the Data Protection Directive regarding consent, in a 2011 report: "(...) these conditions are currently interpreted differently in Member States, ranging from a general requirement of written consent to the acceptance of implicit consent. Moreover, in the online environment—given the opacity of privacy policies—it is often more difficult for individuals to be aware of their rights and give informed consent. This is even more complicated by the fact that, in some cases, it is not even clear what would constitute freely given, specific and informed consent to data processing, such as in the case of behavioural advertising, where internet browser settings are considered by some, but not by others, to deliver the user's consent". See European Commission. COM(2010) 609.

[28] See de Hert and Papakonstantinou (2012) (n 22) at p. 136.

to the processing of personal data of a child (Article 8), the right to be forgotten (Article 17), the right to data portability (Article 18), measures based on profile (Article 20) and processing for historical, statistical and scientific research papers (Article 83). However, perhaps the most intense effect given to consent in data protection law is the administrative sanction provided by Article 79(6)(a), according to which "the supervisory authority shall impose a fine up to 1,000,000 EUR or, in case of an enterprise up to 2 % of its annual worldwide turnover, to anyone who, intentionally or negligently (...) does not comply with the conditions for consent pursuant to Articles 6, 7 and 8".

As a preliminary conclusion, the draft regulation is generous with consent rules. However, consent still represents only one of the six justifications that allow personal data to be processed. In addition, where consent is mentioned in other provisions of the DPR proposal, it also has the nature of an "alternative". Now that the vast majority of concerns regarding consent were met by the draft regulation, it is time for data protection law to find a practical pivotal concept, or cornerstone, which must be directly linked to the object of the right to personal data protection.

12.2.3 Putting Data Processing Based on Consent in Context

Profiling has been defined as "the process of discovering correlations between data in databases that can be used to identify and represent a human or nonhuman subject (individual or group) and/or the application of profiles (sets of correlated data) to individuate and represent a subject or to identify a subject as a member of a group or category".[29] Thus, gathering of data is quintessential for profiling. This procedure is one of the main concerns of privacy advocates nowadays.[30] To meet this concern, the DPR proposal makes a specific reference to "profiling" in Article 20, building on Article 15 DPD, which regulates "automated individuals decisions".

Recital 58 of the Preamble in the DPR proposal explains the conditions under which this special kind of data processing is lawful:

> Every natural person should have the right not to be subject to a measure which is based on profiling by means of automated processing. However, such measure should be allowed when expressly authorized by law, carried out in the course of entering or performance of a contract, or when the data subject has given his consent. In any case, such processing should be subject to suitable safeguards, including specific information of the data subject and the right to obtain human intervention and that such measure should not concern a child.

The extrapolation of these rules to data processing in general explains in a few words the philosophy of data protection law: every natural person should have the right not to be subject to processing of personal data, unless such processing has a lawful basis—which can be a legal provision, consent or other specific condition stipulated by data protection law, and unless the processing is subject to "suitable safeguards".

[29] Hildebrandt (2008a, p. 19).

[30] See, for instance, Zarsky (2010, pp. 53–75); Hildebrandt (2008b).

This means that irrespective of which is the lawful basis for data processing, it must be clear that the individual has some degree of control, pursuant to his or her right to informational self-determination, upon the processing of personal data and that the processing must comply with specific, explicit safeguards so that the fundamental rights of the individual are observed.

For instance, Kightlinger, one of the most vehement critics of consent in European data protection law, argues that under the DPD, the informed consent is never sufficient to ensure that a website operator (he might as well refer to any other type of controller) may collect and use the person's personally identifiable information[31] lawfully and that, as far as the transfer of personal data to third countries is concerned, the consent of the individual plays no role.[32] This happens because the Directive imposes "a panoply of obligations" on operators that have little or nothing to do with a person's consent, including the duty to obtain a license from a DPA, to satisfy the data quality principles, to grant to individuals access to processed data, or to provide information to the individual prior to the processing.[33] He concludes that consent can safely take a "back seat", because it is the job of data protection authorities, not the individual, to protect privacy of personally identifiable information from threats posed by data controllers and possibly from the negative consequences of the individual's own consensual decisions.[34]

While it is true that data protection authorities (DPAs) play an important part in making sure that the data protection provisions are complied with, the supposition that only DPAs are in charge is erroneous. The most obvious counterarguments are the legal remedies and liability rules in Articles 22 and 23 DPD which allow actions for damages in national courts "as a result of an unlawful processing operation or of any act incompatible with the national provisions adopted pursuant to this Directive". Hence, the individual also has an important part in making sure that controllers are engaged in lawful processing operations.[35] In addition, the individual has a few "weapons" accorded to him by data protection law: the rights to intervene directly in the process of processing. So, why shouldn't the focus be on "sharpening" those rights?

Another critique of the central position of consent in conceptualizing data protection can be derived from the idea that, especially in the online world, the reliance on consent for the processing of personal data or the carrying out of an action that would

---

[31] In the American legal system, personal data is often regarded as personally identifiable information. However, the Consumers' Privacy Bill of Rights released in 2012 by the White House opts for the expression "personal data"; see in this regard, Zanfir (2012).

[32] See Kightlinger (2007–2008) (n 7) at p. 21.

[33] Kightlinger (2007–2008) at. p. 20.

[34] Kightlinger (2007–2008) at p. 29.

[35] For instance, in a famous case in Romanian courts, an individual received a 10,000 EUR compensation for moral damages, caused by the publication of details regarding his health condition on the website of the Municipality of Sector 1 of Bucharest as a justification for the individual receiving a public transportation free pass; he based his allegations on the provisions of Law No. 677/2001 which transposes into national law the Data Protection Directive; (See Jud. sect. 1 Bucureş ti, sentinţa civilă din 16.03.2009, irevocabilă).

otherwise constitute a violation to the privacy of the data subject does not always safeguard protection of his privacy.[36] For instance, it was revealed that, in practice, only a fraction of internet users read the privacy notices that precede the collection of their informed consent.[37] As Brownsword argued, such consents are "reduced to a bureaucratic process, where the collection of informed consent is carried out in a casual way, and where we succumb to the temptation to make use of consent as a lazy justification".[38] A probable antidote to the "lazy justification" reality would be, as Kosta construed, asking data controllers to justify their actions not only on the basis of the consent of their users, but also stroking a balance between the controllers' legitimate interests and "the right of the users".[39]

Last, taking into account also that even when data processing is based on consent problems appear in practice, in the sense that "not only consent may be implied or data processed on the basis of opt-out practices, but it may also be traded for perceived immediate economic advantages, or it may be taken contractually or as part of the general terms and conditions of a contract",[40] and that currently "information is automatically processed to an extent not dreamed of when the need for data protection law was first accepted",[41] the next section will look into the object of the right to the protection of personal data with the purpose of identifying safeguards suitable to comply with this right.

12.3 The Object of the Right to the Protection of Personal Data

The right to the protection of personal data has been recognized as such in Article 8 of the Charter after a 30 years history of regulating data protection in Europe.[42] It became clear that, at least in the European Union, this right protects something distinct than private life, as Article 7 of the same Charter expressly protects private life. Having two provisions that share an identical object is illogical. Therefore, what does the right to the protection of personal data protect?

A good way to answer the question is to first categorize the substances of the two rights envisaged. A valuable approach is to see them in terms of "opacity tools" vs. "transparency tools".[43] Opacity tools protect individuals, their liberty and autonomy against state interference and also against interference from other private actors, this being an accurate description of the legal effects of the right to private life

[36] Kosta (2011, p. 315).

[37] Van Alsenoy et al. (2012, p. 31).

[38] Brownsword (2004).

[39] Kosta (2011) (n 36) at p. 315.

[40] See Feretti (2012) (n 3) at p. 476.

[41] See Brownsword (2009) (n 3) at p. 99.

[42] For the beginning of data protection regulation in Europe, see Hondius (1975). For the generational evolution of data protection laws in Europe, see Mayer-Schönberger (1998).

[43] Gutwirth and de Hert (2008).

enshrined in Article 7 of the Charter.[44] Transparency tools limit state powers by devising legal means of control of these powers by the citizens, by controlling bodies or organizations and by the other state powers, which is what Article 8 of the Charter does by organizing the channeling, control and restraint of the processing of personal data.[45]

Following the same line of reasoning, Gomes de Andrade showed that the main difference between the right to privacy and the right to data protection is that the first one is substantive and the other one is procedural. "Substantive rights are created to ensure the protection and promotion of interests that the human individual and society consider important to defend and uphold. Procedural rights operate at a different level, setting the rules, methods and conditions through which substantive rights are effectively enforced and protected".[46] Therefore, even if the enactment of the first data protection rules can be considered a consequence of the affirmation of the right to private life, conceived at the beginning in a narrow understanding, data protection "gradually overflowed this context and assumed a role vis-à-vis all the freedoms enshrined in the European Convention on Human Rights".[47] Data protection, as such, "does not directly represent any value or interest *per se*, it prescribes the procedures and methods for pursuing the respect of values embodied in other rights—such as the right to privacy, identity, freedom of information, security, freedom of religion, etc."[48] These are the grounds for data protection to be considered "a catch-all term for a series of ideas with regard to the processing of personal data; by applying these ideas, governments try to reconcile fundamental but conflicting values such as privacy, free flow of information, the need for government surveillance, applying taxes, etc".[49]

It was acknowledged in the literature that the objective of the data protection regulation in general is to protect individual citizens against unjustified collection, storage, use and dissemination of their personal details.[50] Hence, data protection is pragmatic: it assumes that private and public actors need to be able to use personal information, as it is often necessary for societal reasons.[51]

To answer the question raised earlier, the right to the protection of personal data has as object, just as its name clearly suggests, the protection itself of the personal data being processed, and not private life in general or personal data in particular. As uncommon *a right that protects a protection* sounds, there could be no other way to better express the procedural nature of such a right. It indeed encompasses

[44] Gutwirth and de Hert (2008, pp. 276–278).

[45] Gutwirth and de Hert (2008, pp. 276–278).

[46] Gomes de Andrade (2012, p. 125).

[47] Poullet (2008, p. 41).

[48] Gomes de Andrade (2012) (n 76) at p. 125.

[49] de Hert and Gutwirth (2009, pp. 3–44).

[50] Hustinx (2005, p. 62).

[51] See de Hert and Gutwirth (n 49) at 3.

mechanisms of protection: principles for lawful and fair processing, "interventional" rights of the data subject, data quality rules and accountability rules.

As a preliminary conclusion, the right to data protection, in fact, assumes the inherent nature of processing personal information in the modern society. It is not its purpose *per se* to preclude such processing or to give an absolute right to the individual to object by means of his or her consent to the processing of personal data. Its object is to provide mechanisms of protection or "suitable safeguards" for individuals with regard to the processing of their data. Section 12.4 of this paper will have a look into which are the categories of "suitable safeguards" in data protection law, calling for a deeper analysis of their legal background and an enhanced attention to their future development.

12.4 A New "Cornerstone" for Data Protection Law: The Suitable Safeguards

In order for its protection to be effective, the content of a subjective right, which represents "a prerogative or a bundle of prerogatives"[52] accorded to the subject of the right, must be appropriate for safeguarding the object. The previous section contributed to the identification of the object of the right to the protection of personal data and this section identifies the bundle of prerogatives accorded to the data subject, which are veritable safeguards suited to the protection of personal data—"suitable safeguards".

The most concise and encompassing provision in EU positive law with regard to the protection of personal data is Article 8 of the Charter. Hence, it is sensible to start the search for "suitable safeguards" with this provision, even though most of them were developed since the first enactment of data protection laws in Europe.[53] The second paragraph of Article 8 provides that personal data "must be processed fairly for specified purposes and on the basis of the consent of the person concerned or some other legitimate basis laid down by law. Everyone has the right of access to data which has been collected concerning him or her, and the right to have it rectified".

The first point to be made is that, even though the bases for lawful processing are mentioned in Article 8(2), they should not be included in the category of "suitable safeguards" in the sense analyzed by this paper. They represent more than suitable safeguards, as they allow the processing itself, while the safeguards are the bundle of prerogatives accorded to the data subject so that the procedural object of protecting

[52] Dabin (2007, p. 168).

[53] See generally Nugter (1990). The volume analyzes some of the first data protection laws in Europe – *Bundesdatenschutzgesetz* (Germany, 1977), *Loi relatif a l'informatique, aux fichiers et aux libertes* (France, 1978), *Data Protection Act* (UK, 1984) and *Wet Persoonsregistraties* (The Netherlands, 1989), all of them containing provisions with regard to the specific rights of the data subjects and correlative obligations of the data processors. Information and access rights were omnipresent, while the first European data protection laws contained some variations of the right to object, the right to erasure and the right to correction.

personal data is protected itself. Therefore, pursuant to paragraph 2 of Article 8, the suitable safeguards must first be looked for in the rights of the data subject, on one hand, and in the principles for fair processing and purpose requirements, on the other hand.

The rights of the data subject are already systemized and structured in a well delimited set of prerogatives, and each of them is important for the realization of data protection.

With regard to the principles of fair processing and purpose requirements, it must be observed that Article 6 DPD—under the "Principles relating to data quality" section, is built around the concept of purpose limitation. The only paragraph of Article 6 DPD which does not expressly mention "purpose" is paragraph 1(a), which is a general provision, merely requiring the data processing to be lawful and fair. Thus, purpose requirements are functional and central for fair processing, and they can be converted in a palpable prerogative.

Article 8(3) of the Charter states that "compliance with these rules shall be subject to control by an independent authority". Thus, it refers to a form of accountability. However, accountability in data protection is more complex than the mere control of the data protection authorities. Such a fundamental provision indicates, nevertheless, that accountability plays an important part in the protection of personal data, beyond the general accountability of the "debtors" of correlative obligations stemming from the rights in the Charter. As such, the Charter itself provides a further incarnation of accountability in general in Article 47, which states that everyone whose rights and freedoms guaranteed by the law of the Union are violated "has the right to an effective remedy before a tribunal". The importance of accountability in data protection is highlighted by its extensive regulation in the DPD, under the chapter of "judicial remedies, liability and sanctions", which is further developed and structured in the DPR proposal.

Taking all these considerations into account, the "suitable safeguards" encompassed by the right to the protection of personal data can be structured as such: the rights of the data subject (12.4.1), the purpose requirements (12.4.2) and the mechanisms of accountability (12.4.3). Each of them will be briefly discussed.

12.4.1 Rights of the Data Subject

A core principle of data protection laws in general is that persons should be able to participate in, and have a measure of influence over, the processing of data on them by other individuals or organizations.[54] One of the outcomes of this principle are the consent rules, which were found in the previous sections as being limited with regard to the self-determination of the data subject. However, "the Directive insists on the participation of data subjects even where their consent is not needed",[55] and

[54] See Bygrave (2002) (n 9) at 63.

[55] Simitis (1997, p. 130).

it does so by enforcing a set of specific rights: the right to be informed (Articles 10 and 11), the right to access the processed data and to receive a copy of them (Article 12(a)), the right to object to data processing (Article 14), the right not to be subject to fully automated decisions based on data processing (Article 15), the right to have the data rectified, erased or blocked (Article 12(b)),[56] to which the right to a judicial remedy (Article 22) can be added, although it is more strongly connected with the accountability of the controller.[57]

It has been noted that the purpose of these rights is "to permit the persons concerned to follow and correct processing".[58] Thus, the rights of the data subject are prerogatives which allow the individual to control the way in which his or her personal data are processed, regardless of the legal basis of the processing. Nevertheless, except for the right to a judicial remedy, all of these prerogatives are subject to certain limitations.[59]

Previous literature shows that "the Commission in its draft Regulation has taken bold steps for the improvement of the data subjects' position in contemporary personal data processing conditions"[60] and that the main achievement to this end is that their rights "have been strengthened and data controllers' obligations have been increased respectively".[61] Despite of the enhancement of the provisions regarding the rights of the data subject, these particular safeguards need to be further clarified with regard to their scope and their restrictions.

The DPR proposal contains a chapter dedicated to the "Rights of the data subject" (Chapter 3), which further details and enhances the already existing rights and adds the right to be forgotten and the right to data portability in the panoply of data protection rights. However, none of the two are completely new to data protection law, as both have roots in the DPD, within the right to erasure and the right to receive a copy of the processed data respectively. According to the first draft report on the DPR proposal of the Committee on Civil Liberties, Justice and Home Affairs[62] of the European Parliament, Article 18 is proposed for deletion and its content is moved under Article 15—"the right to access".

The DPR proposal introduces in Article 12 rules regarding the procedures and mechanisms for exercising the rights of the data subject, including means for electronic requests, requiring response to the data subject's request within a defined

---

[56] For a comprehensive analysis of these rights enshrined in the DPD and also in Directive 2002/58 on privacy and electronic communications, see Korff (2005, pp. 71–144).

[57] For instance, the Romanian law transposing Directive 95/46, Law no. 677/2001 for the protection of persons with regard to the processing of personal data and the free movement of such data, enshrines in art. 18 "The right to a judicial remedy", under Chapter IV – "The rights of the data subject in the context of personal data processing".

[58] See Simitis (1997) (n 55) at 131.

[59] See Articles 13(1), 14(a) and 15(2) DPD.

[60] See de Hert and Papakonstantinou (2012) (n 22) at 141–142.

[61] See de Hert and Papakonstantinou (2012).

[62] Committee on Civil Liberties, Justice and Home Affairs (2012).

deadline, and the motivation of refusals.[63] While such specific rules are welcomed, paragraph 3 of this article hampers the efficiency of the rights of the data subject, as it specifically allows the controller to refuse to take action on the request of the data subject, as long as the data subject is informed of the reasons for refusal and on the possibilities for a judicial or administrative remedy.

The rights of the data subject are systemized in the draft Regulation in three categories: (1) information and access (the right of the data subject to be informed—Article 14, and the right of access to data—Article 15), (2) rectification and erasure (the right to rectification—Article 16, the right to be forgotten and to erasure[64]—Article 17, the right to data portability[65]—Article 18) and (3) the right to object and profiling (the general right to object—Article 19, and the right not to be subject to profiling—Article 20).

While the strengthening of the rights of the data subject has been one of the main data protection reform themes, on a closer look their proposed provisions lead to uncertainty and often limit the scope of the rights. For instance, it is true that the right of the data subject to be informed contains, due to the draft regulation, a more consistent set of compulsory details to be provided by the controller to the data subject. However, Article 14(5) provides that this right shall not apply where the data are not collected from the data subject and the provision of such information proves impossible or would involve a disproportionate effort. Similarly, Article 14(5)(c) limits the scope of the right to be informed by excluding from its application the situation of indirect collection of data where it is expressly laid down by law, without requiring further safeguards.

These specifications considerably soften the "teeth" of the provision, as nowadays the cases in which data are collected from other sources than the data subject are numerous. By contrast, the DPD contained a specific provision which covered the information of the data subject when the data were indirectly collected. While the first limitation is also present in Article 11(2) DPD, it is made clear there that it should apply in particular for processing of data for statistical or research purposes. Perhaps the biggest difference between the current provision and the proposed one is the moment of making the information available to the data subject. While Article 11(1) DPD states that the information must be made available "at the time of undertaking the recording of personal data", Article 14(4)(b) of the DPR proposal provides that the information can also be made "within a reasonable period after the collection". This provision obviously hampers the lawful processing of personal data based on consent, when the data is not collected directly from the data subject.[66]

---

[63] See para. 3.4.3.1. from the Explanatory Memorandum of the DPR Proposal.

[64] For a critique of the provision of a right to be forgotten in the data protection reform package see Rosen (2012); See also Ausloos (2012, pp. 143–152); Koops (2012).

[65] For an introductory study about the right to data portability as it is enshrined in the DPR proposal, see Zanfir (2012, pp. 149–163); for a critique of the right to data portability see Swire and Lagos (2013).

[66] For instance, such a situation can easily be imagined in the context of database transactions between data brokers. See Singer (2012).

Another problem of the rights provisions in the DPR proposal is the use of subjective, unclear, criteria for assessing their proper application, such as "the essence of the right to the protection of personal data",[67] "structured and commonly used format"[68] or the "reasonable period" previously mentioned.

The European Data Protection Supervisor (EDPS) formally criticized in its Opinion on the data protection reform package the approach taken by the Commission with regard to the restrictions of the rights of the data subject. The EDPS considers that the scope of possible restrictions has been considerably expanded in comparison to what is currently provided in Article 13 DPD, as all the rights of the data subject can now be restricted due to Article 21 DPR proposal, including the right to object and the measures based on profiling.[69] For instance, the EDPS called for restricting the use of the public interest exemption to clearly identified and limited circumstances including criminal offences or economic financial interests.[70]

The effectiveness of the rights of the data subject is without a doubt a suitable safeguard for fair data processing, the more so as the proposal of a Directive for data protection in criminal matters also provides for a similar set of rights, adapted to the sensitive area of its general scope. The draft Directive recognizes the rights to information, access, rectification, erasure and restriction of processing[71] and it also makes a reference to profiling measures.[72]

12.4.2 Purpose Requirements

Purpose requirements are of paramount importance for processing personal data, as the purpose for processing data is equivalent to a guiding force of the whole "process of processing". Four of the five principles related to data quality enshrined in Article 6 DPD revolve around the purpose of the processing.[73] In addition, the legal definition of "controller" has as point of reference the purpose of the processing.[74]

One of the unanimously recognized data protection principles is the principle of purpose specification. It is considered to be a cluster of three principles: the purposes

[67] Article 17(3)(d) of the DPR proposal.

[68] Article 18(1) of the DPR proposal.

[69] Opinion of the European Data Protection Supervisor (2012), para. 160.

[70] Opinion of the European Data Protection Supervisor (2012), para. 159.

[71] Articles 11 to 16 of the proposal for a Directive on the protection of individuals with regard to the processing of personal data by competent authorities for the purposes of prevention, investigation, detection or prosecution of criminal offences or the execution of criminal penalties, and the free movement of such data.

[72] Article 9 of the draft Directive.

[73] Article 6(1)(b), (c), (d), (e) of the Data Protection Directive.

[74] According to Article 2(d), (d) "'controller' shall mean the natural or legal person, public authority, agency or any other body which alone or jointly with others determines the purposes and means of the processing of personal data".

for which data are collected shall be specified/defined; these purposes shall be lawful/legitimate; and the purposes for which the data are further processed shall not be incompatible with the purposes for which the data are first collected.[75] Moreover, the obligation to connect the processing to a particular purpose predeterminates the selection of the data and confines their use.[76]

The data protection reform package confirms the pivotal role of purpose specification and purpose limitation in data protection law. Both the draft regulation and the draft directive provide that personal data must be "collected for specified, explicit and legitimate purposes and not further processed in a way incompatible with those purposes".[77]

Other common rules are that the processed data must be adequate, relevant, and not excessive in relation to the purposes for which they are processed and that all the reasonable steps must be taken to ensure that the personal data are inaccurate, having regard to the purpose of the processing.[78] The draft regulation adds a very important condition, a proportionality rule, which circumscribes the material scope of lawful data processing by establishing that personal data shall only be processed if, and as long as, the purposes could not be fulfilled by processing information that does not involve personal data.[79] Moreover, the rule of lawful processing pursuant consent in Article 6(1)(a) is directly linked to the "specific purposes" of the data processing.

Another essential requirement related to the processing purpose is that data must be kept in a form which permits identification of data subjects for no longer than is necessary for the purposes for which the personal data are processed.[80]

All of these requirements are currently enshrined in Article 6 DPD, except the express proportionality rule. According to the EDPS, the effectiveness of the purpose limitation principle depends on (1) the interpretation of the notion of 'compatible use' and (2) the possible derogations to the purpose limitation principle, in other words, the possibilities and conditions for incompatible use.[81] Hence, the EDPS calls for additional precision in the proposed Regulation.[82]

Another key issue is the interpretation of "specified", "explicit" and "legitimate" purpose, taking into account that the three conditions are cumulative. Interpreting these conditions *stricto sensu* is vital for the efficiency of the purpose requirements. For instance, a general purpose such as "public interest" must not be considered as fulfilling the "explicit" requirement. From this point of view, the position taken by the Commission in recital 44 of the draft proposal is subject to critique, as it specifically allows political parties to "compile data on people's political opinions"

[75] See Bygrave (2002) (n 9) at 61.

[76] See Simitis (1997) (n 55) at 129.

[77] Article 5(b) of the draft regulation and Article 4(b) of the draft directive.

[78] Article 5(c), (d) of the draft regulation and Article 4(c), (d) of the draft directive.

[79] Article 5(c) of the draft regulation, second thesis.

[80] Article 5(e) of the draft regulation and Article 4(e) of the draft directive.

[81] EDPS Opinion (n 58), para. 116.

[82] EDPS Opinion (n 58), para. 117.

for "reasons of public interest", if the "operation of the democratic system requires so" in a Member State. It is difficult to find a valid argument which legitimizes a database of political partisans necessary for the operation of the democratic system.

A criterion for the "explicit" requirement could be that the purpose of the processing should allow the quantitative assessment in time of the data processing.[83] As for the meaning of "legitimate", previous literature underlined that this notion "denotes a criterion of social acceptability, such that personal data should only be processed for purposes that do not run counter to predominant social mores".[84]

12.4.3 Mechanisms of Accountability

The draft regulation introduces expressly a principle of accountability in Article 5(f), stating that personal data must be "processed under the responsibility and liability of the controller, who shall ensure and demonstrate for each processing operation the compliance with the provisions of this Regulation". A similar provision is enshrined in the draft directive, in Article 4(f). However, even though such a principle was not expressly recognized in the DPD, certain provisions, such as the ones related to the judicial remedies and the control competences of the data protection authorities, indicated a certain degree of accountability of the controllers. Moreover, Article 6(2) DPD states that "it shall be for the controller to ensure that paragraph 1 is complied with", which can be seen as an approximate definition of accountability, as paragraph 1 contained all the principles relating to data quality.

It has been shown that, in broad terms, a principle of accountability would place upon data controllers the burden of implementing within their organizations specific measures in order to ensure that data protection requirements are met.[85] At the same time, from the data subject's point of view, a principle of accountability would enable her to efficiently protect her right to data protection in front of or even against the competent authorities.[86] Hence, accountability translates into two types of mechanisms.

On the one hand, such mechanisms include anything from the introduction of a Data Protection Officer to implementing Data Protection Impact Assessments or employing a Privacy by Design system architecture.[87] On the other hand, they include

[83] For instance, personal data related to the students of a University are processed with the purpose of keeping track of their academic results; hence, the period of time needed for this processing equals to the period of the students' enrollment. If all or some of their personal data need to be processed for statistical purposes after this period, the legal safeguards for this situation must be observed.

[84] See Bygrave (2002) (n 9) at 61.

[85] See de Hert and Papakonstantinou (2012) (n 22) at 134.

[86] de Hert and Papakonstantinou (2012).

[87] de Hert and Papakonstantinou (2012).

rules regarding remedies, liability and sanctions. Articles 22, 23, and 24 of the DPD have been consistently developed both in the draft regulation and the draft directive.[88]

The DPR proposal excels in expressly providing procedural rights for the data subject: the right to lodge a complaint with a supervisory authority (Article 73)—which is also extended to any body, organization or association which aims to protect data subjects' rights, the right to a judicial remedy against a supervisory authority—which is provided also for legal persons[89] (Article 74), the right to a judicial remedy against a controller or processor (Article 75), the right to compensation and liability (Article 77), and even common rules for court proceedings[90] (Article 76).

Also, the administrative sanctions provided for in the draft regulation are severe. Article 79 provides that each supervisory authority shall be empowered to impose administrative sanctions, which can amount up to one 1 million EUR or, in case of an enterprise, up to 2 % of its annual worldwide turnover. As a general rule, pursuant to Article 79(2), the administrative sanction shall be in each individual case "effective, proportionate and dissuasive", a formula which will need further clarification.

This particular safeguard needs attention taking into account at least the fact that the administrative sanctions as provided for in the draft regulation have four thresholds—from a warning in written to the 1 million EUR fine, each threshold having its conditions, which amount in the case of the most serious one to 15 different hypotheses (Article 79(6) from (a) to (o)). Thus, accountability of the controller is taken very seriously in the future of data protection law in Europe.

12.5 Conclusion

This article explored a conclusion drawn from a "sketch" of the legal philosophy of data protection: every natural person should have the right not to be subject to processing of personal data, unless such processing has a lawful basis—which can be a legal provision, consent or other specific condition stipulated by data protection law, and unless the processing is subject to "suitable safeguards". Thus, it put consent rules into context and highlighted in section 12.2 that while they are an important part of data protection law, focusing on them is not productive for the achievement of the goal of the right to the protection of personal data, which has a highly procedural object. Instead, the focus should be on a set of rules that apply to all the types of data processing flowing from the six lawful bases recognized by data protection law and also to both spheres recognized in EU for the general data protection rules (the general framework and the *criminal matters* sphere).

[88] See Articles 50 to 55 from the draft directive.

[89] This provision must refer to legal persons in their controller or representative of a controller capacity, as the DPR proposal makes it very clear that its provisions only apply to natural persons.

[90] Since the Treaty of Amsterdam, an explicit base for harmonization of civil procedural law is to be found in Article 65 of the EC Treaty (currently Article 81 TFEU); See Eliantonio (2009).

Section 12.3 clarified what the object of the right to the protection of personal data is, in order to identify the suitable safeguards which match the achievement of its goal. It found that this right assumes the inherent nature of processing personal information in the modern society and that it is not its purpose *per se* to preclude such processing or to give an absolute right to the individual to object by means of his or her consent to the processing of personal data. In fact, its object is to provide mechanisms of protection or "suitable safeguards" for individuals with regard to the processing of their personal data: it "protects the protection of personal data". This paper aimed at correlating the object of the right to the protection of personal data to its content.

The last section identified three types of "suitable safeguards", the bundle of prerogatives that constitute what it was identified as the content of the right to the protection of personal data, that need equal attention from lawmakers and privacy professionals and that need to be further developed and clarified: the rights of the data subject, the purpose requirements and the accountability mechanisms. Each of them enjoys broad improvements in the EU's data protection reform package. However, section 12.4 showed that they are far from being clear and that they need further systematization and development.

After making a thorough analysis of consent in data protection law in her thesis, one of the conclusions Kosta reached was that "the role of consent in this era is reduced, as the control of the individual over his personal information is overcome by the facilitation of everyday activities in electronic communications and especially the internet, to the extent that the privacy of the individual is not infringed".[91] If we accept that the role of consent in data protection is reduced, then the right to the protection of personal data needs, both in theory and in practice, to rely on other specific and well defined prerogatives of the data subject so that its purpose is achieved. The proposal of considering a systematization of these prerogatives under the concept of "suitable safeguards" is one possible solution of this problem, a solution which could also contribute to a functional redress system[92] for data protection in the European Union.

Acknowledgments This work was supported by the strategic grant POSDRU/CPP107/DMI1.5/S/78421, Project ID 78421 (2010), co-financed by the European Social Fund—Investing in People, within the Sectoral Operational Programme Human Resources Development 2007–2013. The author would like to thank the Tilburg Institute for Law, Technology and Society for providing valuable support for her research during her research visit there.

[91] Kosta (2011) (n 36) at 318.

[92] The preliminary results of the EU Fundamental Rights Agency project on "Data protection: Redress Mechanisms and Their Use", presented at the Computers, Privacy and Data Protection Conference in Bruselles, January 23–25, 2013, show that „data protection cases are few and dispersed between a variety of different courts" in the Member States and that „in most jurisdictions data protection does not form an important area for the specialization and development of judicial expertise".

References

Volumes

Bygrave, Lee A. 2002. *Data protection law. Approaching its rationale, logic and limits*. The Hague: Kluwer Law International.
Dabin, Jean. 2007. *Le Droit Subjectif*. Paris: Dalloz.
Hondius, Frits W. 1975. *Emerging data protection in Europe*. Amsterdam/New York: North-Holland Publishing Co./American Elsevier Publishing Co.
Korff, Douwe. 2005. *Data protection laws in the European Union*. Federation of European Direct Marketing and Direct Marketing Association.
Manson, Neil C., and Onora O'Neill. 2007. *Rethinking informed consent in bioethics*. Cambridge University Press.
Nugter, Adriana C. M. 1990. *Transborder flow of personal data within the EC*. Dordrecht: Springer.

Chapters of Volumes

Brownsword, Roger. 2009. Consent in data protection law: Privacy, fair processing and confidentiality. In *Reinventing Data Protection?* ed. Serge Gutwirth, Yves Poullet, Paul de Hert, Cecile de Terwangne, and Sjaak Nouwt, 83–110. Heidelberg: Springer.
Bygrave, Lee A., and Dag W. Schartum. 2009. Consent, proportionality and collective power. In *Reinventing data protection?* ed. Serge Gutwirth, Yves Poullet, Paul de Hert, Cecile de Terwangne, and Sjaak Nouwt, 157–173. Heidelberg: Springer.
de Hert, Paul, and Serge Gutwirth. 2009. Data protection in the case law of Strasbourg and Luxemburg: Constitutionalism in action, in *Reinventing Data Protection?* ed. Serge Gutwirth, Yves Poullet, Paul de Hert, Cecile de Terwangne, and Sjaak Nouwt, 3–44. Heidelberg: Springer.
Gutwirth, Serge, and Paul de Hert. 2008. Regulating profiling in a democratic constitutional state. In *Profiling the European citizen*, ed. Mirelle Hildebrandt, and Serge Gutwirth, 271–303. Dordrecht: Springer.
Hildebrandt, Mirelle. 2008a. Defining profiling: A new type of knowledge? In *Profiling the European citizen*, ed. Mirelle Hildebrandt, and Serge Gutwirth, 17–45. Dordrecht: Springer.
Mayer-Schönberger, Viktor. 1998. Generational development of data protection in Europe. In *Technology and privacy: The new landscape*, ed. Philip E. Agre, and Marc Rotenberg, 219–242. Cambridge, MA: The MIT Press.
Poullet, Yves. 2008. *Pour une troisième génération de réglementation de protection des données, dans Défis du droit à la protection à la vie privée. In coll. Cahiers du Centre de Recherches Informatique et Droit, 31*. Bruxelles: Bruylant.
Simitis, Spiros. 1997. *Data Protection in the European Union—The quest for common rules. In Collected courses of the Academy of European Law*. Vol. VIII-1, 95–141. European University Institute: Kluwer Law International.
Zarsky, Tal. 2010. Responding to the inevitable outcomes of profiling: Recent lessons from consumer financial markets, and beyond. In *Data protection in a profiled world*, Yves Poullet, Serge Gutwirth, and Paul de Hert, 53–75. Dordrecht: Springer.

Articles

Ausloos, Jef. 2012. The right to be forgotten—Worth remembering? *Computers Law and Security Review* 28:143–152.
Brownsword, Roger. 2004. The cult of consent: fixation and fallacy. *King's Law Journal* 15:223–252.

Curren, Liam, and Jane Kaye. 2010. Revoking consent: a blind spot in data protection law? *Computer Law and Security Review* 26:273–283.
de Hert, Paul, and Vagelis Papakonstantinou. 2012. The proposed data protection regulation replacing directive 95/46: A sound system for the protection of individuals. *Computer Law & Security Review* 28:130–142.
Eliantonio, Mariolina. 2009. The future of National Procedural Law in Europe: Harmonisation vs. Judge made standards in the field of administrative justice. *Electronic Journal of Comparative Law* 13.3:1–11.
Feretti, Federico. 2012. A European perspective on data processing consent through the re-conceptualization of European data protection's looking glass after the Lisbon treaty: Taking rights seriously. *European Review of Private Law* 2:473–506.
Gomes de Andrade, Nuno Norberto. 2012. Oblivion, the right to be different from oneself. Reproposing the right to be forgotten. *Revista de Internet, Derecho y Politica* 13:122–137.
Hildebrandt, Mirelle. 2008b. Profiling and the rule of law. 1. *Identity in the Information Society* 1:55–70.
Kightlinger, Mark F. 2007–2008. Twilight of the idols? EU internet privacy and the postenlightenment paradigm. *Columbia Journal of European Law* 14:1–62.
Koops, Bert Jap. 2012. Forgetting footprints, shunning shadows. A Critical Analysis of the Right to be Forgotten in Big Data Practice. *Tilburg Law School Legal Studies Research Paper Series* 8.
Le Métayer, Daniel, and Sarah Monteleone. 2009. Automated consent through privacy agents: Legal requirements and technical architecture. *Computer Law & Security Review* 25(2):136–144.
Rosen, Jeffrey. 2012. The right to be forgotten. 64 *Stanford Law Review Online* 88.
Swire, Peter, and Yanni Lagos. 2013. Why the right to data portability likely reduces consumer welfare: Antitrust and privacy critique. *Maryland Law Review* 72(2):335. http://ssrn.com/abstract=2159157. Accessed 26 Feb 2013.
Traung, Peter. 2012. The proposed new EU general data protection regulation. *CRi* 2:33–49.
Zanfir, Gabriela. 2012. The right to data portability in the context of the EU data protection reform. *International Data Privacy Law* 2(3):149–163.

Theses

Kosta, Eleni. Unraveling consent in European Data Protection legislation. A prospective study on consent in electronic communications. Doctoral Thesis, submitted on June 1, 2011, Faculty of Law, K. U. Leuven, Interdisciplinary Center for Law and ICT.

Official Reports/Opinions

Article 29 Working Party. 2011. Opinion 15/2011 on the definition of consent, WP 187.
Committee on Civil Liberties, Justice and Home Affairs. 2012. Draft report on the proposal for a regulation of the European Parliament and of the Council on the protection of individual with regard to the processing of personal data and on the free movement of such data (General Data Protection Regulation), (COM 2012. 0011– C7-0025/2012–2012/0011(COD)). December 17, 2012.
European Commission. 2010. COM(2010) 609 final, A comprehensive approach of data protection in Europe (4 November 2010), p. 8–9.
European Commission. 2012a. COM(2012) 10 final, Proposal for a Directive of the European Parliament and of the Council on the protection of individuals with regard to the processing of personal data by competent authorities for the purposes of prevention, investigation, detection or prosecution of criminal offences or the execution of criminal penalties, and the free movement of such data, 25.1.2012.

European Commission. 2012b. COM(2012) 11 final, Proposal for a regulation of the European Parliament and of the Council on the protection of individuals with regard to the processing of personal data and on the free movement of such data (General Data Protection Regulation), Brussels, 25 January 2012.

European Data Protection Supervisor. Opinion of the European Data Protection Supervisor on the data protection reform package, issued on March 7, 2012.

Other Sources

Hustinx, Peter. 2005. *Data protection in the European Union. Privacy & Informatie* 2:62.

Singer, Natasha. 2012. You for sale: Mapping, and sharing, the consumer genome. *New York Times,* 16th June. http://www.nytimes.com/2012/06/17/technology/acxiom-the-quiet-giant-of-consumer-database-marketing.html?_r=1&pagewanted;=all. Accessed 28 Feb 2013.

Van Alsenoy, Brendan, Eleni Kosta, and Jos Dumortier. 2012. D6.1—Legal requirements for privacy-friendly model privacy policies. *The IWT SBO SPION Project.*

Whitely, Edgar A., and Nadja Kanellopoulou. 2010. Privacy and informed consent in online interactions: Evidence from expert focus groups. International Conference on Information Systems, St. Louis, Missouri.

Zanfir, Gabriela. 2012. EU and US data protection reforms. A comparative view, in *7th edition of The International Conference "The European Integration, Realities and Perspectives" Proceedings.* http://papers.ssrn.com/sol3/papers.cfm?abstract_id=2079484. Accessed 26 Feb 2012.

Chapter 13
"All my mates have got it, so it must be okay": Constructing a Richer Understanding of Privacy Concerns—An Exploratory Focus Group Study

Anthony Morton

13.1 Introduction

In a 2010 UK survey[1], 76.4 % of respondents were either *very concerned* or *somewhat concerned* about their privacy while using the Internet. In the same survey, 44 % had experienced an invasion of their privacy[2] *very frequently* or *somewhat frequently*. A 2011 survey by the European Commission on attitudes towards data protection and electronic identity found that although three out of four respondents accepted the need to provide personal information as part of everyday life, 70 % were concerned about how organisations use their personal information, believing they had "*only partial, if any, control of their own data*".[3] Although privacy surveys have attracted some criticism[4], they highlight a genuine concern amongst people of the impact of technology-underpinned services on their privacy. This concern is not misplaced. In the last five years, services as diverse as street-level mapping[5], smartphones and smartphone applications[6], video-gaming[7], social networking[8], targeted advertising[9] and peer-to-peer file-sharing[10] have attracted adverse publicity

---

[1] Coles-Kemp et al. 2010.

[2] For this survey question, an invasion of privacy included offline intrusions, e.g. unsolicited telephone calls, in addition to online intrusions, e.g. e-mails.

[3] European Commission (EUROPA) 2011.

[4] Harper and Singleton 2001.

[5] Barnett 2008; BBC 2008; Mills 2007.

[6] Sarno 2010; Panzarino 2011; Leavitt 2011.

[7] Quinn and Arthur 2011.

[8] BBC 2010, 2011.

[9] Fiveash 2007; Ashford 2011.

[10] Mennecke 2007; NBC 2009; Federal Trade Commission 2010.

A. Morton (✉)
Department of Computer Science, University College London,
Gower Street, London WC1E 6BT, UK
e-mail: anthony.morton.09@ucl.ac.uk

S. Gutwirth et al. (eds.), *Reloading Data Protection*,
DOI 10.1007/978-94-007-7540-4_13,

or criticism, for collecting or leaking personal information. Such services, although offering benefits such as easier navigation and travel, entertainment, social contact, relevant advertising and access to media content, may explicitly request personal information, collect it covertly or accidentally[11], or distribute it without the user's knowledge.[12]

Investigation of an individual's privacy concerns has traditionally focused on their general level of privacy concern, or their perception of organisations' collection, use, management, control and securing of personal information, by asking them to respond to a selection of statements about government and/or organisations' information handling practices. However, responding to statements about information handling practices in the abstract is problematic. When asked to consider one of Westin's statements—"*Most businesses handle the personal information they collect about consumers in a proper and confidential way*"[13]—a survey participant may reasonably think, "*It depends on the organisation. I trust organisation X, but not organisation Y, as I don't believe it will look after my personal information carefully*". Furthermore, when providing personal information to an organisation, the nature of the technology platform involved is omitted from most general privacy concern surveys. For example, a customer may be comfortable conducting financial transactions using a bank's website, but not using a smartphone application—even when interacting with the same bank. Finally, such surveys do not take into account environmental influences, such as the experiences of an individual's friends, or media stories concerning similar technology-underpinned services. Peoples' attitudes to disclosing their personal information is complex, as they may state they value their information privacy, but are usually prepared to trade personal information for benefits.[14]

A more holistic approach to the construction of privacy concern is required, which encompasses the technical, organisational and environmental factors individuals take into account when choosing to use a technology-underpinned service requiring the provision of personal information. Existing privacy concern indexes, such as Westin's, only provide measurement of an individual's general level of privacy concern. An individual's privacy concern is likely to be constructed from their concerns about the technology-mediated interaction they are having with a specific organisation, others' views of the organisation and/or technology, and their personality, life experiences and innate desire, or otherwise, to protect their privacy.

To emphasise the importance of considering a broad range of factors in determining privacy concern, this paper henceforth refers to the socio-technical construct of

---

[11] Farrell 2010.

[12] Johnson 2008; El Emam 2010.

[13] This statement was one of those used by Westin to derive his Privacy Segmentation Index, and Core Privacy Orientation Index used in his studies between 1995 and 2003—quoted in Kumaraguru and Cranor (2005).

[14] Beldad et al. 2011.

a *technology service*—proposed by Morton & Sasse[15]—in place of the more cumbersome phrase *technology-underpinned service.* A *technology service* consists of a technology platform[16] and the organisation providing it. The use of the *technology service* construct emphasises the need, when attempting to understand peoples' privacy concerns, of not only considering the hardware and software in the technology platform, but the motivation, principles, culture and privacy practice of the organisation providing the *technology service.*

When deciding to use a technology service[17], an individual's desire to achieve their goal(s) usually results in them having to balance relinquishing some aspect of their information privacy in exchange for benefits[18] (e.g. saving credit card details on an e-commerce website to achieve their goal of saving time). In essence, "[...] *individuals will exchange personal information as long as they perceive adequate benefits will be received in return—that is, benefits which exceed the perceived risks of information disclosure*" (p. 327).[19] However, individuals do not always rationally consider the risks and consequences—including long-term ones—of information disclosure[20], and are often unable to predict the nature of the information to be managed.[21] Nevertheless, the phrase, "*perceived risks of information disclosure*" does encapsulate the meaning of privacy concern. If an individual believes the party requesting the information is not capable of looking after their personal information properly, they will perceive a high degree of risk. Privacy concern—in the context of a technology-mediated interaction—can therefore be thought of as an individual's perceived risk of disclosing personal information; the higher the perceived risk, the higher the individual's level of privacy concern. Beldad views online information privacy as a response to the risks of disclosing personal data, influenced by the amount and type disclosed.[22] This suggests privacy concern, like privacy, is highly contextual, depending, in part, on an individual's expectations of the privacy behaviour of the technology service under consideration—it cannot be measured in the abstract.

An individual's privacy-sensitive decision making process is likely to affected by incomplete information, bounded rationality (their ability to understand the available information and use it to make a rational privacy-sensitive decision), and psychological factors.[23] However, an individual will usually make some effort to consider

---

[15] Morton and Sasse 2012.

[16] Morton and Sasse use the term *technology lens* for the technology platform to highlight that a poorly implemented or designed technology platform may lead an individual to have a distorted view of the organisation, no matter how benign its motivation.

[17] For simplicity it is assumed an individual actively makes a decision to use a technology service, rather than its use being mandatory or unavoidable (e.g. closed-circuit television), or it existence being unknown.

[18] Sheehan and Hoy 2000.

[19] Culnan and Bies 2003.

[20] Acquisti 2004; Acquisti and Grossklags 2005.

[21] Laufer and Wolfe 1977.

[22] Beldad et al. 2011.

[23] Acquisti and Grossklags 2005.

information about a technology service to assist, or justify, their privacy-sensitive decision making, unless they are solely focused on the benefits it offers. Generally speaking, levels of risk increase when there is insufficient information to assess the true level of risk. Similarly, an individual's level of privacy concern will rise if there is incomplete information about a technology service's ability to safeguard the personal information they need to provide. To reduce the discomfort with a lack of information about the collection and usage of their personal data, people will engage in '*information-seeking behaviours*'[24], with Beldad suggesting "[a]*n online privacy statement is often the only source of information*" (p. 222).[25] However, people are also likely to seek information from other sources, such as the attributes of the technology service which are important to them (e.g. security mechanisms, brand name, professionalism of website design etc.) and the advice of friends and colleagues.

Perfect information about a technology service is not possible, and even if it was available, bounded rationality would be likely to prevent an individual from correctly processing it. Fortifying notice-and-consent, such as clearer privacy policies—although welcome—assumes "[i]*ndividuals can understand all facts relevant to true choice at the moment of pair-wise contracting between individuals and data gatherers*" (p. 32).[26] An individual is therefore likely to look for certain attributes of the technology service they are considering using, which they consider to be important. These may include: its professionalism; design; ease of use; perceived security protections; nature of the information requested; perceived ethics of the providing organisation; evidence of sound information handling practices; and links to trusted third parties (e.g. online payment systems). Environmental cues, such as friends' experiences, reviews by existing users and changing social privacy norms[27] are also likely to influence an individual's privacy concern.

The absence of, or incomplete information about, the technology service attributes or environmental cues an individual seeks for reassurance in their decision to engage with it, are likely to lead to an increase in privacy concern. For example, a missing or unclear privacy policy on a website may cause an individual's level of privacy concern to increase, fearing the providing organisation will sell their personal information to a third-party without their permission. Similarly, a lack of information from friends and colleagues about their experiences with a particular technology service may also increase an individual's level of privacy concern.

If it is assumed the technology service attributes and environmental cues an individual seeks for reassurance are underpinned by their innate level of privacy concern, it can be seen that an individual's privacy concern when engaging with a technology

[24] Beldad et al. 2011.

[25] Beldad et al. 2011.

[26] Nissenbaum 2011.

[27] Mark Zuckerberg, the CEO of Faceboook's observation that "*People have really gotten comfortable not only sharing more information and different kinds, but more openly and with more people. That social norm is just something that has evolved over time*" is a good example of this—TechCrunchTV 2010.

service can be constructed from: (1) their innate level of privacy concern; (2) environmental cues, which may be general or related to the technology service under consideration; and (3) attributes of the technology service under consideration. The last of these components (attributes of the technology service) will be highly contextual (i.e. relevant to a particular technology service), with the second one partially influenced by context (i.e. media stories about technology services similar to the one under consideration). Furthermore, each of the three privacy concern components is also likely to be influenced by an individual's personality and attributes (e.g. age, gender, educational level, computer experience etc.). For example, an individual may have been told by friends of their negative privacy experiences with a technology service, but discounted these views because they believe their friends are "*not particularly Internet-savvy*" and "*probably didn't tick the right boxes*".

As the first stage in developing this richer approach to the construction of peoples' privacy concern, the rest of this chapter describes an exploratory study—using focus groups and an online survey—to explore what people consider when deciding to use a technology service offering benefits, but requiring personal information. Section 13.2 situates the proposed approach to constructing privacy concerns in the context of existing work in trust, and the measurement of peoples' privacy concerns, and proposes a hypothetical model based on this work. Sections 13.3, 13.4 and 13.5 describe the research objectives addressed by the study, and provide a description of the research method used. The results from the qualitative and quantitative analyses of the focus group transcripts are discussed in Sects. 13.6–13.9. The paper concludes in Sects. 13.10 and 13.11 with a discussion of the limitations of the study and the next steps for the research, which is to create a richer representation of individuals' privacy concerns, linking this with organisational privacy practice and privacy by design.

13.2 Related Work

Westin—who defines *privacy* as, "*the claim of individuals, groups, or institutions to determine for themselves when, how and to what extent information about themselves is communicated to others*" (p. 7)[28]—categorised people in his surveys of their privacy concern as: (1) *Privacy Fundamentalists*—who are protective of their privacy, distrustful of organisations collecting personal data, and believe in privacy regulation; (2) *Privacy Pragmatists*—who consider the consequences of providing private information *vs.* the benefits received; and (3) *Privacy Unconcerned*—who are least protective of their privacy, believing any benefits they receive for disclosing personal information outweigh its potential misuse.[29]

[28] Westin 1967.

[29] Kumaraguru and Cranor 2005. Perri 6 observes a fourth group is now also recognised—*privacy fatalists*—"*who believe that there is little that they or anyone else can do to ensure proper use of personal information*" (p. 2)—6 et al. 1998.

One-dimensional measurements of peoples' general level privacy concern, such as Westin's categorisation, do not explain the specific dimensions of that concern.[30] To address this, Smith et al. (1996) created a multidimensional scale called, concern for information privacy (CFIP), constructed from individuals' concerns about organisations' information handling practices in the context of offline direct marketing. CFIP is constructed from four factors relating to the handling of information by organisations: (1) collection; (2) errors; (3) unauthorised secondary use; and (4) improper access to information. Stewart & Segars observed, "[...] *the theoretical and operational assumptions underlying the structure of constructs such as CFIP should be reinvestigated in light of emerging technology, practice, and research*" (p. 37)[31], and empirically validated CFIP. They concluded that CFIP was a second-order factor mediating the relationship between computer anxiety and behavioural intention. They also suggested that growing awareness amongst consumers of explicit and implicit information collection and processing by organisations are likely to impact the nature of CFIP—in essence, the effect of environmental cues, such as media stories and the experiences of friends and colleagues.

Using CFIP as the foundation, Malhotra et al. (2004) created the more parsimonious Internet Users' Information Privacy Concerns (IUIPC) scale, specifically aimed at the Internet environment. IUIPC consists of ten items measuring three factors: (1) information collection—identified by Smith et al. (1996); (2) control over personal information; and (3) awareness of an organisation's privacy practices. Although these privacy concern scales recognise an individual's perception of an organisation's information handling practices is an important constituent in their level of privacy concern, they do not explain the influence of external factors, an individual's innate privacy concern, and the specific attributes of the technology service (e.g. perceived security protections, professionalism, design, ease of use, perceived brand and ethics of the providing organisation, service etc.), which an individual seeks for reassurance.

If, as posited earlier, privacy concern is assumed to be the "*perceived risks of information disclosure*"[32], trust in the technology platform and providing organisation's privacy behaviour is key to people feeling comfortable disclosing personal information, as '[t]*rust is only required in situations that are characterized by risk and uncertainty*' (p. 384).[33] Social exchange theory posits that if the benefits of a social transaction with another party outweigh the perceived costs (or risks), an individual will enter into it; trust therefore plays a critical role in this process as it reduces perceived costs and is a precondition for self-disclosure.[34]

The relationship between privacy concern, trust and behavioural intention were explored by Liu et al. in the context of e-commerce, who found that "*privacy has a strong influence on whether an individual trusts an EC* [(e-commerce)] *business.*

---

[30] Malhotra et al. 2004.

[31] Stewart and Segars 2002.

[32] Culnan and Bies 2003, p. 327.

[33] Riegelsberger et al. 2005.

[34] Metzger 2004.

In turn, this will influence their behavioral intentions to purchase from or visit the site again" (p. 300).[35] Privacy, in their *privacy–trust–behavioural intention model* consists of the dimensions of notice, access, choice and security, matching the Fair Information Practices set out by the US Federal Trade Commission for e-commerce.[36] This suggests a technology service implementing and following fair information practices, and making this behaviour visible to an individual considering using it, is more likely to engender trust than one which does not.

Like privacy, trust's multi-dimensional nature makes it impossible to arrive at a unitary definition. To address this, McKnight & Chervany[37] developed a typology of three trust constructs: (1) *dispositional trust*; (2) *interpersonal trust*; and (3) *institutional trust. Dispositional trust* is essentially the general level of trust an individual has, consisting of *faith in humanity* and *trusting stance*.[38] Rotter[39] was the first to develop a scale for this construct of an individual's generalised trust in others[40]—effectively an innate level of trust—which an individual carries with them and applies to each situation. An individual's upbringing and culture will mould their persona and hence their disposition to trust.[41] Peoples' disposition to trust has been found to be positively related to their enthusiasm to embrace new technology[42]—their *Personal Innovativeness with respect to Information Technology (PIIT)*—a construct developed by Agarwal & Prasad, which they define as, "*the willingness of an individual to try out any new information technology*".[43]

Tan & Sutherland[44] include dispositional trust in their multidimensional model of trust, to emphasise the importance of this personality-based trust on consumers' trusting behaviour. They suggest that interpersonal trust and institutional trust are founded upon dispositional trust, observing, "[i]*f the individual typically finds it hard to trust in general, they are not likely to find the internet a comfortable place to conduct business* [...]" (p. 47)[45] Similarly, it is likely an innately private person will not feel comfortable providing personal information to technology services. An organisation making its privacy policy available will have little impact on the views of Privacy Fundamentalists, or those who believe any information disclosure is risky.[46]

Institutional trust is split into: (1) *situational normality*—things appear normal; and (2) *structural assurances*—contracts, regulations and warranties are in place and

---

[35] Liu et al. 2005.

[36] Federal Trade Commission 2000.

[37] McKnight and Chervany 2001.

[38] McKnight and Chervany 2001.

[39] Rotter 1967.

[40] Rotter refers to this as *interpersonal trust*.

[41] Tan and Sutherland 2004.

[42] McKnight et al. 2002.

[43] Agarwal and Prasad 1998.

[44] Tan and Sutherland 2004.

[45] Tan and Sutherland 2004.

[46] Beldad et al. 2011.

evident.[47] In the context of a technology service, an example of *situational normality* is an e-commerce website appearing professional and following a familiar shopping basket and checkout paradigm; an example of a *structural assurance* is the website adhering to distance selling legislation.

An individual's assumptions and expectations of a technology service, which form part of their initial level of trust in it, may be influenced by *trust signals* emitted by the technology service, allowing the individual to determine if trust should be given[48], and hence whether personal information should be disclosed. Trust signals include *trust symbols* (e.g. evidence of HTTPS and trusted third-party seals) and *trust symptoms* (e.g. user reviews, usability and professionalism of web site design).[49] These trust signals originate from the technology service and mainly influence *interpersonal trust*, allowing an individual (*trustor*) to decide if trust should be given to the *trustee*. If the sources of trust signals are the attributes of a technology service related to its information privacy practice, e.g. use of technical security controls, stated privacy policy and control over personal information provided to its users, absent or weak trust signals are likely to increase an individual's level of perceived risk of information disclosure, leading to increased privacy concern and decreased trust.

In addition to the trust signals emitted by a technology service, individuals' trust in a technology service may be influenced by environmental cues, such as the experiences of friends, advertising material (e.g. television and poster advertisements) and social privacy norms. For example, an individual's level of trust in a particular technology service is likely to be increased if their friends have used it with no perceived problems—hence the quote in the title of this paper. These environmental cues may not be directly related to the technology service under consideration. Environmental cues such as media reports and experiences of using similar technology services are likely to be an important constituent of peoples' privacy concern, as individuals generalise broadly from their experiences.[50]

Gefen et al. (2003), in their study of trust and technology acceptance in the context of online shopping, suggest the decision to purchase from an e-vendor has two antecedents: (1) their trust in the technological aspects of the website interface (influenced by its perceived ease of use); and (2) the trust the consumer has in the vendor (essentially *interpersonal trust*), reflecting the two main components of a technology service, the technology platform and providing organisation. Using three terms from Riegelsberger et al.'s (2005) framework—an organisation's *internalised norms* (e.g. policies, privacy behaviour etc.), *benevolence* (e.g. an easy faulty product return process) and *ability* (e.g. professionalism of its website)—an individual may trust the *norms and benevolence* of an organisation, but not its *ability* to provide a secure technology platform to protect users' information from unauthorised intrusion.[51] For example, an individual may believe an organisation possesses strong information

[47] McKnight and Chervany 2001.

[48] Riegelsberger et al. 2005.

[49] Riegelsberger et al. 2005.

[50] Camp et al. 2002.

[51] Beldad et al. 2011.

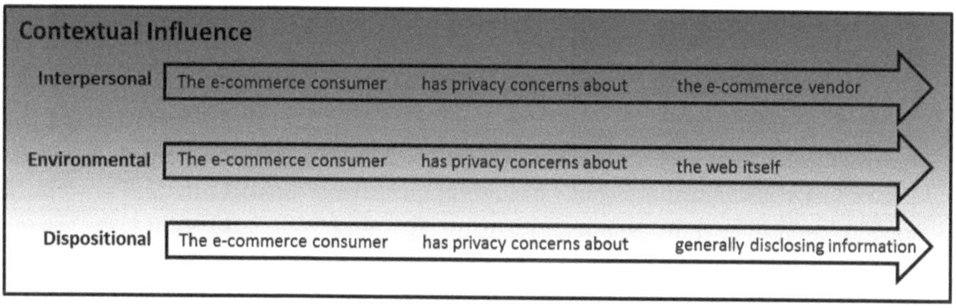

Fig. 13.1 Grammar of the privacy concern model (for an e-commerce website)

ethics and intention to safeguard collected information[52]—causing privacy concern to lessen—but may also suspect the organisation is unable to realise those ethical principles due to a poorly implemented technology platform—increasing the individual's level of privacy concern.

Research has shown a relationship between an individual's degree of *agreeableness* measured by the Five-Factor Model of personality (the Big Five) and their level of privacy concern measured by CFIP—albeit with a restricted sample of respondents.[53] Other studies have shown a relationship between an individual's personality traits and their level of privacy concern.[54] Peoples' attributes (e.g. age, gender, experience, cultural background and intellectual capability) also influence their adoption of new technology.[55]

If an individual's personality influences their level of privacy concern, trust and technology adoption, it is also likely to influence the environmental cues and technology service attributes an individual looks for to lessen their privacy concern. For example, one personality type may place a high degree of importance on technical security controls (e.g. the HTTPS browser 'padlock') and the existence of a privacy policy, whilst another may only consider the advice of friends or social norms.

McKnight & Chervany represent their trust construct using three sentences in a *grammar of trust*, with each one constructed as an action sentence with a subject, verb, and direct object.[56] If an individual's level of privacy concern is influenced by: (1) their innate level of privacy concern; (2) environmental cues; and (3) the attributes of the technology service, a *grammar of privacy concern* (Fig. 13.1) can be constructed using a similar approach, with an individual's privacy concern constructed from:

1. **Dispositional privacy concern**. An individual's innate concern about disclosing any information to other parties. *Dispositional privacy concern* is essentially the construct captured by traditional privacy surveys such as Westin's, and therefore

[52] Beldad et al. 2011.

[53] Korzaan and Boswell 2008.

[54] Iris et al. 2008.

[55] Agarwal and Prasad 1999; Venkatesh and Morris 2000.

[56] McKnight and Chervany 2001.

represents only a partial understanding of the nature of an individual's privacy concern.

2. **Environmental privacy concern**. An individual's level of privacy concern created by environmental cues, such as media reports, anecdotes from friends and family, and social privacy norms. *Environmental privacy concern* may also be lessened by structural contracts and regulations (e.g. applicable data protection legislation) being in place and evident.
3. **Interpersonal privacy concern**. An individual's level of privacy concern about the party they are transacting with. The level of privacy concern will be increased or lessened by the existence or absence of technology service attributes an individual considers important and therefore looks for.

The shading in Fig. 13.1 represents the increasing influence of context on the three components of privacy concern. The philosophy of *privacy as contextual integrity*[57] posits that the transfer of personal information between two entities (e.g. a consumer and an e-commerce website) should be tied to the widely accepted norms of particular contexts, so that information collection and dissemination is appropriate to each context and the roles of the entities, and in line with expectations. It is the violation of these norms and expectations which is one of the principal factors leading to peoples' perception that their privacy has been invaded. The collection and dissemination of personal information by increasingly powerful technologies and digital media serve to subvert these norms and expected information flows.[58] Contextual integrity is constructed from: (1) informational norms; (2) appropriateness of collection and dissemination; (3) roles of the entities involved; and (4) principles of transmission.[59] Given this construction of contextual integrity, Fig. 13.1 shows that interpersonal privacy concern will be more influenced by context (e.g. a specific transaction with a particular e-commerce website), than dispositional privacy concern.

Perri 6 describes research by Brunel University, which suggests a "*more nuanced approach to segmentation*" (p. 39) than Westin's—based on a repertoire of behaviours—and argues people take different privacy stances in different contexts, and very few can be simply categorised as fundamentalist, unconcerned or pragmatic.[60] The construction of privacy concerns shown in Fig. 13.1 addresses this by recognising that although an individual's *dispositional privacy concern* is an important factor underpinning their privacy concern, the contextual influences at the *interpersonal privacy concern* layer, and to some extent, at the *environmental privacy concern* layer, are extremely important.

Organisations may be able to influence peoples' level of interpersonal information privacy concern through information handling practices which avoid substantive harm, and the use of trust signals[61], but are unlikely to be able to significantly, or

[57] Nissenbaum 2004.

[58] Nissenbaum 2004.

[59] Barth et al. 2006.

[60] 6 et al. 1998.

[61] Riegelsberger et al. 2005.

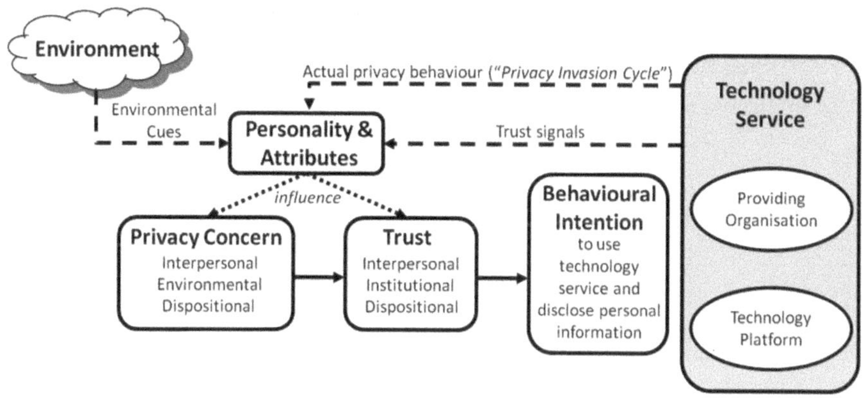

Fig. 13.2 Hypothesised extended model of privacy concern, trust and behavioural intention

quickly, influence an individual's dispositional privacy concern. Pragmatically, this approach to privacy concern seems reasonable. An individual with a very high level of dispositional privacy concern is unlikely to provide their personal information, irrespective of the experiences of others, or the perceived privacy behaviour of the other party. Similarly, an individual with a moderately high level of dispositional privacy concern may be dissuaded from providing their personal information by the experiences of friends who have had their information passed to third parties—increasing the individual's environmental privacy concern.

Using the *privacy–trust–behavioural intention model* of Liu et al. (2005) and the idea of trust signals proposed by Riegelsberger & Sasse[62], a hypothesised extended model of privacy concern, trust and behavioural intention is proposed (Fig. 13.2). The model illustrates how trust signals emitted by a technology service—albeit modified by an individual's personality and attributes—influence interpersonal trust (e.g. trust in the ability of the technology service to provide the products ordered) and interpersonal privacy concern (e.g. the level of concern about the technology service's privacy practices). Similarly, environmental cues will be modified by an individual's personality and attributes, influencing their environmental privacy concern.

An individual will have assumptions and expectations of a technology service's likely privacy behaviour, with their "*privacy perceptions often reflect*[ing] *their trust in the organisation, technology and thus expectations for privacy protection*" (p.52) (also please add superscripted reference to footnote 64). If an individual's experience of the technology service's actual privacy behaviour does not match these expectations and assumptions, because of a malicious or incompetent organisation, error, or a badly designed technology platform leaking sensitive information, the individual is likely to feel their privacy has been invaded, have an emotive reaction, and reject the technology and providing organisation—Adams & Sasse call this the *Privacy Invasion Cycle*[63], and this concept has been included in the hypothesised model

[62] Riegelsberger et al. 2005.

[63] Adams and Sasse 2001.

shown in Fig. 13.2. The invasion of an individual's privacy is likely to result in a decrease in their trust as a result of an increase in their privacy concern about the technology service, suggesting the construction of peoples' privacy concern is likely to be a dynamic process.

13.3 Research Objectives & Method

The principal aim of this exploratory study was to understand the factors—organisational, technological and environmental—which people consider when deciding to use a technology service. More specifically, it was to explore the factors which increase or decrease their *interpersonal privacy concern* and *environmental privacy concern* (Fig. 13.1). The secondary aim was to investigate if peoples' attributes (age, computer experience, gender etc.) influence the organisational, technological and environmental factors they consider to be important (Fig. 13.2). These aims resulted in two research objectives:

1. To investigate the organisational, technological and environmental factors people consider when deciding to use a technology service.
2. To investigate if there is a relationship between individuals' attributes (e.g. age, gender, computer experience etc.), willingness to adopt new technologies and general privacy concern (e.g. their Westin category), and the organisational, technological and environmental factors they consider.

If people look for very disparate organisational, technological and environmental factors when faced with different types of technology services, the hypothesised model (Fig. 13.2) is unlikely to be feasible. Therefore a third research objective was defined:

3. To investigate if the factors individuals consider are broadly common to all technology services.

To address these three objectives, a research method was required—richer than online surveys—which facilitated open-ended investigation of these factors, without unduly influencing study participants. Focus groups were selected as the research method as they are suited to the investigation of complex behaviours and motivations[64], such as technology adoption and privacy-sensitive decision making. Focus groups also allow participants to query each other, explain themselves and comment on each other's experiences.[65] There has been some use of focus groups in understanding privacy concerns about technologies, and acceptance of new technologies[66]; this latter area being focused on informational privacy in healthcare.[67] There has however, been

[64] Morgan and Krueger 1993.

[65] Kitzinger 1995; Morgan 1996.

[66] Zhang et al. 2010; Hundley and Shyles 2010; 6 et al. 1998.

[67] Skinner et al. 2003; Snell et al. 2012.

Table 13.1 Focus group scenarios and composition

Focus group no	Scenario	Group composition
G1	Photograph sharing web site	Technical PhD students and postdoctoral researchers
G2	Social networking discounts	Undergraduate students
G3	Supermarket RFID ordering	Technology outsourcing business development and administrative staff
G4	Smartphone assistant	Postgraduate students
G5	Smart metering	IT support and development, IT business development, IT project management, retirees and administrative staff
G6	Landmark identification web site	Extended family group consisting of retirees, middle managers, administrative staff and tradesmen

promising work by the VOME project[68], which has run interactive sessions with users discussing citizen-centric privacy by design.[69]

In addition to the focus groups, an online survey was used to collect quantitative data prior to participants' attendance at each focus group, although completion of the survey was not a pre-requisite for attendance. The principal objective of the survey was to provide data for quantitative analysis to investigate research objective 2. The survey was split into four sections: (1) eight questions concerning the participant; (2) one question to ascertain the participant's willingness to adopt new technologies; (3) three questions to ascertain the participant's general level of privacy concern; and (4) two questions based on Sheehan's study of privacy concerns[70], which were not used in the study.

13.4 Focus Group Procedure

Six focus groups—considered to be an adequate number[71]—took place, capturing the views of 35 individuals. To ensure participants represented a broad range of experiences and ages, opportunistic sampling with participant peer recruitment was used for four of the groups, with the other two groups consisting of volunteers from a UK university's participant pool (Table 13.1).

At the start of each focus group the researcher provided an overview of the objectives of the session and briefly described the concept of a technology service. This

[68] Visualisation and Other Methods of Expression (VOME) is a project involving researchers from the Information Security Group at Royal Holloway, University of London, Salford and Cranfield Universities. It has explored how users engage with the concepts of information privacy. For further information about VOME see http://www.vome.org.uk.

[69] VOME 2012.

[70] Sheehan 2002.

[71] Morgan 1996.

was to encourage participants to think more widely than the technology described in the scenario, and also consider the organisation providing it.

Once the focus group had read the scenario randomly allocated to it (see Appendix for the six scenarios), it was shown the following three questions:

1. What things would you consider when deciding to use, or not use, this technology service?
2. How would you go about deciding if the benefits offered by this technology service were worth the potential loss of some of your privacy?
3. How would you decide whether to trust this technology service to look after your privacy?

Each focus group lasted approximately one hour, with 15–20 min spent discussing each question. Use of the same questions and procedure for each focus group facilitated investigation into the similarity of the themes discussed across the focus groups. The groups were designed to encourage participants to interact with each other, rather than the researcher, allowing "*structured eavesdropping*" (p. 301).[72] The researcher attempted to restrict their contribution to reading the three questions out aloud, and asking further probing questions when required.

13.5 Qualitative Analysis of Focus Groups

A thematic analysis—similar to that described by Braun & Clarke[73], albeit without producing a thematic map—was undertaken for the qualitative analysis of the focus group transcripts.

Each focus group was recorded by the researcher, and transcribed by a professional typing agency. The researcher listened to the audio recording of each session twice, correcting any errors in the transcripts, and ensuring anything in the transcript which identified the focus group or its members was redacted. This ensured the data was "*transcribed to an appropriate level of detail, and the transcripts* [...] *checked against the tapes for 'accuracy*'" (p. 96).[74] This process of active reading and re-reading, and becoming familiar with the data, assisted with generating initial ideas for base-level codes.

Once the transcripts had been checked they were loaded into ATLAS.ti, and participants' comments—*quotations* in ATLAS.ti—coded by the researcher in a systematic fashion with an initial set of base-level codes. The entire transcript from each focus group was coded to ensure '[e]*ach data item has been given equal attention in the coding process.*' (p. 96).[75] At the end of the initial coding phase 39 base-level codes had been created, excluding the ATLAS.ti super-codes and non-substantive codes used to facilitate subsequent quantitative analysis. The 39 base-level codes

[72] Powney J., quoted in Kitzinger 1995.

[73] Braun and Clarke 2006.

[74] Braun and Clarke 2006.

[75] Braun and Clarke 2006.

Table 13.2 Demographic profile of online survey participants

Demographic characteristic		Percentage of respondents
Gender ($n = 27$)	Male	51.9
	Female	48.1
Age (years; $n = 27$)[a]	Under 18	3.7
	18–24	7.4
	25–34	33.3
	35–44	25.9
	45–54	14.8
	55–64	11.1
	Over 65	0.0
	Rather not say	3.7
Education level[a] ($n = 26$)[b]	Doctoral	7.7
	Postgraduate	19.2
	Undergraduate	38.5
	Diploma level	19.2
	School leaver	15.4

[a] The total of the percentages in Table 13.2 for this survey item does not equal 100 % because of rounding

[b] $n = 26$ as one online survey participant was still attending school

were then collated into candidate themes by considering whether a code could be combined with others into an overarching theme.

Although Braun & Clarke suggest quotations may be coded '[...] *in as many different 'themes' as they fit into* [...]' (p. 89)[76], the researcher coded each quotation to a single base-level code, and hence theme. This encouraged the researcher to consider carefully what each participant was actually alluding to in their comment, and also facilitated the reconciliation of totals during quantitative analysis. Each quotation was also coded with a non-substantive reference for the participant who spoke it, e.g. G5P7, for participant 7 in group 5. This enabled cross-referencing of focus group quotations with the results from the online survey, during the quantitative analysis.

13.6 Online Survey Results

13.6.1 Survey Participant Demographics

Prior to attending the focus groups, 27 of the 35 focus group participants completed the online survey, with response rates ranging from 56 to 100 % within each focus group. Table 13.2 shows the demographic profile of online survey participants, and Table 13.3 their level of computer experience and daily computer use.

[76] Braun and Clarke 2006.

Table 13.3 Computer experience and use profile of online survey participants

Demographic characteristic		Item statistics ($n = 27$)
Computer experience (years)	Mean	19.9
	Standard deviation	6.8
Computer use at home and work per day (hours)	Mean	6.7
	Standard deviation	3.1

13.7 Survey Participants' General Attitude to Technology and Privacy

Four of the questions in the online survey were asked to determine participants' general level of privacy concern, and their attitude towards adopting new technologies, the results of which are discussed briefly below:

- **Technology Privacy Concern**: Participants were asked to respond to two statements concerning their perception of the impact on their privacy of: (1) existing technology, which had been around for at least three years; and (2) emerging technology, which had appeared in the last year. These two statements represented a survey participant's *technology privacy concern* (TPC), which can range from 2 to 8, where 2 represents two '*Not concerned at all*' responses and 8 represents two '*Very concerned*' responses. Ignoring the results from the two survey participants who selected '*Don't know*', resulted in a sample size of 25, with a mean TPC score of 6.0 ($\sigma = 1.55$), suggesting a relatively high level of TPC amongst survey participants, which may be caused by the relatively high percentage (40.7 %) of Privacy Fundamentalists in the survey group.
- **Westin's Privacy Segmentation Index**: Survey participants were asked to respond to the same three statements used by Westin between 1995 and 2003 to determine peoples' Privacy Segmentation Index/Core Privacy Orientation Index[77], with their responses used to place them into one of three privacy categories using the same criteria as Westin (Table 13.4). In Westin's surveys from 1996 to 2003, between 55 and 64 % of participants were categorised as Privacy Pragmatists, with the remaining participants split approximately equally between Privacy Fundamentalists and Privacy Unconcerned.[78] The high percentage of Privacy Fundamentalists (40.7 %) amongst survey participants in this exploratory study may be caused by the large percentage (65.4 %) educated to undergraduate degree level or above. Previous studies have found a relationship between higher levels of education and increased privacy concern.[79]

[77] This index was called the *Privacy Segmentation Index* for Westin's surveys between 1995 and 1999, and the *Core Privacy Orientation Index* for the surveys since mid-2000 (Kumaraguru and Cranor 2005).

[78] Kumaraguru and Cranor 2005.

[79] Phelps et al. 2000; Sheehan 2002.

Table 13.4 Percentage of online survey participants in each Westin category ($n = 27$)

Westin category	Percentage of online survey participants
Privacy fundamentalists	40.7
Privacy unconcerned	18.5
Privacy pragmatists	40.7

The total of the percentages in Table 13.4 does not equal 100 % because of rounding.

Table 13.5 Responses to: "Here are some predictions about how technology will impact peoples' privacy in the next five years. Which of the following statements comes closest to the way you feel?" ($n = 27$)

Statement	Percentage of survey participants	Coding
Technology will make peoples' privacy worse	59.3	3
Technology will make peoples' privacy better	7.4	2
Despite advances in technology, peoples' privacy will remain about the same as it is today	18.5	1
Don't know	14.8	N/A

- **Future Impact of Technology on Privacy**: Table 13.5 shows the percentages for the responses from survey participants to the same question used by Westin in 1996.[80] Almost 60 % of survey participants believed technology would make people's privacy worse over the next five years.

To test if survey participants' responses to this question were consistent with their responses to the TPC statements, they were coded as shown in the far right-hand column in Table 13.5, and a Pearson two-tailed correlation test (with '*listwise*' exclusion, so $n = 22$) performed between these and the TPC scores. Correlation was significant at 0.599 ($p = 0.003$), indicative of consistency between the responses from each participant to these two survey questions about privacy concern in relation to technology.

- **Willingness to adopt new technologies**: To assess survey participants' willingness to adopt new technologies, a score for each participant was calculated by taking the mean of scores for the four statements in Agarwal & Prasad's PIIT scale.[81] A mean score of 7 represented someone with the highest level of willingness to adopt new technologies, and a mean score of 1 represented someone with the lowest. The overall mean PIIT score for focus group survey participants was 4.82 ($\sigma = 1.18$), suggesting a reasonable level of comfort using new technology. To assess the reliability of the measure, Cronbach's α was calculated to be 0.825—indicating good internal consistency.

[80] Kumaraguru and Cranor 2005.

[81] Agarwal and Prasad 1998.

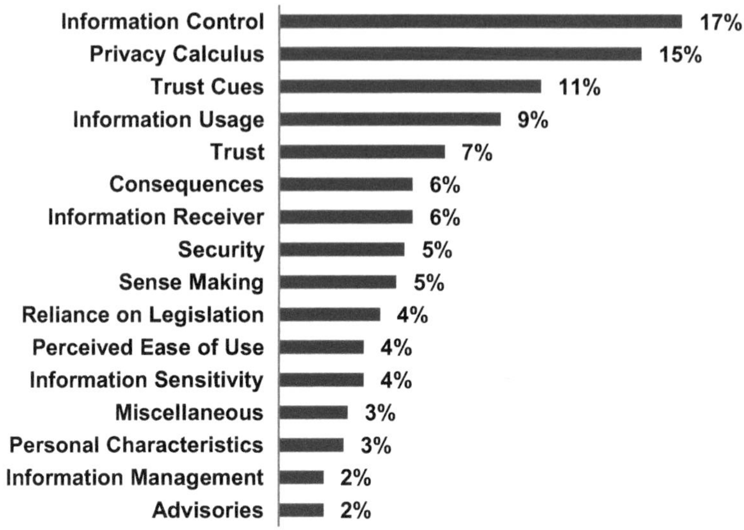

Fig. 13.3 Percentage of coded quotations from focus groups for each identified theme ($n = 599$)

13.8 Focus Groups—Qualitative Analysis

From the analysis of the six focus group transcripts, 599 quotations were coded and allocated to one of 39 base-level codes, grouped into 16 themes.[82] Fig. 13.3 shows the percentage of coded quotations allocated to each of the 16 themes.

Quotations from the focus group transcripts not directly related to the research objectives, such as copyright, peoples' desire for an online presence, the societal impact of technology and general criminal activity, were placed in a *Miscellaneous* theme—representing 3 % of coded quotations. The three themes most frequently discussed in the groups: (1) *Information Control*; (2) *Privacy Calculus*; and (3) *Trust Cues*—representing 43 % of all coded quotations—are discussed in the next three sections.

13.8.1 Information Control

The *Information Control* theme includes four base-level codes used for quotations relating to: (1) organisations passing personal information to third parties (34 %); (2) the ability to opt-in or opt-out of aspects of a technology service (27 %); (3) individual control over access to personal information (24 %); and (4) information control provided by a technology service (15 %).

Organisations passing personal information to third parties was the most common concern within the *Information Control* theme. This was either because of a perception that some organisations act unethically, or individuals being unaware

[82] See Appendix for a description of the types of quotations covered by each theme.

they had given permission to share personal information. Passing of information to third parties without notice, and the secondary use of that information, is an important factor in people feeling their privacy has been invaded.

In common with the three out of four of the respondents in the European Commission survey[83], focus group participants recognised organisations' commercial objective to collect and sell consumers' personal information. One focus group participant observed, "*Well, they're probably selling the data to the retailers, aren't they? It's a revenue stream from that*" (G5). A participant in another group remarked, "*Well obviously the company wants to get as much money as they possibly can, and this site would be for free, would be free sign up, so they have to get the money through the links we share*" (G2).

As two of the groups' discussions progressed (G3 & G4), participants became aware of the potential consequences—collated under the *Consequences* theme—of information being passed to third parties for secondary use. A participant in the group discussing the Smartphone Assistant scenario said:

> So how would that affect insurance companies for example? Because I've thought about that as well, so if there's all this data about what I'm buying, where I'm going, you know, let's say I have diabetes and I went and bought sweets all the time that would get recorded and then I'd have an issue with my diabetes, and insurance would go like, 'Well, you know, she's doing all of that stuff and she's not helping herself.'

The ability to opt-in or opt-out of aspects of a technology service was discussed in all groups, particularly in the context of passing information to third parties. Under the UK Data Protection Act 1998, if a data holder wishes to pass a data subject's information to a third-party, permission must be sought from the data subject. Three of the groups referred to the framing of such questions[84] (G1, G5 & G6); with participants generally suspecting commercial pressures lead organisations to offer consumers the choice to opt-out, rather than the more acceptable opt-in. Many in the group discussing the Photograph Sharing Web Site scenario felt organisations deliberately obfuscated the opt-in/opt-out process, with one participant expressing anger about having to opt-out of receiving e-mails from third parties: "[...] *when it's a little tiny little tick that you have to find it in some buried place within the website that actually really annoys me.*" Individuals' effort to control their privacy and their selection of services—particularly with respect to opting out—was also raised in the group discussing the Smart Metering scenario, with one participant remarking, "*Now you have to actively protect your privacy by, you know, looking for the boxes of 'I don't want to be contacted'*, [...] *'I don't want my details to be sent out'*".

Individual control over access to personal information, which relates to the control people wish to exercise over access by others to their personal information, was touched upon by five groups (G1, G2, G4, G5 & G6). Unsurprisingly, the group discussing the Landmark Identification Website scenario contributed the majority (63 %) of the quotations within this base-level code, as this scenario was potentially the most personally invasive. One participant in this group reflected the group's

[83] European Commission (EUROPA) 2011.

[84] Bellman et al. 2001.

consensus, stating, "*I wouldn't be altogether comfortable knowing that other people can just take a photograph of you and find out all sorts of information about you.*"

Information control provided by a technology service relates to the amount of control a technology service provides a user—specifically the data subject—to manage the disclosure of personal information. Participants raised concerns about technology services requesting: unnecessarily mandatory data items (G4); information perceived as irrelevant by the user (G5); and information considered sensitive or intrusive by the user during initial interaction with a technology service (G1 & G4). Three groups (G1, G3 & G6) stressed their need for a technology service to notify them about which information was being shared, and with whom. One participant (G1) observed:

> You don't really know what you are going to be sharing. They never really say, 'If you join the service we will then take all of these eight items'. That doesn't really exist.

Providing users with accurate feedback about which information is being collected, was taken further by a participant in another group (G4) who said:

> It's always you're in or you're out, [it's] never the option to, 'I would still like to join your service providing these things are not recorded or done for me' and then put them all back in the other person's court. 'Do you still think I'm valuable enough to be a customer for you?'

This approach would allow users to decide which personal information they are comfortable to provide, given the context and benefit received—leaving the organisation to determine if they still wish to provide them with the benefits offered.

13.8.2 *Privacy Calculus*

Westin's Privacy Pragmatist, who is someone prepared to forgo some of their privacy in exchange for some sort of benefit, exemplifies the concept of *privacy calculus*. Pragmatically, privacy calculus is the cost/benefit[85] analysis in which an individual considers the benefits against the potentially unforeseen consequences of information disclosure. The privacy calculus concept is founded on Laufer & Wolfe's idea of a "*calculus of behaviour*", in which people consider the consequences of engaging in a particular behaviour.[86] They cite the example of an individual submitting to personality testing, and disclosing personal information, in the belief the outcome will be beneficial—this individual will not consider this as an invasion of privacy.[87] However, they are likely to consider it privacy-sensitive, and ask themselves, "*Can I trust this person to safeguard the personal information I have passed to them?*"

---

[85] Dinev and Hart (2006) refer to these polarities as '*risk beliefs*' and '*confidence and enticement beliefs*' respectively.

[86] Laufer and Wolfe 1977.

[87] Laufer and Wolfe 1977.

The notion of privacy calculus is therefore effectively a trade-off in which an individual weighs up the cost to their privacy against the benefits of disclosing private information; Beldad et al. (2011) observe:

> [w]hen the expected benefits from the information disclosure do not outweigh the value attached to the personal data to be disclosed, information withholding or incomplete information disclosure could be forthcoming (p. 226).

The *Privacy Calculus* theme includes two base-level codes: (1) the benefits received by an individual for using a technology service (72 %); and (2) an individual's thought processes when considering the benefit offered by a technology service, against the amount and type of personal information requested (28 %). During coding it was often difficult to decide which of these two base-level codes a quotation should be coded against; this was mitigated by including both within the *Privacy Calculus* theme.

The benefits received by an individual for using a technology service were placed on a continuum by the focus groups, with tangible benefits at one end: cash payments (G1 & G2); cheaper products and services (G3 & G6); and cost savings (G3 & G5). At the other end of this continuum are intangible benefits: recommendations (G1 & G4); removal of the effort of visiting shops (G3 & G4); assistance with lifestyle (G3); ability to save credit card details (G1 & G4); social benefits (G5); and socialising (G1 & G2). Beldad et al. also categorise benefits as *tangible* or *intangible*, with tangible benefits (e.g. cash, vouchers or gift items), and intangible benefits (e.g. convenience, joining a social networks and personalised services).[88] However, focus group participants differentiated between discount vouchers or credits, and cash payments, when considering their privacy. One participant in the group discussing the Social Networking Assistant scenario—which offered credits which could only be spent with participating companies—stated:

> [...] if they gave me cash, physical cash, I wouldn't mind, but because I don't have the choice of where I spend the credits, it has to be targeted on certain sites, my privacy concerns would be dominant in a situation like this.

This suggests certain benefits, such as discount vouchers and two-for-one offers (G1, G2 & G6) are actually situated in the middle of the benefits continuum, with their relevance to an individual's goal(s) at a particular point in time increasing their attractiveness. For example, the group discussing the Landmark Identification Web Site scenario initially rejected the idea, but one participant observed:

> I think if I was at a landmark that costs thirty, forty pounds each to get into, and all of a sudden I was offered two-for-one tickets, but by accepting that offer and using that offer, there isn't an opt-out button to receive mailings, for example, from a company. At the time I'd probably take it.

An individual's thought processes when considering the benefit offered by a technology service against their perceived loss of privacy accounted for 28 % of the coded quotations under the *Privacy Calculus* theme. However, all groups at some point during their discussions referred to the privacy-sensitive decision making

[88] Beldad et al. 2011.

process they adopt. The relevance of a technology service's benefits to an individual's goals has already been alluded to, and this appeared to be a particularly important factor in participants' privacy calculus. One participant in the group discussing the Landmark Identification Web Site scenario exemplified this by observing:

> The tipping point for me would be, would I get benefit, do I think I would use this enough? [...] I'd make a conscious decision, 'Will I use this software, is it of benefit to me, are there savings there, generally across the board? Yes or no?' Yes, I would use it, and, to a certain extent, take this on board.

Another participant in the same group explicitly referred to the degree of privacy invasion versus the benefit received:

> For me it comes back to a decision about trade-off. So, am I happy to be bombarded with emails? Yes. Am I happy to be bombarded with emails, but any stranger could identify me? Probably not. And if I'm not willing then I don't want the offer, because it's not worth it.

13.8.3 Trust Cues

Although the focus groups referred to the importance of *trust symptoms*, such as other users' reviews (G2 & G6), findings from personal research (G1 & G2), friends' recommendations (G1, G2, G3, G4 & G6), and magazine reviews (G4), the groups' discussions also highlighted the importance of other prompts from the wider environment—*environmental cues*—in the construction of an individual's trust in a technology service and therefore a willingness to provide it with personal information. The *Trust Cues* theme therefore encompasses not only coded quotations relating to *trust symbols* and *trust symptoms* (45 %)[89], but also environmental cues, which account for the remaining 55 % of coded quotations in this theme.

Environmental cues include: (1) social privacy norms—specifically participants perception of peoples' information sharing behaviour; (2) technology norms—particularly the increasing capabilities of technology to collect and process information; and (3) other external cues, such as advertisements (G1), media stories of hacking and loss of credit card details (G1, G2 & G5), payment for goods and services through recognised methods, e.g. Verified by Visa (G1), and use of the technology service by other people (G6). With reference to this last cue, participants admitted there was comfort in 'following the crowd', with a participant in the group discussing the Landmark Identification Website scenario admitting:

> [...] you hear of more and more people using it, so you think, 'Well it must be okay.' So, it goes back to the fear of the unknown, and whilst you don't know much about it, the more people that use it, the more comfortable you become with it.

Social privacy norms and technology norms appear to define a 'privacy floor' for participants in terms of acceptable levels of information sharing behaviour. One participant in the group discussing the Landmark Identification Website scenario observed:

[89] Riegelsberger et al. 2005.

> So, it's just a totally different world now, and I think people are more willing to accept it, if they've come through that generation. I think there's much more willingness to accept what's out in the public domain and what you're going to share with people [...]

Examples from the focus groups of societal and technology norms used by participants in their privacy-sensitive decision making were:

- **Societal norms**—the need to share information as part of modern life (G2 & G6); peoples' apparent comfort with sharing personal information (G1 & G6); increasing availability of personal information (G2); and the need to enter personal details to gain access to discounts and services (G3 & G6).
- **Technology norms**—the increasing levels of surveillance, e.g. CCTV (G1 & G4); behavioural tracking by websites and supermarket loyalty cards (G1, G3 & G6); unsolicited e-mails and targeted advertising (G4 & G6); and the relentless progress of technology (G3).[90]

13.9 Focus Group Theme Similarity

There was insufficient data across all theme/focus group combinations to perform a statistical test to determine if there was a broad similarity between the focus groups, in terms of the number of times each theme was discussed (research objective 3 in Sect. 13.3). An approach was therefore required to facilitate visual inspection of the qualitative data in the transcripts.

The percentages for each of the 15[91] themes were calculated as the number of quotations for that theme, divided by the total number of quotations in each focus group. A frequency table was created, and the 25 and 75 % percentiles calculated to define three categories based on the percentage of a focus group's coded quotations relating to each theme: (1) *H*—between 9 and 22 %; (2) *M*—between 3 and 9 %; and (3) *L*—less than 3 %; these categories were used to label each theme/focus group combination in Table 13.6. Table 13.6 is divided into three sections—shown by the outlined cells—based on the number of *H*, *M* and *L* categories, so that the most common category in each section—reading from left to right—is *L*, *M* and *H*. This overview grid suggests a broad degree of communality of themes raised and discussed across the focus groups despite the use of different scenarios.

The two most popular themes—*Information Control* and *Privacy Calculus*—have an *H* category in all six focus groups. Despite the broad commonality of the remaining 13 themes across the focus groups—there are some exceptions—most noticeably the *H* categories in five of the scenario/theme combinations in the *Somewhat Discussed* group, and the single *L* category in the *Frequently Discussed* group. Transcripts from focus groups where there were unexpected *H* or *L* categories in the scenario/theme combinations were therefore re-examined.

---

[90] In many respects this acceptance of technology norms, i.e. the inevitability of technological progress and increasing collection and processing of personal information, echoes the views of the *privacy fatalists*—6 et al. (1998).

[91] The *Miscellaneous* theme was excluded, resulting in a sample size of 582 quotations.

Table 13.6 Percentage of quotations—as a category—for each theme across all focus groups ($n = 582$ quotations)

Technology service scenario	Advisories	Information management	Personal characteristics	Information sensitivity	Perceived ease of use	Reliance on legislation	Sense making	Security	Information receiver	Consequences	Trust	Information usage	Trust cues	Privacy calculus	Information control
Photograph sharing web site (Group 1)	L	L	L	M	M	H	M	L	M	M	H	M	H	H	H
Social networking discounts (Group 2)	M	M	L	L	M	M	M	M	M	L	M	M	H	H	H
Supermarket RFID ordering (Group 3)	L	L	L	L	L	L	L	H	M	H	M	H	M	H	H
Smartphone assistant (Group 4)	M	L	L	L	M	L	M	M	L	L	M	H	M	H	H
Smart metering (Group 5)	L	L	M	M	M	M	M	M	M	M	H	H	L	H	H
Landmark identification web site (Group 6)	L	L	M	M	L	L	M	L	M	M	L	M	H	H	H
	Rarely discussed			Somewhat discussed								Frequently discussed			

75% of themes in the *Frequently discussed* section have an *H* category
58% of themes in the *Somewhat discussed* section have an *M* category
72% of themes in the *Rarely discussed* section have an *L* category

13.9.1 *Photograph Sharing Web Site Scenario (Group 1)*

Participants in this group discussed how they felt the law protects their data and financial transactions, along with their distrust of organisations' motives—resulting in an *H* category for the following two themes:

- **Reliance on Legislation**—The discussion in this group began with concerns about copyright of users' photographs—coded under the *Miscellaneous* theme—and frequently returned to how the law protects consumers. The group discussed terms and conditions, data protection, consumer fraud and financial protection.
- **Trust**—Like the Smart Metering scenario, participants generally mistrusted organisations' motives—with one participant observing, "*these big companies have all been shown to act in very dubious ways and that's I think that's what actually scares people the most*"—or trusted particular brands based on their experience—with another participant remarking, "*I trust Amazon, or I am happy to give them the information I have* [...] *given them*".

13.9.2 Supermarket RFID Ordering Scenario (Group 3)

Participants in this group discussed the collection and use of data about shopping habits, and the overall security of the system—resulting in an *H* category for the following two themes:

- **Consequences**—Participants were worried about the potential financial consequences of their shopping habits, with one participant fearing that details about products purchased could be sold to the UK National Health Service (NHS), leading them to say, "*you've got a non-healthy diet, because of that and you're more likely to develop diabetes. Therefore we're going to charge you more money in tax, because you're more likely to use our hospitals*".
- **Security**—Participants frequently discussed their concerns regarding the security of the system, due to the sanctity of the home, concerns about the authorities (e.g. police) checking up on them, and disquiet about the security of information captured by the system.

13.9.3 Smart Metering Scenario (Group 5)

Participants in this group frequently returned to their overall mistrust of energy suppliers, resulting in an *H* category for the *Trust* theme. Two comments encapsulated the group's opinions—"*Do you know any electricity suppliers we trust?*", and "*I don't trust any of these companies really, deep down*". This appeared to be primarily caused by participants' previous experiences of incorrect utility meter readings, with one observing for estimated bills, "*they have actually estimated it for the future* [...] *and half the time it's wrong*". This mistrust in energy suppliers' competence appeared to be generalised to a suspicion that energy suppliers would probably misuse the detailed electricity consumption data from smart meters. Participants also mistrusted energy suppliers' motives—with one stating "*My concern would also be the likelihood is that they will benefit more than I*".

Despite this focus group's mistrust of energy suppliers this scenario was the only one to have an *L* category for the *Trust Cues* theme. This may have been because participants knew—or at least suspected—that the UK Government will make smart meter installation mandatory in the future. When asked, "*What things would you consider when deciding to use or not use this technology service?*", one participant responded, "*I think whether it's optional or not, whether you have got a choice or whether it's part of the contract to have this meter fitted*". If smart metering does become mandatory, environmental cues are likely to have minimal effect on peoples' adoption behaviour.

13.10 Personal Characteristics and Themes Discussed

A quantitative analysis of focus group transcripts was carried out to investigate the hypothesised relationship between participants' personal characteristics (i.e. their attributes and general attitudes to technology adoption and privacy) and the themes discussed in the focus groups. As this required data for each focus group participant from the online survey, only quotations made by the 27 participants who completed the survey were used in the quantitative analysis.

Where required, the personal characteristics from the online survey were converted into categorical variables[92] using the criteria shown. Pearson's chi-square (χ^2) test was used to determine if there was a relationship between the personal characteristics captured by the online survey and the themes raised in the focus groups. Pearson's chi-square test is used as a test of independence of two categorical variables (e.g. a personal characteristic and a theme discussed in the focus groups). The chi-square statistical test calculates the deviations between the actual frequencies observed in each combination of categorical variables and the frequencies which might be expected due to chance. The sum of the standardised deviations between the observed and expected frequencies for each combination of categorical variable results in Pearson's chi-square (χ^2) statistic.

An important criterion for Pearson's chi-square test to be valid is that the expected frequency in each combination of categorical variables is greater than 5. However, in thematic analysis there are likely to be themes with relatively few coded quotations attributed to them, therefore quotations relating to themes representing less than 5 % of the total 599 coded were removed, resulting in nine themes covering 477 quotations. As chi-square tests could only be carried out for participants responding to the relevant survey question used in each chi-squared test, the maximum data set size used for the chi-square test was 420 quotations—70 % of total quotations coded—across nine themes.

To transform the focus group transcript data into a format allowing quantitative analysis, it was exported from ATLAS.ti as an XML file and processed with a Microsoft Excel Visual Basic module developed by the researcher. This created a Microsoft Excel worksheet, which could be imported into SPSS, with each row

[92] These are marked with an asterisk in Table 13.7.

containing a coded quotation, its theme, and information from the survey pertaining to the participant who made the quotation, e.g. age, computer experience, Westin category, etc.

The results in Table 13.7 show the chi-square figure in six of the nine Pearson chi-square tests as significant, supporting an association between certain participant characteristics and the themes discussed in the focus groups. Conventionally, a Pearson chi-square test is considered statistically significant, i.e. two categorical variables are not independent, if the value of $p < 0.05$ (the column headed "p (sig.)" in Table 13.7). The results suggest participants' intrinsic attributes (e.g. age and gender) are not related to the number of quotations within each theme discussed in the focus groups, but there is evidence to support an association with educational level and computer experience. The one exception to this latter category was the amount of time a participant used a computer each day, which did not appear to have a significant association with the themes raised and discussed in the focus groups.

The standardised residuals for the six chi-square tests, which supported a significant association between the personal characteristic category and the theme discussed in the focus groups, were used to understand which themes contributed significantly to the overall association, thus:

- **Educational Level**—Participants in the *Lower* education level category made significantly less quotations relating to the *Trust* theme ($z = -2.1$) than expected—this was the only theme with a significant effect for this categorical variable.
- **Computer Experience (in years)**—There was no theme about which participants made significantly more or fewer quotations.
- **Personal Innovativeness in Information Technology (PIIT)**—Participants in the *Late Adopters* category (PIIT score $< \mu$) made significantly fewer quotations relating to the *Information Receiver* theme ($z = -2.1$) than expected—this was the only theme with a significant effect for this categorical variable.
- **Technology Privacy Concern**—Participants with a *High* TPC score (TPC $>= \mu + \sigma$) made significantly more quotations relating to the *Security* theme ($z = 2.5$), and significantly fewer quotations relating to the *Information Receiver* theme ($z = -2.0$) than expected. Participants with a *Low* TPC score (TPC $<= \mu - \sigma$) made significantly more quotations relating to the *Trust Cues* theme ($z = 2.5$), than expected.
- **Westin Category**—Participants in Westin's *Privacy Unconcerned* category made significantly more quotations relating to the *Trust Cues* theme ($z = 3.2$), than expected; participants in the *Privacy Fundamentalists* category made significantly fewer ($z = -2.3$). Participants in the *Privacy Fundamentalists* category also made significantly more quotations relating to the *Security* theme ($z = 2.2$) than expected.

Table 13.7 Pearson chi-square test results for personal characteristic categories and themes discussed across all focus groups

Personal characteristic category	Categorical variables and criteria	n[a]	df	Chi-square value[b]	p (sig.)	Possible assoc'n?
Gender	Male Female	420	8	12.997	0.112	No
Age*	Under 35 35 and older	420	8	9.442	0.306	No
Educational level*	Lower (< undergraduate) Higher (≥ undergraduate)	420	8	17.727	0.023	Yes
Computer experience*	Low (years < μ) High (years ≥ μ)	420	8	17.715	0.023	Yes
Daily computer use*	Low (hours < μ) High (hours ≥ μ)	420	8	7.742	0.459	No
Personal innovativeness in information technology (PIIT)*	Late adopter (PIIT score < μ) Early adopter (PIIT score ≥ μ)	420	8	21.817	0.005	Yes
Technology privacy concern*	Low (TPC $\leq \mu - \sigma$) Medium ($\mu - \sigma <$ TPC $< \mu + \sigma$) High (TPC $\geq \mu + \sigma$)	411[c]	16	35.813	0.003	Yes
Westin category	Privacy unconcerned Privacy pragmatist Privacy fundamentalist	420	16	36.332	0.003	Yes
Future impact of technology on privacy (Westin)	Better Same Worse	366[c]	16	45.092	< 0.001	Yes

[a] Number of focus group quotations in chi-square analysis
[b] Pearson chi-square test
[c] This was less than the maximum possible 420 quotations as some survey participants responded with "*Don't know*" to this survey item

- **Future Impact of Technology on Privacy (Westin)**—This result should be treated with some caution, as the expected frequency in 29.6 % of the cells in the contingency table was less than 5. This was a large contingency Table (3 × 9), and "[i]*n larger tables the rule is that all expected counts should be greater than 1 and no more than 20 % of expected counts should be less than 5*" (p. 695).[93] For this chi-square test all expected counts were ≥ 4 and $p < 0.001$, suggesting the possibility of a relationship. This personal characteristic also had a significant effect on the largest number of themes. Participants who thought technology would make peoples' privacy better over the next five years made significantly fewer quotations relating to the *Trust* theme ($z = -2.3$) than expected. Participants who thought technology would make peoples' privacy worse over the next five years made significantly fewer quotations relating to the *Trust Cues* theme ($z = -2.0$) than expected. Participants who thought the effect of technology over the next five years would be about the same as it is today, made significantly more quotations relating to the *Trust Cues* theme ($z = 3.3$), and significantly fewer quotations relating to the *Information Usage* theme ($z = -2.5$), than expected.

13.11 Limitations of Study

An obvious limitation of this study was the small sample size of 35 people who took part in the focus groups. However, as this was an exploratory study involving focus groups, and the "[. . .] *common rule of thumb is that most projects consist of four to six focus groups*" (p. 144)[94], an average of six participants in each focus group is reasonable.

Although the sample size for the online survey was also small ($n = 27$), the chi-square tests used to find a relationship between participants' personal characteristics and the themes discussed across the focus groups, used up to 420 coded quotations, split across nine themes. Despite the creation of the themes being data-driven 'from the ground up', significant statistical relationships were found between specific attributes of people (e.g. computer experience, measures of general privacy concern and PIIT), and the number of times specific themes were discussed across the focus groups. However, the chi-square tests only investigated the relationship between two categorical variables—the focus group theme and personal attribute—for all quotations across all focus groups. Such a test cannot provide a probability that an individual with a particular attribute will be the one to raise a particular topic in the group. The analyses also did not differentiate between those quotations which were the first time a particular theme was discussed, and those that were related to further discussions on the same theme.

A major advantage of focus groups—their ability to encourage group level discussion—is potentially one of their major limitations. Participants may behave

---

[93] Field 2009.

[94] Morgan 1996.

differently if faced with the technology service assigned to their focus group in a different context (e.g. using it alone to achieve a specific goal). The *privacy paradox*, in which peoples' stated privacy behaviour is not the same as their actual behaviour, is a well-known phenomenon.[95] However, in the context of a focus group, particularly when discussing a specific technology service, people may be more truthful about their privacy behaviour in front of others who may challenge them and ask for justification of their views.

In focus group discussions people may be reminded by other participants of factors they would not normally consider, and therefore there is a danger of dominant personalities steering a group's discussion—both these biases were mitigated to some extent by the study's design. Firstly, the use of a standard set of three questions, with an approximately similar amount of time allotted to each question, ensured discussion remained focused, and was not hijacked by particular participants. Secondly, the use of an online survey gave participants the chance to provide their views of privacy in a different and solitary context—the data from these two different research methods still resulted in statistically significant relationships.

13.12 Conclusions and Further Work

Information Control was the most frequently discussed theme in the focus groups, with 17 % of all coded quotations. Not only does this lend credence to the idea that people principally seek informational self-determination when engaging with technology services, but also echoes one of the factors—control over collection and usage personal information—in the IUIPC scale.[96] When empirically validating the CFIP scale, Stewart & Segars observe[97]:

> [A] central concern that seems to underlie consumer attitudes, and is perhaps the common theme captured by the higher-order concept of CFIP, is the issue of control. Consumers desire levels of personalization and customization but also want some sense of control over how this service occurs. (p. 46)

The control-based privacy paradigm is a recurring theme in privacy literature[98], and is supported by empirical studies[99], but has attracted some criticism.[100] Furthermore, although definitions of privacy, such has Westin's[101], consider control as an important dimension of privacy, due in part to the importance of individual autonomy in Western culture[102], Laufer & Wolfe suggest "*the privacy phenomenon is conceptually different*

[95] Norberg et al. 2007.

[96] Malhotra et al. 2004.

[97] Stewart and Segars 2002.

[98] Westin 1967; Altman 1976; Fried 1968.

[99] Sheehan and Hoy 2000; Malhotra et al. 2004.

[100] Allen 2000; Tavani 2007.

[101] Westin 1967.

[102] Laufer and Wolfe 1977.

from control/choice" (p. 39), and that control/choice is actually a mediating variable in the privacy system.

An individual's information control is more than the disclosure or non-disclosure of information, but a decision making process in which an individual considers the future consequences of engaging in a particular behaviour—the "*calculus of behavior*".[103] Laufer & Wolfe suggest new technologies affect this calculus, so an "*individual is often unable to predict the nature of that which has to be managed*" (p. 37).[104] Their idea of a calculus of a behaviour underpins Culnan & Bies' observation that this "*social exchange perspective also applies to a consumer context*" (p. 327)[105], i.e. consumers carry out a similar cost-benefit analysis, or what they refer to as a "*privacy calculus*"—the second most discussed theme in the focus groups. Although this implies people consider to some degree, the risks and benefits of providing personal information, the significant percentage (72 %) of coded quotations in the *Privacy Calculus* theme relating to the benefits offered by a technology service, indicates people are principally focused on the benefits they believe they will receive for disclosing personal information.[106] The fact that the *Consequences* theme only accounted for 6 % of all coded quotations, supports the idea that people do not always consider the medium and long-term consequences of disclosing personal information.

The third most discussed theme in the focus groups—*Trust Cues*—not only included coded quotations relating to *trust symbols* and *trust symptoms*[107], but also environmental cues. Of the coded quotations within the *Trust Cues* theme, 55 % related to environmental cues, including the advice of friends, social and technology norms, and media stories, indicating the possible existence of another component of peoples' privacy concern: *environmental privacy concern*.

Although there was insufficient data to statistically support the research objective of investigating if the factors individuals consider are common to all technology services, analysis of 582 of the total of 599 coded quotations does—*prima facie*—support this. For the four most frequently discussed themes: (1) *Information Control*; (2) *Privacy Calculus*; (3) *Trust Cues*; and (4) *Information Usage*, there was a high incidence of these themes representing more than 9 % of the total coded quotations in each of the focus groups. Furthermore, cogent reasons could be found where less than 9 % of coded quotations in each focus group were related to these particular themes. There was also a reasonably evident grouping of the themes into the other two groups: (1) *somewhat discussed* (i.e. between 3 and 9 % of the coded quotations in each group); and (2) *infrequently discussed* (i.e. between 0 and 3 % of the coded quotations in each group). These exploratory findings suggest it may be feasible to abstract the technology service attributes and environmental cues people typically look for—across disparate technologies.

---

[103] Laufer and Wolfe 1977.

[104] Laufer and Wolfe 1977.

[105] Culnan and Bies 2003.

[106] Acquisti 2004.

[107] Riegelsberger et al. 2005.

Table 13.8 Relationship between significant residuals and personal characteristic categories

	High level of privacy concern		Low level of privacy concern	
	High technology privacy concern[a]	Privacy fundamentalists	Low technology privacy concern[a]	Privacy unconcerned
Trust cues		Less	More	More
Trust	*Less*		Less	
Information usage		*More*	Less	*Less*
Information receiver	Less		Less	*Less*
Security	More	More		

[a] In Table 13.8 those in the *High technology privacy concern* category includes participants who believe privacy will get worse in response to Westin's question on the future impact of technology on privacy, and those whose TPC score was $\geq \mu + \sigma$; all other participants were placed in *Low technology privacy concern* category. The two categories in Table 13.8 use the highest standardised residuals for the Westin and TPC categories

Significant statistical relationships were found between the themes raised and discussed in the focus groups and: (1) the attributes of educational level and computer experience; (2) personal innovativeness in information technology (PIIT)[108]; and (3) general privacy concern (including those measured using Westin's categories). For those cases where a significant statistical relationship was found, examination of the standardised residuals helps to explain the relationship. For example, those in the Westin's Privacy Fundamentalists category made significantly more quotations relating to the *Security* theme than expected, but significantly fewer quotations relating to *Trust Cues* theme than expected. Those categorised as Privacy Unconcerned made significantly more quotations relating to the *Trust Cues* than expected.

Table 13.8 shows the relationship between five themes where there were significantly more or less comments made in the focus groups than expected (i.e. $z > \pm 1.96$)[109], and two different types of users: (1) those with a high level of privacy concern; and (2) those with a low level of privacy concern. The results in Table 13.8 suggest there is potentially a type of person with a high level of general privacy concern, who will attach more importance to a technology service's security, and how their personal information might be used, than the advice of friends. Similarly there may be people who are generally unconcerned about privacy and likely to be influenced in their adoption of technology by social privacy norms or the advice of others.

This suggests it may be feasible, with further research, to identify a richer set of *privacy concern types* for groups of people, representing the technology service attributes and environmental cues which each group consider important and therefore look for. This will assist in understanding how *interpersonal privacy concern* and *environmental privacy concern* are constructed.

[108] Agarwal and Prasad 1998.

[109] Table 13.8 also shows those relationships where there is standardised residual between 1.7 and 1.96 in *faint text*. As this table is the result of quantitative analysis of qualitative data it is considered unrealistic to have an absolute cut-off at 1.96.

The exploratory study did not explore the impact of individuals' personality on their level of privacy concern, i.e. *dispositional privacy concern*. However, the significant statistical relationships between peoples' innovativeness and general level of privacy concern, and the themes raised and discussed in the focus groups suggests certain aspects of peoples' personality is likely to determine the technology service attributes and environmental cues they consider important.

Morton & Sasse[110] propose a layered approach—the Privacy Security Trust (PST) Framework—to assist practitioners with effective privacy practice for both the technology platform and providing organisation within a technology service. The layers within their framework are: (1) information security; (2) information management; (3) information principles; (4) information use; and (5) information privacy culture. It is trust signals from each of the PST Framework layers in a technology service's privacy practice, which will be contextual and assist in the construction of an individual's *interpersonal privacy concern*. For example, the *Information Principles Layer* in the PST Framework should encapsulate fair information practices, echoing the CFIP scale of privacy concern with its emphasis on peoples' concerns about organisations' information privacy practices.[111]

Morton & Sasse suggest the trust signals originating from the organisation's privacy practice may become distorted by a badly designed or implemented technology platform leaking personal information. The fact information control was the most commonly discussed topic in the focus groups, highlights the importance of providing users with feedback and control of their personal information, implemented in the technology platform using the tenets of *privacy by design*[112], and seamlessly linked to the organisation's *Information Management Layer* as defined in the PST Framework.

The qualitative analysis of the focus group transcripts suggests individuals are likely to seek out specific technology service attributes, whose absence, or inadequate implementation, will increase their level of *interpersonal privacy concern*. Similarly, the focus group results suggest individuals also take environmental cues into account, which may increase or decrease their level of *environmental privacy concern*. Finally, the quantitative analysis of the focus groups transcripts and survey data suggest certain individual characteristics influence the technology service attributes and environmental cues people consider important.

Acknowledgements Special thanks are owed to all participants who took time to participate in the focus groups, which formed part of this study. Anthony Morton is funded by a PhD scholarship—part of a UK Engineering and Physical Sciences Research Council (EPSRC) grant (EP/G034303/1)—awarded to the Centre for Secure Information Technologies (CSIT) at Queen's University Belfast.

[110] Morton and Sasse 2012.

[111] Smith et al. 1996.

[112] Cavoukian 2009.

Appendix

Focus Group Technology Service Scenarios

Photograph Sharing Web Site (discussed by Group 1) A photograph sharing web site continually runs a software application, which uses facial recognition technology coupled with information from popular social networking sites, to label ('tag') people in all uploaded photographs. If the picture contains a landmark that the software recognises by searching images on the Internet, this is also labelled. For example, if the picture contains identifiable people and a landmark, the picture will be labelled as '*Mr. Fred Smith and Mrs Jane Jones by Big Ben in London*'. If the picture has meta-data within it, the date and time are extracted and appended to the picture's title, e.g. '*Mr. Fred Smith and Mrs. Jane Jones by Big Ben in London on June 21st at 2:30pm*'.

Organisations that manage landmarks, such as Legoland, Woburn Abbey, Tower Bridge, and Edinburgh Castle etc., can subscribe to a service to be sent photographs of people visiting their landmarks, which they show on their web sites. Users registered with the photograph sharing web site, can sign up to a service to get '*2 for 1*' offers on landmarks similar to the one they have been photographed at, with the coupon being e-mailed to all of the people identified by the software application in the photograph who are registered with the photograph sharing web site (whether they have signed up for the '*2 for 1*' offer or not).

Social Networking Discounts (discussed by Group 2) A social networking site for which users must register and create a profile containing their personal information. Registered users can:

- Link to each other by sending invitations.
- Post status messages about themselves.
- Post messages on other users' pages.
- Send private messages to each other.
- Upload photographs.
- Create and join groups with other users.
- Link to content on the Internet they consider worth looking at.

Users can gain 'social networking credits', which they may use as discounts on products sold on affiliated web sites. The amount of credits is based on a user's amount of use of the social networking site, the number of links with other users they have, and the amount of information about themselves they have entered into their profile.

Supermarket RFID Ordering (discussed by Group 3) A supermarket uses RFID (radio frequency identification) chips, which are not disabled at the supermarket checkout, in the product tags on their food goods.

The supermarket is trialling a new automatic ordering service for shoppers who are registered on their home delivery web site. For a single payment of £ 25 the registered customer is given a small RFID reader unit, which is placed near their food cupboards and wirelessly connects with the household's broadband router.

This RFID reader unit continually scans the product tags of goods in the cupboards and e-mails a message to the customer, at a selected frequency (monthly or weekly), containing a list of items no longer in the cupboard (and therefore assumed to be used). The customer can click the *Buy Now* button in the e-mail and replacement goods are delivered to the house at the customer's chosen delivery time. Goods purchased using this e-mail automatically attract a discount and also get priority for delivery times.

Smartphone Assistant (discussed by Group 4) A smartphone assistant software application, which monitors an individual's location and provides information about things nearby which may be of interest, including:

- Events (e.g. concerts, theatre, films etc.)
- Places to visit (e.g. museums, parks etc.)
- Shops selling products an individual might be interested in
- Restaurants
- Clearance sales

To ensure the application provides relevant content, individuals must register and enter information about themselves, their interests and lifestyle. The developers of the application provide these details to other companies to allow them to provide targeted advertisements to the registered users' smartphones. If an individual visits a retail outlet, which is part of the scheme, a coupon code flashes up on the screen that can be used to receive a discount at that retail outlet. If an individual has clicked on an advert on their smartphone and ordered goods online they also receive a discount.

Smart Metering (discussed by Group 5) An electricity company offers its customers the opportunity to have a smart meter installed, which sends back details of electricity consumed by taking readings of electricity consumption every half-hour. The readings are sent via the customer's broadband connection to both the electricity infrastructure provider and the electricity supplier. Customers who have a smart meter installed are given a discount on their electricity bill, every quarter.

If a customer has agreed to have a smart meter installed, they are sent updates via e-mail telling them which appliances are inefficient and therefore costing money to run. The electricity supplier, through its relationship with retailers, can offer discounts on household appliances with better energy efficiency. Customers are sent e-mails with adverts offering these appliances.

Landmark Identification Web Site (discussed by Group 6) A smartphone software application which allows individuals to use the camera built into their mobile phone to take a picture of a landmark and request identification of it.

The software uses images from the Internet to identify the landmark, its name being displayed on the smartphone, along with links to relevant web sites providing more information (e.g. opening hours, special events). This information may also include special offers relating to the landmark, such as 2-for-1 tickets, discounted food, private 'behind the scenes' tours etc.

Table 13.9 Types of focus group quotations covered by each theme

Theme name	Types of quotations coded under theme
Advisories	Information and warnings provided by a technology service concerning privacy and data handling
Information management	Individual's perception of how an organisation manages peoples' information once they are in possession of it
Personal characteristics	How an individual's age and personal experience is perceived to affect their use of technology, views on privacy, trust etc
Miscellaneous	General quotations not related to the research questions
Information sensitivity	An individual's view of the information they consider sensitive within the context of a technology service
Perceived ease of use	The effort an individual has to make to use a technology service; design; and whether use is mandatory
Reliance on legislation	An individual's reliance on legislation to protect them, e.g. data protection, consumer protection etc
Sense making	How individuals avoid/minimize privacy invasion, and use previous experiences or similar situations to understand a technology service
Security	The technology service's security, and organisations' physical and information security
Consequences	The impact on an individual's personal security, finances or behavior of using a technology service
Information receiver	Organisations' ability to provide the technology service, its objectives and its characteristics
Trust	Technological and organisational trust
Information usage	Use of information by organisations for location tracking, behavior profiling, and targeted advertising
Trust cues	How individuals use news stories, reviews, third parties etc. to aid their decision to engage with a technology service; social and technological norms; and trust symbols and trust symptoms
Privacy calculus	Individuals' views of the benefits a technology service offers, and the decision process individuals undertake when considering the potential benefits *vs.* private information that has to be provided
Information control	The information control offered by a technology service (e.g. opt-in/opt-out, feedback and control), organisations passing information to third parties without authorisation, and how individuals control information disclosure

The software is free, but to download it and continue using it, you must register and provide links to your profile on social networking sites you use such as Facebook, LinkedIn etc.

This same smartphone software also allows individuals to take a picture of a person on the street, and using facial recognition technology coupled with information from popular social networking sites, provide the name of the person in the picture.

A link is provided to the web site(s) so the user can find out any other publicly available information about the person, such as address, job title etc. (where this can be found).

Qualitative Analysis Themes

During qualitative analysis of focus group transcripts 39 base-level codes were created—grouped into 16 themes—shown in Table 13.9 with a description of the types of quotations coded within each theme.

References

6, Perri, Kristen Lasky, and Adrian Fletcher. 1998. *The future of privacy—Public trust and the use of private information.* Vol. 2. 2 vols. Demos. http://www.demos.co.uk/files/thefutureofprivacyvolume2.pdf.

Acquisti, Alessandro. 2004. Privacy in electronic commerce and the economics of immediate gratification. Presented at the Proceedings of the 5th ACM conference on Electronic commerce—EC '04, New York, NY, USA, 2004, 21, doi:10.1145/988772.988777.

Acquisti, A., and J. Grossklags. 2005. Privacy and rationality in individual decision making. *IEEE Security and Privacy Magazine* 3 (1): 26–33. doi:10.1109/MSP.2005.22.

Adams, A., and M. Angela Sasse. 2001. Privacy in multimedia communications: Protecting users, not just data. In *People and Computers XV: Interaction without Frontiers*, eds. A. Blandford, J. Vanderdonckt and P. Gray, 49–64. London: Springer.

Agarwal, R., and J. Prasad. 1998. A conceptual and operational definition of personal innovativeness in the domain of information technology. *Information Systems Research* 9 (2): 204–215. doi:10.1287/isre.9.2.204.

Agarwal, Ritu, and Jayesh Prasad. 1999. Are individual differences germane to the acceptance of new information technologies? *Decision Sciences* 30:361–392.

Allen, Anita L. 2000. Privacy-as-data control: Conceptual, practical, and moral limits of the paradigm. *Connecticut Law Review* 32:861.

Altman, Irwin. 1976. Privacy—A conceptual analysis. *Environment and Behavior* 8 (1): 7–29. doi:10.1177/001391657600800102.

Ashford, Warwick. US woman sues google over gmail scanning. *ComputerWeekly.com*, August 11, 2011. http://www.computerweekly.com/news/2240105327/US-woman-sues-Google-over-Gmail-scanning.

Barnett, Emma. Google street view: Survey raises privacy concerns. *Telegraph.co.uk*, March 12, 2010, sec. Technology. http://www.telegraph.co.uk/technology/google/7430245/Google-Street-View-survey-raises-privacy-concerns.html.

Barth, Adam, Anupam Datta, John C. Mitchell, and Helen Nissenbaum. 2006. Privacy and contextual integrity: Framework and applications. In *Security and Privacy, 2006 IEEE Symposium On*, p. 15.

BBC. Google's street view under fire. *BBC News*, July 9, 2008. http://news.bbc.co.uk/1/hi/sci/tech/7498613.stm.

BBC. Google buzz 'breaks privacy laws'. *BBC News*, February 17, 2010. http://news.bbc.co.uk/1/hi/technology/8519314.stm.

BBC. Facebook sorry over tagging launch. *BBC News*, June 8, 2011. http://www.bbc.co.uk/news/technology-13693791.

Beldad, Ardion, Menno de Jong, and Michaël Steehouder. 2011. A comprehensive theoretical framework for personal information-related behaviors on the internet. *The Information Society* 27 (4): 220–232. doi:10.1080/01972243.2011.583802.

Bellman, Steven, Eric J. Johnson, and Gerald L. Lohse. 2001. To opt-in or opt-out? It depends on the question. *Communications of the ACM* 44 (2): 25–27.

Braun, Virginia, and Victoria Clarke. 2006. Using thematic analysis in psychology. *Qualitative Research in Psychology* 3 (2): 77–101.

Camp, L. Jean, Helen Nissenbaum, and Cathleen McGrath. 2002. Trust: A collision of paradigms. *Financial Cryptography* 2339:91–105.

Cavoukian, Ann. 2009. *Privacy by Design*. Ontario: Office of the Information and Privacy Commissioner. http://www.ipc.on.ca/images/Resources/privacybydesign.pdf.

Coles-Kemp, Lizzie, Lai Yee-Lin, and Margaret Ford. 2010. *Privacy on the internet: Attitudes and behaviours*. VOME (Royal Holloway—Information Security Group). http://www.vome.org.uk/wp-content/uploads/2010/03/VOME-exploratorium-survey-summary-results.pdf.

Culnan, Mary J., and Robert J. Bies. 2003. Consumer privacy: Balancing economic and justice considerations. *Journal of Social Issues* 59 (2): 323–342.

Dinev, Tamara, and Paul Hart. 2006. An extended privacy calculus model for e-commerce transactions. *Information Systems Research* 17 (1): 61–80. doi:10.1287/isre.1060.0080.

El Emam, K., E. Neri, E. Jonker, M. Sokolova, L. Peyton, A. Neisa, and T. Scassa. 2010. The inadvertent disclosure of personal health information through peer-to-peer file sharing programs. *Journal of the American Medical Informatics Association* 17 (2): 148.

European Commission (EUROPA). Data protection: Europeans share data online, but privacy concerns remain—new survey. *europa.eu*, June 16, 2011. http://europa.eu/rapid/pressReleasesAction.do?reference=IP/11/742&format=HTML&aged=0&language=EN&guiLanguage=en.

Farrell, Nick. Google admits it sniffed out people's data. *TechEye.net*, May 17, 2010. http://www.techeye.net/security/google-admits-it-sniffed-out-peoples-data.

Federal Trade Commission. 2000. *Privacy online: Fair information practices in the electronic marketplace—A Report to Congress*, May 2000. http://www.ftc.gov/reports/privacy2000/privacy2000.pdf.

Federal Trade Commission. 2010. Widespread data breaches uncovered by FTC probe. *Federal Trade Commission*, February 22, 2010. http://www.ftc.gov/opa/2010/02/p2palert.shtm.

Field, Andy. 2009. *Discovering statistics using SPSS*. 3rd ed. SAGE Publications Ltd.

Fiveash, Kelly. 2007. MoveOn tells facebook to stop shining beacon. *The Register*, November 21, 2007. http://www.theregister.co.uk/2007/11/21/facebook_moveon_privacy_beacon/.

Fried, Charles. 1968. Privacy. *The Yale Law Journal* 77 (3): pp. 475–493.

Gefen, David, Elena Karahanna, and Detmar W Straub. 2003. Trust and TAM in online shopping: An integrated model. *MIS Quarterly* 27 (1): pp. 51–90.

Harper, J., and S. Singleton. 2001. *With a grain of salt: What consumer privacy surveys don't tell us*. Competitive Enterprise Institute & The Cato Institute, June 2001. http://www.slis.indiana.edu/faculty/hrosenba/www/l574/pdf/harper_privacy-surveys.pdf.

Hundley, Heather L., and Leonard Shyles. 2010. US teenagers' perceptions and awareness of digital technology: A focus group approach. *New Media & Society* 12 (3): 417–433. doi:10.1177/1461444809342558.

Johnson, M. Eric. 2008. Information risk of inadvertent disclosure: An Analysis of file-sharing risk in the financial supply chain. *Journal of Management Information Systems* 25 (2): 97–124.

Junglas, Iris A, Norman A Johnson, and Christiane Spitzmüller. 2008. Personality traits and concern for privacy: An empirical study in the context of location-based services. *European Journal of Information Systems* 17 (4): 387–402. doi:10.1057/ejis.2008.29.

Kitzinger, Jenny. 1995. Introducing focus groups. *BMJ: British Medical Journal* 311 (7000): 299–302.

Korzaan, Melinda L., and Katherine T. Boswell. 2008. The influence of personality traits and information privacy concerns on behavioral intentions. *The Journal of Computer Information Systems* 48 (4): 15–24.

Kumaraguru P. and L. F. Cranor. 2005. Privacy Indexes: A Survey of Westin's Studies. *Technical Report CMU-ISRI-05-138*, Institute for Software Research International, School of Computer Science, Carnegie Mellon University, December.

Laufer, Robert S., and Maxine Wolfe. 1977. Privacy as a concept and a social issue: A multidimensional developmental theory. *Journal of Social Issues* 33 (3): 22–42.

Leavitt, Lydia. 2011. Mobile app makers probed over privacy concerns. *TG Daily*, April 5, 2011. http://www.tgdaily.com/business-and-law-features/55204-mobile-app-makers-probed-in-privacy-investigation.

Liu, Chang, Jack T. Marchewka, June Lu, and Chun-Sheng Yu. 2005. Beyond concern—A privacy-trust-behavioral intention model of electronic commerce. *Information & Management* 42 (2): 289–304. doi:10.1016/j.im.2004.01.003.

Malhotra, Naresh K., Sung S. Kim, and James Agarwal. 2004. Internet users' information privacy concerns (IUIPC): The construct, the scale, and a causal model. *Information Systems Research* 15 (4): 336–355. doi:10.1287/isre.1040.0032.

McKnight, D. Harrison, and Norman L. Chervany. 2001. "What trust means in e-commerce customer relationships: An interdisciplinary conceptual typology." *International Journal of Electronic Commerce* 6 (2): 35–59.

McKnight, D. Harrison, Vivek Choudhury, and Charles Kacmar. 2002. Developing and validating trust measures for e-commerce: An integrative typology. *Information Systems Research* 13 (3): 334–359.

Mennecke, Thomas. 2007. Pfizer P2P security breach. *Slyck News*, June 20, 2007. http://www.slyck.com/story1496_Pfizer_P2P_Security_Breach.

Metzger, Miriam J. 2004. Privacy, trust, and disclosure: Exploring barriers to electronic commerce." *Journal of Computer-Mediated Communication* 9 (4). doi:10.1111/j.1083-6101.2004.tb00292.x.

Mills, Elinor. 2007. Google's street-level maps raising privacy concerns. *USA Today*, June 4, 2007. http://www.usatoday.com/tech/news/internetprivacy/2007–06-01-google-maps-privacy_N.htm.

Morgan, D. L. 1996. Focus groups. *Annual Review of Sociology* 22:129–152.

Morgan, David. L, and Richard A. Krueger. 1993. When to use focus groups and why. In *Successful focus groups: Advancing the state of the art*, Vol. 1. Sage Publications, Inc, pp. 3–19.

Morton, Anthony, and M. Angela Sasse. 2012. Privacy is a process, not a PET: A theory for effective privacy practice. In *Proceedings of the 2012 Workshop on New Security Paradigms*, 87–104. NSPW '12. New York, NY, USA: ACM, 2012. doi:10.1145/2413296.2413305.

NBC. 2009. New warnings on cyber-thieves. *TODAY Investigates*. United States: NBC, February 26, 2009. http://www.today.com/id/26184891/vp/29405819#29405819.

Nissenbaum, Helen. 2004. Privacy as contextual integrity. *Washington Law Review* 79 (1): 119.

Nissenbaum, Helen. 2011. A contextual approach to privacy online. *Daedalus* 140 (4): 32–48.

Norberg, Patricia A., Daniel R. Horne, and David A. Horne. 2007. The privacy paradox: Personal information disclosure intentions versus behaviors. *Journal of Consumer Affairs* 41 (1): 100–126. doi:10.1111/j.1745–6606.2006.00070.x.

Panzarino, Matthew. 2011. It's not just the iphone, android stores your location data too. *TNW—The Next Web*, April 21, 2011. http://thenextweb.com/google/2011/04/21/its-not-just-the-iphone-android-stores-your-location-data-too/.

Phelps, Joseph, Glen Nowak, and Elizabeth Ferrell. 2000. Privacy concerns and consumer willingness to provide personal information. *Journal of Public Policy & Marketing* 19 (1): 27–41. doi:10.2307/30000485.

Quinn, Ben, and Charles Arthur. 2011. PlayStation network hackers access data of 77 million users. *The Guardian*, April 26, 2011. http://www.guardian.co.uk/technology/2011/apr/26/playstation-network-hackers-data?intcmp=239.

Riegelsberger, Jens, M. Angela Sasse, and John D. McCarthy. 2005. The mechanics of trust: A framework for research and design. *International Journal of Human-Computer Studies* 62 (3): 381–422.

Rotter, Julian B. 1967. A new scale for the measurement of interpersonal trust. *Journal of Personality* 35 (4): 651–665. doi:10.1111/j.1467–6494.1967.tb01454.x.

Sarno, David. 2010. Apple collecting, sharing iphone users' precise locations [Updated]. *Los Angeles Times*, June 21, 2010, sec. Business. http://latimesblogs.latimes.com/technology/2010/06/apple-location-privacy-iphone-ipad.html.

Sheehan, Kim Bartel. 2002. Toward a typology of internet users and online privacy concerns. *The Information Society* 18 (1): 21–32. doi:10.1080/01972240252818207.
Sheehan, Kim Bartel, and Mariea Grubbs Hoy. 2000. Dimensions of privacy concern among online consumers. *Journal of Public Policy & Marketing* 19 (1): 62–73.
Skinner, Harvey, Sherry Biscope, Blake Poland, and Eudice Goldberg. 2003. How adolescents use technology for health information: Implications for health professionals from focus group studies. *Journal of Medical Internet Research* 5 (4). doi:10.2196/jmir.5.4.e32.
Smith, H. Jeff, Sandra J. Milberg, and Sandra J. Burke. 1996. Information privacy: Measuring individuals' concerns about organizational practices. *MIS Quarterly* 20 (2): 167–196.
Snell, K., J. Starkbaum, G. Lauß, A. Vermeer, and I. Helén. 2012. From protection of privacy to control of data streams: A focus group study on biobanks in the information society. *Public Health Genomics* 15 (5): 293–302. doi:10.1159/000336541.
Stewart, Kathy A., and Albert H. Segars. 2002. An empirical examination of the concern for information privacy instrument. *Information Systems Research* 13 (1): 36–49.
Tan, Felix B., and Paul Sutherland. 2004. Online consumer trust: A multi-dimensional model. *Journal of Electronic Commerce in Organizations (JECO)* 2 (3): 40–58.
Tavani, Herman T. 2007. Philosophical theories of privacy: implications for an adequate online privacy policy. *Metaphilosophy* 38 (1): 1–22. doi:10.1111/j.1467–9973.2006.00474.x.
TechCrunchTV. Crunchies Awards 2010. *Mike Arrington Interrogates Mark Zuckerberg*. Las Vegas, January 9, 2010. http://static-cdn1.ustream.tv/swf/live/viewer:55.swf?vid=3848950&vrsl=c:170.
Venkatesh, Viswanath, and Michael G. Morris. 2000. Why don't men ever stop to ask for directions? Gender, social influence, and their role in technology acceptance and usage behavior. *MIS Quarterly* 24 (1): 115–139.
VOME. 2012. Citizen-centric privacy by design. http://www.vome.org.uk/wp-content/uploads/2012/06/citizen-centric-privacy-by-design.pdf.
Westin, Alan F. 1967. *Privacy and freedom*. [1st ed.]. New York, NY, USA: Atheneum.
Zhang, Harry, Claudia Guerrero, David Wheatley, and Young Seok Lee. 2010. Privacy issues and user attitudes towards targeted advertising: A focus group study. In *Human Factors and Ergonomics Society Annual Meeting Proceedings*, 54:1416–1420.

Chapter 14
Data Mining and Its Paradoxical Relationship to the Purpose Limitation Principle

Liana Colonna

14.1 Introduction

Fair information principles have played a significant role in structuring privacy statutes around the world. These principles seek to ensure the legitimate collection and use of personal information. They are designed to empower the individual by giving him/her rights to control the processing of his/her data. Through exercising control over personal data, an individual is ostensibly able to decide when and how information about him/her is revealed to the general public. This helps preserve, among other things, a zone of solitude where an individual can develop his/her sense of self.[1]

One of the problems, however, with the current reliance on fair information principles is that they are increasingly challenged by the technological reality. One technology that seems particularly disruptive is that of data mining. As early as 1998, commentators have noted that there is quite a paradoxical relationship between data mining and some data protection principles.[2] This paper seeks to explore this so-called paradoxical relationship further and to specifically examine how data mining calls into question the purpose limitation principle: the idea that personal data must be collected for specified, explicit and legitimate purposes and not further processed in a way incompatible with those purposes.[3]

The outline of the paper is as follows: first, it will explore the theoretical foundation of the purpose limitation principle, which is deeply rooted in the notion of privacy as control. Then it will explain the advancing technology of data mining in order to highlight the specific attributes of the technology that run counter to the purpose

[1] Westin 1967.
[2] Cavoukian 1998; de Hert and Bellanova 2008.
[3] Article 6(1)(b) of EU Data Protection Directive 95/46/EC; Article 5(b) CoE Convention for the Protection of Individuals with regard to Automatic Processing of Personal Data; Article 3 of EU Framework Decision 2008/977/JHA.

L. Colonna (✉)
Liane Colonna/Castenfors, Skeppargatan 55, 11459 Stockholm, Sweden
e-mail: liane.colonna@juridicum.su.se

S. Gutwirth et al. (eds.), *Reloading Data Protection*,
DOI 10.1007/978-94-007-7540-4_14,

limitation principle. Next the paper will elaborate upon the contradictory relationship that exists between the purpose limitation principle and data mining. The paper will conclude with a discussion about how to cope with the paradoxical situation such as by moving away from the extensive reliance on the fair information principles towards more of an abuse-centered regulatory approach that focuses on preventing the misuse of personal data instead of delivering notice and obtaining prior authorization to process personal data.

14.2 Privacy as Control

The theoretical foundation for the purpose limitation principle is steeped in the notion of privacy as informational control. In *Privacy and Freedom*, Westin sets forth one of the most prominent articulations of this theory.[4] In this landmark book, Westin analyzes the nature and functions of privacy, its roles in society, and the challenges posed to privacy by advancing technologies.[5] Westin's work has subsequently informed most scholarly discussions of privacy and most privacy regulations enacted around the world since its publication in 1967.

Westin defines privacy as "the claims of individuals, groups, or institutions to determine for themselves when, how and to what extent information about them is communicated to others."[6] This formulation of privacy can be considered an early expression of the fair information principles. Essentially, the notion of privacy as control, and the fair information principles in turn, seek to secure privacy through processes that are fair and provide individuals with the capacity to make fully informed decisions about their personal information.[7]

As part of the overall theory put forth in the book, Westin elaborates upon four central functions of individual privacy that reflect the value of the concept within society, the first of which is personal autonomy.[8] He explains that autonomy is at risk when an individual's inner zone is penetrated and his/her ultimate secrets are discovered because an individual's "psychological armor, would leave him (her) naked to ridicule and shame and would put him (her) under the control of those who knew his secrets."[9] He also explains that autonomy is threatened when individuals do not have space to develop their individuality, which requires time for sheltered experimentation and the testing of ideas.[10]

---

[4] Westin 1967; *see also*, Fried 1968 ("Privacy is ... the control we have over information about ourselves."); Miller 1971 ("the basic attribute of an effective right of privacy is the individual's ability to control the circulation of information relating to him.").

[5] Westin 1967.

[6] Westin 1967.

[7] Mulligan and King 2012.

[8] Westin 1967, p. 33.

[9] Westin 1967, p. 33.

[10] Westin 1967, p. 34.

The second function of individual privacy described by Westin is emotional release, which means, among other things, that individuals must have moments to be "off stage": tender, angry, irritable, lustful, or dream-filled.[11] The third function of individual privacy is self-evaluation. That is, every individual needs space to integrate his/her experiences into a meaningful pattern and to exert his individuality on events.[12] The fourth function of privacy is to afford limited and protected communication which entails both the need for the individual to share confidences with those he trusts and to "set necessary boundaries of mental distance in interpersonal situations ranging from the most intimate to the most formal and public."[13]

These functions of the private life are maintained through what Westin describes as four basic states of individual privacy.[14] The first, and most complete, state of privacy is solitude; here, he explains, the individual is separated from the group and freed from the observation of other persons.[15] In the second state of privacy, intimacy, the individual acts as a part of a small unit that seeks to achieve a close, relaxed, and frank relationship among each other.[16] The third state of privacy, anonymity, occurs when "the individual is in public places or performing public acts but still seeks, and finds, freedom from identification and surveillance."[17] The fourth and most subtle state of privacy is reserve: the creation of a psychological barrier against unwanted intrusion.[18]

It is important to note that many legal scholars reject the notion that privacy can be entirely understood as control over information by the individual. First, it is contended that the notion of privacy as control is too vague, too narrow and/or too broad (Solove 2008). It is too vague in that there is neither a clear understanding of the scope of what constitutes personal information over which an individual should exercise control nor a clear understanding of what precisely individual control actually entails.[19] It is too narrow because it excludes aspects of life having little to do with personal information that should, at least arguably, be considered private such as, for example, a woman's decision to have an abortion.[20] It is too broad when "personal information" is defined in an expansive manner.[21] That is, not every piece of personal information is necessarily private.[22]

---

[11] Westin 1967, p. 35.

[12] Westin 1967, p. 36.

[13] Westin 1967, p. 38.

[14] Westin 1967, pp. 31–32.

[15] Westin 1967, pp. 31–32.

[16] Westin 1967, pp. 31–32.

[17] Westin 1967, pp. 31–32.

[18] Westin 1967, pp. 31–32.

[19] Solove 2008.

[20] Allen 1988.

[21] Solove 2002.

[22] Solove 2002; *see also*, Inness 1996.

Second, it is contended that privacy as control improperly rests on the notion that personal information can be construed as property.[23] Schwartz explains that "(p)rivacy-control . . . encourages a property approach to personal information that transforms data into a commodity."[24] The problem here is that once privacy is equated with property it receives a price tag, which seemingly contradicts the notion of privacy as a human right that should not be traded away to the highest bidder.[25]

Theorists further reject the privacy-control paradigm because it is too focused on the individual.[26] It has been argued, for example, that the control-over information conception incorrectly assumes that privacy is a subjective matter of individual prerogative and fails to take account of the fact that privacy is also an issue of what society deems appropriate to protect.[27] More specifically, Regan contends that the notion of privacy as control does not recognize that privacy is important for a myriad of social interests such as the full functioning of democratic institutions.[28]

Additionally, it is contended that the notion of privacy as control incorrectly assumes that individuals have the ability to exercise control over their data in the first place. Schwartz questions whether individuals are able to exercise meaningful choices with regard to their information, given the disparities in knowledge and power in bargaining over the transfer of their information.[29] He argues that control over personal information is illusory because people are not only uninformed about all the ways their data can be used but also because they are often powerless to make demands on data controllers. In other words, he suggests that people do not necessarily have the education or the empowerment to always exercise meaningful control over their personal data.

Finally, the privacy-control paradigm is rejected, at least partly, because it fails to address many of the concerns raised by modern technology. Privacy as control does not account for situations, such as participation in social media, where "individuals feel pressured to reveal their private information because others have done so and a social or economic stigma will attach if they stay quiet."[30] Nissenbaum contends a more contextual understanding of privacy is necessary.[31] That is, she argues that privacy is not necessarily control but rather respect for context-relative informational norms.[32]

---

[23] Schwartz 2000.

[24] Schwartz 2000.

[25] *See generally*, Cohen 2000.

[26] Nissenbaum 2010, p. 71 (raising the question, "(h)as a person who intentionally posts photographs of himself to a Web site such as Flickr lost privacy?").

[27] Nissenbaum 2010, p. 71 (raising the question, "(h)as a person who intentionally posts photographs of himself to a Web site such as Flickr lost privacy?"); *citing* Schoeman 1992.

[28] Regan 1995.

[29] Schwartz 1999.

[30] Peppet 2012.

[31] Nissenbaum 2010, p. 71 (raising the question, "(h)as a person who intentionally posts photographs of himself to a Web site such as Flickr lost privacy?").

[32] Nissenbaum 2010, p. 71 (raising the question, "(h)as a person who intentionally posts photographs of himself to a Web site such as Flickr lost privacy?").

14.3 The Purpose Limitation Principle

14.3.1 Understanding the Concept

At the outset, it must be made clear that it is the European Union (EU) that applies the phrase "purpose limitation." Other regimes use similar but different terminology. For example, the OECD Guidelines refers to "purpose specification."[33] It is not obvious whether "purpose specification" and "purpose limitation" have the same meaning and so as to avoid any confusion: this paper is written from a European perspective.[34]

The purpose limitation principle is expressly laid down in Article 6(1)(b) of the EU Data Protection Directive 95/46/EC which, requires that personal data must be "collected for specified, explicit and legitimate purposes and not further processed in a way incompatible with those purposes."[35] More specifically, it requires a data controller to provide a specific and lawfully-stated reason to the data subject as to why it is collecting his/her data. This explanation should be made explicit no later than at the time of the data collection. The principle further requires that the data controller only use the data for the purposes, which were stated to the data subject at the time the data was collected.

There are, however, a number of different ways that the subsequent use of the data can be extended beyond the fulfillment of the initial purposes stated to the data subject. For example, data can be further processed if the subsequent use is compatible with the initial purpose. Because there is no definition of what exactly constitutes a "compatible purpose" this determination is made on a case-by-case basis. Additionally, derogations from this principle are permitted when such derogations are authorized by law and constitute a necessary measure in a democratic society.[36]

---

[33] Guidelines on the Protection of Privacy and Transborder Flows of Personal Data, adopted Sept. 23, 1980 (explaining "(t)he purposes for which personal data are collected should be specified not later than at the time of data collection and the subsequent use limited to the fulfillment of those purposes or such others as are not incompatible with those purposes and as are specified on each occasion of change of purpose.").

[34] Cate 2006 (raising the following serious of questions: "What is the difference between 'collection limitation,' 'purpose specification,' and 'use limitation,' all three of which appear in the OECD Guidelines, and how do they compare with 'purpose limitation' as that term is used to describe the EU directive? Does the latter include all three of the former?").

[35] Article 6(1)(b) of Directive 95/46/EC; *see also*, Article 3 of Framework Decision 2008/977/JHA and Article 5(b) of the Council of Europe Convention for the Protection of Individuals with regard to Automatic Processing of Personal Data, ETS 108 (1981).

[36] *See generally*, Article 9 of the Council of Europe Convention for the Protection of Individuals with regard to Automatic Processing of Personal Data, ETS 108 (1981) (stating, "(d)erogation from (the purpose limitation principle) shall be allowed when such derogation is provided for by the law of the Party and constitutes a necessary measure in a democratic society in the interests of: protecting State security, public safety, the monetary interests of the State or the suppression of criminal offences (sic); protecting the data subject or the rights and freedoms of others."); *see also*, *Rotaru v Romania*, ECtHR, judgment of 4 May 2000 (Para. 55) (adding added another three conditions, notably that violations of privacy should also be precise, foreseeable and proportionate).

The Article 29 Working Party has recently provided guidance on the purpose limitation principle. In its opinion on purpose limitation, it provides practical examples for understanding the principle.[37] One particularly helpful example concerns a situation where a customer contracts with an online retailer to deliver a box of organic vegetables to the customer's house each week.[38] The Working Party explains, "(a)fter initial 'collection' of the customer's address and banking information, these data are 'further processed' by the retailer each week for payment and delivery. This obviously complies with the principle of purpose limitation and requires no further analysis."[39] If, however, the online retailer seeks to use the customer's email address and purchase history to send the individual personalized offers or to forward the customer's data to a business colleague then the Working Party explains that the compatibility is not obvious and needs further analysis.[40]

The purpose limitation principle is deeply embedded in the theory of privacy as informational control, which presupposes that an individual can only have full authority over his/her personal information when he/she understands the purpose of the data processing before turning it over to a data controller. By requiring that an individual be told why his/her personal data is gathered and in which ways the data can be processed and especially by who, the purpose limitation principle seeks to place the individual in a position to decide whether a specific information transfer is worth it to him/her.[41] The idea is to fully inform the individual about the data processing so that he/she can either provide consent or object to the processing. Theoretically, by affording the individual an opportunity to consent or object after being made aware of the purpose of processing, the individual is able to safeguard his/her privacy by, for example, limiting his/her exposure to unwanted public observation and ridicule.[42]

Essentially, the core value of the purpose limitation principle lies in its ability to provide limits on how a data controller will use a piece of personal information so that an individual can assess and foresee the dangers of giving up this information in the first place.[43] In this respect, the principle can be understood as a shield to protect an individual against unwanted intrusions into his/her private life. It can also be understand as a sword to ground a cause of action against a data controller who violates its promises to the individual.

---

[37] Article 29 Data Protection Working Party: Opinion 03/2013 on purpose limitation (2 April 2013) available online at http://idpc.gov.mt/dbfile.aspx/Opinion3_2013.pdf.

[38] Article 29 Data Protection Working Party: Opinion 03/2013 on purpose limitation (2 April 2013) available online at http://idpc.gov.mt/dbfile.aspx/Opinion3_2013.pdf .

[39] Article 29 Data Protection Working Party: Opinion 03/2013 on purpose limitation (2 April 2013) available online at http://idpc.gov.mt/dbfile.aspx/Opinion3_2013.pdf .

[40] Article 29 Data Protection Working Party: Opinion 03/2013 on purpose limitation (2 April 2013) available online at http://idpc.gov.mt/dbfile.aspx/Opinion3_2013.pdf .

[41] *See generally*, *Rotaru v Romania*, ECtHR, judgment of 4 May 2000; *see also*, *Leander v. Sweden*, ECtHR, judgment of 26 March 1987.

[42] Westin, Alan. 1967. *Privacy and freedom*. New York: Atheneum.

[43] ECtHR, *Peck v. the United Kingdom* (Para. 62); *see also*, Solove 2011.

Take, for example, the situation where an individual relinquishes personal information to a private entity, say a social media company like Facebook. The individual has agreed that this transfer of information is worth the perceived privacy risks: he/she is comfortable providing his/her name, email address, a few photos etc. to the private company so that he/she can connect better with his/her friends. At the same time, however, he/she expects that this information will not be further processed without his/her consent or for incompatible purposes unless, of course, there is a very good reason to do so. In other words, if the man or woman knew that the data collected by the social media company might be used by his/her local bank to determine his/her creditworthiness and to deny him/her a loan then he/she might not decide to join the club after all (or, he/she might have a cause of action against the company).[44]

14.3.2 A Level of Abstraction

It is important to mention, before proceeding to an analysis of how technology challenges the purpose limitation principle, that this principle can be criticized as fairly abstract and thus problematic even before concerns about technology are raised.[45] One study found that the principle is open to divergent applications. Specifically, it concluded, "different EU Member States apply different tests ... ranging from the 'reasonable expectations' of the data subject, to 'fairness' or the application of various 'balance' tests."[46]

The study further discovered that, in a few Member States, the principle is subject to quite sweeping exemptions, in particular for public-sector controllers. In other EU Member States, the study concluded, purposes are sometimes defined in excessively broad terms, thus undermining the principle itself.[47] For example, if the initial purpose for collection of the personal data is marked broadly as "for any law enforcement purpose" than the police and judicial authorities are able to process (and further process) the data for a variety of purposes.[48] These purposes can range from the prevention to the investigation to the detection to the prosecution of specific criminal offences without ever technically deviating from the purpose limitation principle.[49]

It is also problematic that the EU Data Protection Directive does not define what is meant by "compatible use." As such, the term has been left open for interpretation by

[44] White 2012.

[45] *See* Cate 2006.

[46] Korff and Brown 2010.

[47] Korff and Brown 2010 (explaining that " ... UK law refers to 'policing purposes' in one breath (and thus allows data obtained for one police purpose to be used for any such purpose), where German law strictly distinguishes between 'countering immediate threats', 'general and specific prevention', and 'investigation and prosecution of [suspected] criminal offences.'").

[48] Colonna 2012.

[49] Colonna 2012.

the various Member States raising the question of whether the notion of "compatible use" can be considered synonymous with "any use." Indeed, the European Data Protection Supervisor has noted that "the notion of 'compatible use' is interpreted differently in various Member States" and it has called for additional precision in the expression.[50]

14.4 Data Mining

14.4.1 A Solution to Information Overload

Every day, quintillions of bytes of data are created. The amount of data is so growing so fast that scientists are no longer talking in terms of megabytes, gigabytes, terabytes or even petabytes, but have, instead, had to create new terms such as "zettabyte" to describe it.[51] This data comes from everywhere: transactional records captured by payment providers, location information tracked by mobile phone companies, posts to social media sites, clinical diagnoses captured by hospitals, to name just a few. Because much of this data is in digital form "they can be stored, shared, searched, combined, and duplicated with extraordinary speed and at very little cost."[52] It is also significant that this data is often embedded with metadata, data about data, which can reveal a tremendous amount about a person's life and habits.[53]

The ability to generate data has to a large extent outstripped the human ability to do useful things with it.[54] As a result, some data collected in large repositories have become data tombs, "data stores that are effectively write-only; data is deposited to merely rest in peace, since in all likelihood it will never be accessed again."[55] The promise of data mining is to turn these "data tombs" into "golden nuggets"

---

[50] EDPS Opinion on the data protection reform package (March 7, 2012).

[51] A "zettabyte" is equivalent to about 250 billion DVDs. *For more, see* Arthur 2011; *see also,* Kuner et al. 2012.

[52] Kuner et al. 2012.

[53] Kuner et al. 2012 (explaining that metadata is "data about when and where and how the underlying information was generated."); *see also,* Biersdorfer 2006 (explaining that (m)etadata, a term created by the fusion of an ancient Greek prefix with a Latin word, has come to mean "information about information" when used in technology and database contexts. The Greek meta means behind, hidden or after, and refers to something in the background or not obviously visible, yet still present. Data, the Latin term, is factual information used for calculating, reasoning or measuring.): *see also,* Government Surveillance 2012 (explaining that "(m)etadata (the records of who people call and e-mail, and when, as distinct from the content of conversations) can now be amassed on a vast scale, and run through powerful software that can use it to create a fairly complete portrait of a person's life and habits—often far more complete than just a few recorded conversations.).

[54] Han and Kamber 2001.

[55] Fayyad and Uthurusamy 2002.

of knowledge.[56] And, as inexpensive and widely distributed computing capacity continues to soar, the promise becomes more and more likely to be fulfilled.[57]

14.4.2 A Semantic Muddle

At the moment, there is no widely held understanding of what precisely data mining means.[58] The nebulousness of the term data mining is demonstrated by pointing to a few of the terms that carry a similar or slightly different meaning to it such as knowledge mining from databases, knowledge extraction, data archaeology, data/pattern/predictive analysis, data fishing, data dredging, big data analysis and analytics.[59] These terms are, among others, used rather interchangeably. The problem with the free flow and ad hoc use of these terms is that confusion arises because all of the words do not necessarily mean the same thing: important perceptions about the technology are lost when the different meanings are blurred together.

Because the term data mining is the most popular term applied in the media, in the scientific research field, in the business community and even in the legal discourse, it will be the term adopted here.[60] However, the term is applied broadly and perhaps imprecisely. While debating the terminological haze of data mining is important, the goal here is to focus on those attributes of this advancing technology that separate it from more traditional forms of data analysis tools, such as structured queries or statistical analysis software, in order to pinpoint what has changed in the world of data processing that might require legal attention be paid the purpose limitation principle.

14.4.3 A Difference in Kind and in Degree

Generally, there are two main approaches to data mining described in the literature.[61] The first is verification-driven data mining where information is extracted in

[56] Han and Kamber 2001; *see also*, Symeonidis and Mitkas 2005 (stating, "(t)he human quest for knowledge and the inability to perceive the—continuously increasing—data volumes of a system has led to what we today call data mining."); Schermer 2011, p. 45 (stating "(o)ver the past decades (data mining) as evolved from an experimental technology to an important instrument for private companies and institutions to help overcome to problem of information overload.").

[57] Kuner et al. 2012.

[58] For an example of the lack of agreement on what "data mining" actually means compare the definition of "data mining" in the 2007 US Data Mining Reporting Act (defining "data mining" in such a manner that requires the reporting of "pattern-based" tools but not "linked based" tools), with the definition of "data mining" provided by the US General Accountability Office (GAO)(defining data mining in its May 2004 report entitled "Data Mining: Federal Efforts Cover a Wide Range of Uses", more broadly as "the application of database technology and techniques—such as statistical analysis and modeling—to uncover hidden patterns and subtle relationships in data and to infer rules that allow for the prediction of future results.").

[59] Han and Kamber 2001.

[60] Han and Kamber 2001.

[61] *See e.g.*, Sivanandam and Sumathi 2006.

the process of validating a hypothesis postulated by a user.[62] Notwithstanding the application of this approach to large data sets, it is similar to conventional data analysis methods.[63] Because this approach is not "opportunistic" or "bottom up" or "data driven," it is questionable whether verification-driven data mining can be considered "true" data mining.

The second approach is discovery-driven data mining. Here, a variety of different tools are applied to automatically extract information about the data with limited guidance from the user.[64] Discovery driven data mining can further be subdivided to include descriptive data mining and predictive data mining.[65] Descriptive data mining "provides analysts with a better understanding of the information at their disposal, while uncovering hidden traits and trends within the dataset."[66] It focuses on, among other things, the intrinsic structure, relations, interconnectedness of the data and aspires to improve the overall comprehension of the dataset.[67] Predictive data mining, as the name suggests, is used to make predictions based on inferences found in the available data such as the possibility that people who buy diapers might also buy beer.[68]

The fact that data mining does not necessarily begin with a question to be falsified or verified by the logical analysis of the data set is a significant departure from earlier forms of data processing.[69] In an almost "voodoo-science" way, data mining can be used to analyze multiple dimensions of a dataset and to use the data itself to generate hypotheses.[70] Yoo et.al. explain, that according to the conventional scientific method,

---

[62] Sivanandam and Sumathi 2006.

[63] Jackson 2002.

[64] Sivanandam 2006.

[65] Symeonidis and Mitkas 2005.

[66] Zarsky 2011; *see also*, Schermer 2011 (explaining, "(t)he goal of descriptive data mining is to discover unknown relations between different data objects in a database. Descriptive data mining algorithms try to discover knowledge about a certain domain by determining commonalties between different objects and attributes. By discovering correlations between data objects in a dataset that is representative of a certain domain, we can gain insight to it.").

[67] Symeonidis and Mitkas 2005.

[68] Zarsky 2011 (explaining that "(i)n a predictive process, the analysts use data mining applications to generate rules based on preexisting data. Thereafter, these rules are applied to newer (while partial) data, which is constantly gathered and examined as the software constantly searches for previously encountered patterns and rules. Based on new information and previously established patterns, the analysts strive to predict outcomes prior to their occurrence (while assuming that the patterns revealed in the past pertain to the current data as well."); *see also*, Schermer 2011 (explaining "(a)s the name implies, the goal of predictive data mining is to make a prediction about events based on patterns that were determines using known information."); Whitehorn 2006.

[69] Custers 2013 (explaining that "… traditional statistical analysis usually begins with an hypothesis that is tested against the available data. Data mining tools usually generate hypotheses themselves and test these hypotheses against the available data.").

[70] Calders and Custers 2013 (explaining, "Unlike in statistics, where the data is collected specially with the purpose of testing a particular hypothesis, or estimating the parameters of a model, in data mining one usually starts with historical data that was not necessarily collected with the purpose of analysis, but rather as a by-product of an operational system.").

a hypothesis is built and then data is collected to test the hypothesis: this involves a process of reasoning from the general (i.e., a hypothesis) to the specific (i.e. data). Unlike with the conventional scientific method, the data mining method involves an exploration of a dataset without a hypothesis in order to discover hidden patterns from data: this involves a process of producing the general (i.e., knowledge or an evidence-based hypothesis) from the specific (i.e., data).[71] The automation of the scientific inquiry means that data mining is not limited by the creativity of humans to come up with a relevant hypothesis.[72]

It is important to remember, however, that the data that are initially discovered through the mining process must subsequently be validated by applying the result to a new subset of data. Many variables may appear related during the initial exploration of the data but this could occur purely through chance alone.[73] While the results may be significant, further study is required to ensure that it is not just chance connecting the variables but some other factor.[74] In this respect, it is key to keep in mind that data mining is a dynamic and iterative process that involves human judgment.

Another interesting feature of data mining is, unlike in earlier forms of data processing where the data is collected especially with the purpose of testing a particular hypothesis, in data mining, "one usually starts with historical data that was not necessarily collected with the purpose of analysis, but rather as a byproduct of an operational system."[75] A considerable amount of this data comes from the public sphere as opposed to the "intimate" sphere.[76] In this context, data mining can be referred to as secondary data analysis because the data were not collected to answer the questions now posed.[77] It is useful to think of secondary analysis as the research equivalent of recycling. For example, the secondary analysis of telephone records might not concern the analysis of billing questions for which the data was initially collected but rather the analysis of calling patterns such as: call length, time-of-day or from where-to-where, etc.[78]

Because data mining often relies upon data that was collected by someone else for some other purpose, it is more likely to have missing values and noise (random error in the data) than the data found in statistics, for example.[79] It has been remarked

---

[71] Yoo et al. 2012.

[72] de Hert and Bellanova 2008, *but see also*, Wiley 2008.

[73] Seifert 2006.

[74] Schermer 2011 (explaining that "(i)n unguided descriptive data mining we look for correlations in the data without using a pre-defined working hypothesis. Dependent on the size of the dataset and the 'confidence interval' used to determine correlations, our data mining exercise will yield certain results. While these results might indeed be significant, there is also a chance they are completely random. So, the results we find (and the hypothesis we formulate on the basis of these results) need to be validated to exclude the possibility that the correlation is in fact totally random.").

[75] Calders and Custers 2013.

[76] Nissenbaum 2010.

[77] Imprecise Causality In Mined Rules 2003.

[78] Imprecise Causality In Mined Rules 2003.

[79] Han and Kamber 2001.

that "(d)ata mining results are inherently soft or fuzzy as the data is generally both incomplete and inexact."[80] The quality of data mining results, therefore, relies on the data preparation process and on the excellence of the underlying datasets.[81]

Another distinguishing feature of data mining is that, unlike traditional query and reporting tools, which only reveal explicit information found in the database, data mining reveals implicit information.[82] That is, instead of simply extracting "what" is in a database, data mining can turn data into something new and more useful.[83] The "novel" information that arises through data mining can consist of, among other things, a higher form of knowledge, a more compact summary, a more abstract description or a more useful form such as a prediction.[84] In this respect, data mining can be understood to be the "silkworm" of data processing: it takes a raw material, data, and transforms it into something transcendent that was not there before. If a system can only perform data or information retrieval then it should be more appropriately categorized as either a database system or an information retrieval system or a deductive database system.[85]

Relatedly, data mining is able to confront the visualization and understanding of large data sets more efficiently than previous forms of data processing.[86] Here, data mining is closely linked with the science of information visualization.[87] The knowledge that is discovered in data mining can be expressed in high-level languages, visual representations, or other expressive forms so that it is easily understood and directly usable by humans.[88]

It is also important to mention that despite the expressiveness of the end results, some aspects of data mining can be very opaque and highly automated, relying on "black boxes." Zarsky explains that data mining can rely upon non-interpretable processes where "the rationales for actions premised upon the predictions the data

---

[80] Imprecise Causality In Mined Rules 2003.

[81] Berti-Equille 2007, p. 101.

[82] Taipale 2003.

[83] Taipale 2003.

[84] Taipale 2003 (explaining that "Data mining is the process of looking for new knowledge in existing data. The basic problem addressed by data mining is turning low-level data, usually too voluminous to understand, into higher forms (information or knowledge) that might be more compact (for example, a summary), more abstract (for example, a descriptive model), or more useful (for example, a predictive model). At the core of the data mining process is the application of data analysis and discovery algorithms to enumerate and extract patterns from data in a database.").

[85] Han and Kamber 2001.

[86] Symeonidis and Mitkas 2005, (explaining that data mining "... confronts the visualization and understanding of large data sets efficiently.").

[87] For one explanation of the difference between the two fields *see* Bertini and Lalanne 2009, p. 12 (explaining that "(w)hile information visualization (infovis) targets the visual representation of large-scale data collections to help people understand and analyze information, data mining, on the other hand, aims at extracting hidden patterns and models from data, automatically or semi-automatically.").

[88] Han and Kamber 2001.

mining process provides are not necessarily explainable to humans."[89] When non-interpretable processes are applied, "the software makes its decisions based upon multiple variables (even thousands!) that were learned throughout the data analysis the software makes its decisions based upon multiple variables that were learned throughout the analysis process, but are not easily explained in text."[90] In these contexts, the role of the analyst is minimized. Sometimes it is difficult to provide an answer as to why a specific result was reached beyond stating, "this is what the algorithm found based on previous similar cases in the past."[91]

Furthermore, and perhaps obviously at this point, data mining takes on a much larger scale than previous forms of data processing. It is usually conducted on huge volumes of data, which often includes sensitive and private information about individuals or companies, and it seeks to extract *value* from such *volume*.[92] With traditional techniques, the usefulness of retrieving data from a database is diminished when the database is really big. For example, if an analyst ran a database query against a large table such as "find customers who live in New York", it might be possible to locate millions of responses and thus, the usefulness of the search is eviscerated by the large volume of data. Enter data mining: instead of returning a list of names of people who live in New York, the result of the search could be the creation of useful model of the data in question.[93] A data analysis system that does not handle large amounts of data can at most be categorized as a machine learning system, a statistical data analysis tool, or an experimental system prototype.[94]

Another unique feature of data mining is that the data to be mined is usually of a high dimensionality such as micro-array data, which may have tens of thousands of dimensions.[95] Statistics, for example, does not consider the dimensionality of data. Furthermore, the data to be mined can also be very complex, such as social network data, whereas most of the data in older forms of processing are flatted, simple data files in text or binary format.[96]

Lastly, data mining involves very complex interdependencies between humans and technology that have hitherto not existed. Data mining is not a simple, stand-alone

---

[89] Zarsky 2011.

[90] Zarsky 2011.

[91] Zarsky 2011.

[92] Lloyd-Williams 1997.

[93] Lloyd-Williams 1997.

[94] Han and Kamber 2001.

[95] *For more, see* Keim 2002 (explaining that "(o)ne-dimensional data usually has one dense dimension. A typical example of one-dimensional data is temporal data ... Two-dimensional data has two distinct dimensions. A typical example is geographical data where the two distinct dimensions are longitude and latitude ... Many data sets consists of more than three attributes and therefore, they do not allow a simple visualization as 2-dimensional or 3-dimensional plots. Examples of multidimensional (or multivariate) data are tables from relational databases, which often have tens to hundreds of columns (or attributes). Since there is no simple mapping of the attributes to the two dimensions of the screen, more sophisticated visualization techniques are needed.").

[96] Keim 2002.

technology but rather it is a complicated socio-technical system.[97] The advanced algorithms and mathematical models applied in data mining are connected to a world of interlinked databases and information networks and they rely (at least partly) on humans to build, train and apply them.

14.5 Data Mining and its Paradoxical Relationship to the Purpose Limitation Principle

At this point, it should be relatively clear that data mining is a rather complex technology. It should also be clear that the purpose limitation principle is a good concept in theory but vague and ridden with exceptions in reality: in the words of Schwartz, it promises a lot and delivers too little.[98] The ambition of this section is to demonstrate, for the reasons set forth below, that whatever concerns already exists about the purpose limitation principle, it is going to be made worse as the technology of data mining continues to advance.

First, data mining challenges the purpose limitation principle because of its reliance on huge amounts of data. This reliance on "big data" encourages the large-scale collection and retention of data by governments and private companies alike. These entities are creating data silos because of an almost instinctive feeling that the information stored in such silos might have great value at some point in the future. This, in turn, encourages an obfuscation of the purpose limitation principle: instead of stating a clear and specific purpose, data controllers are incentivized to create broadly defined purposes in order to bolster their freedom to process the data.

The result of this obfuscation of the purpose limitation principle is that the individual does not fully understand what happens to his/her data beyond the initial transfer. He/she almost certainly does not grasp that, in many situations, his/her personal data are being systematically collected for later use in the information marketplace.[99] This creates what van den Hoven calls informational inequality: an unfairness inflicted upon data subjects that results because of a lack of openness, transparency, participation and notification on the part of the data controller.[100] Ostensibly, this imbalance of power over information can lead to manipulation, profiling and discrimination in society.[101]

Second, it is impossible to reconcile data mining with the purpose limitation principle because no one knows beforehand what the results of a data-mining search will turn up. An individual cannot possibly foresee how his/her data will be used because the very point of data mining is to provide a window to the unforeseeable.[102]

[97] Nissenbaum 2010.

[98] Schwartz 1999.

[99] Nissenbaum 2010.

[100] Van den Hoven 2007, p. 462; *see also*, Nissenbaum 2010.

[101] Gandy 2009.

[102] *See*, Rosenzweig 2010 (explaining that "the purpose and use limitations, if fully applied, would significantly degrade the analytical utility of many knowledge discovery systems.").

This means that a data controller cannot conceive beforehand the specific purpose of the data collection (and then restrict the use to that purpose) because it will not be obvious until *after* the processing what data will be of value or what relationships will emerge.[103] Because the first prerequisite to data protection is not fulfilled—purpose specification—other data protection requirements, including use limitation, adequacy, relevance and proportionality cannot be met since the individual has no sense of the boundaries within which his/her personal data may be processed and therefore, cannot enforce his/her rights.[104]

A data miner can, of course, inform the individual ex post facto that his/her personal data has been subjected to data mining and that some "new" information about him/her has been extrapolated through the process. It can provide a specific and legitimate purpose for how it would like to use the new information and allow the individual an opportunity to consent or reject to the processing. The problem with requiring that purpose limitation be required *after* the data mining is that it misses the theoretical point that the individual is not supposed to possess on-and-off control over his/her personal data. By pushing the purpose limitation principle aside to data mine, the control the individual has over his/her data becomes illusory.

To illustrate this point, it is helpful to examine two data mining scenarios, even if they are both somewhat prosaic and oversimplified. In the first scenario, transactional data revealing an individual has recently purchased certain fertilizers is combined with watch-list data and mined to reveal that the individual is a suspected terrorist.[105] In the second scenario, customer-loyalty-card information is combined with demographic information and mined to reveal that a woman is pregnant.[106] In both of these situations, the "new" information that is derived from data mining can easily get passed on to various authorities, without the individual even knowing the data exists, who might then use it for any number of purposes: for example, as evidence to obtain a wiretap warrant to collect electronic surveillance about the suspected terrorist or, to send diaper coupons to the woman, uncovering to her family the she is pregnant.[107] Gunasekara explains rather eloquently that the "new" personal information that is derived from data mining becomes part of " . . . the swelling river of data whose channels are, in the private and public sectors, ever changing and difficult to follow, much less control."[108]

---

[103] Cavoukian 1998 (explaining that a good data mining program cannot, in advance, delineate what the primary purpose will be because the "discovery model" upon which data mining is based, does not need an hypothesis, and without an hypothesis, establishing the specific purpose of data collection or data processing is a much more complex task); *see also* de Hert and Bellanova 2008.

[104] Article 29 Data Protection Working Party: Opinion 03/2013 on purpose limitation (2 April 2013) available online at http://idpc.gov.mt/dbfile.aspx/Opinion3_2013.pdf (explaining that "(s)pecification of purpose is an essential first step in applying data protection laws and designing data protection safeguards for any processing operation.").

[105] Gunasekara 2009.

[106] Duhigg 2012.

[107] Duhigg 2012.

[108] Gunasekara 2009.

It is further worth mentioning that the "new" information that results from data mining often takes the form of an abstraction of the underlying dataset such as a predictive or a descriptive model. These abstractions are the consequence of a dynamic and iterative process of retrieving, excluding, comparing reorganizing, digging, pulling etc. huge amounts of data of which any one data unit can technically be identified, singled out and used again and again in the process. In this respect, data mining has a jig-saw nature where the idea is to piece together units of data in order create a clearer picture of something: little by little, pieces of data are placed together and a picture emerges.[109]

Lyon raises concerns over the constant circulation of fragments of personal data within computer systems beyond any agent's authority.[110] These fragments of data circulate and accumulate in ways that not only make it impossible for an individual to control the use and reuse of the data but also raise larger questions about discrimination and social justice.[111] Likewise, Amoore describes "data derivatives" as fragmented elements, which are easily moved, shared, traded and exchanged with complete indifference to underlying data set and data subject.[112] She explains that the "data derivative" is not centered on who the data subject is or even on what a particular data set says about a data subject but it is instead focused on what can be imagined and inferred about data subjects.[113]

The fragmentation of data that takes place in data mining is hard to reconcile with the purpose limitation principle because it may not be obvious whether a unit of data is related to any particular data subject, which calls into question the application of data protection law in the first place. This is because it is only when data can be linked to an actual individual that the principle of purpose limitation even applies. For example, there could be data about two, one-way-ticket flights bought on the same credit card, from two different telephone numbers which incidentally were the same numbers dialed by three of the 9/11 hijackers used in the creation of some descriptive or predictive model.[114] Yet, it may not be obvious how to connect this data to an identifiable human being and therefore, no obligations under data protection law.

The relationship between the purpose limitation principle and data mining becomes even more antithetical when one considers that data mining invariably involves taking data collected in one context and using the data in a different context in a way never contemplated at the time of its initial collection.[115] For example, public-transit

[109] *See generally*, Jonas 2009.

[110] Lyon 1994.

[111] Lyon 1994; *see also*, Bennett 2011.

[112] Amoore 2011.

[113] Amoore 2011.

[114] This example is based off a similar example provided by Louise Amoore during her presentation "Risk based security practices—Risk and the war on terror" held at the Amsterdam Privacy Conference (October 9, 2012).

[115] Calders and Custers 2013.

card information can be mined for police inquiries[116] or social-media data can be mined to help politicians get their political supporters to the poll boxes.[117] Here, the purpose limitation principle provides an important safeguard against the data from being "lost in translation" when it migrates from one context to another context without any granular understanding of how the data was originally classified.[118] It provides the individual the opportunity to step in, exercise his/her control and to say, "STOP! I do not want you to that with my data!" The problem is, however, as explained above, there is no moment to afford the individual a chance for meaningful intervention.

14.6 Moving Forward

The purpose limitation principle has provided an important baseline for protecting personal data. Over time, however, technological advances such as data mining have caused the principle to lose much of its meaning. The strict application of this principle simply does not work well in a world where personal data is frequently transferred from one context to another context, often without the individual's knowledge or consent, and then mined to give raise to new information about the individual that is useful for organizational decision-making.[119] If one considers the fact that a data controller cannot meaningfully inform the individual of a specific and legitimate purpose for the data processing in advance of data mining, the purpose limitation principle becomes a conception existing only on paper and individual control of his/her data just a chimera of the law.

One way of addressing the new kind of privacy threats posed by technological innovations such as data mining is to move away from the rigid and inflexible fair information principles such as the purpose limitation principle and to adopt an alternative paradigm to data protection. Cate fiercely critics the fair information principles and suggests that, at least in the commercial context, data protection laws should regulate information flows only when necessary to protect individuals from harmful uses of information.[120] That is, "data protection law should be designed to prevent tangible harms to individuals and to provide for appropriate recovery for those harms if they occur."[121]

---

[116] *See generally*, Lyon 2008 (explaining, "In the case of Oyster cards in the UK, data that begin life in the commercial sphere of public transit, are increasingly required in police inquiries. Such data may also stay in the same context but as their uses grow, they may acquire some dangerous characteristics; internal citations omitted).

[117] Duhigg 2013.

[118] Ramasastry 2006, p. 757 (aptly using the phrase "lost in translation" to explain the problems that arise when data migrates from commercial data brokers to government entities in counter-terrorism data mining programs).

[119] Tavani 1999, p. 137.

[120] *See* Cate 2006.

[121] Cate 2006.

Cate's suggestion is related to Sweden's "abuse-centered regulatory approach" which is an approach that seeks to enhance the efficacy of data-protection rules by simplifying and focusing them on preventing the misuse of personal data.[122] Pursuant to Sweden's approach, unstructured processing of personal data is allowed *unless* it constitutes a misuse of the privacy of an individual.[123] The exemption of unstructured processing of personal data from several obligations arising from the Swedish data protection legislation marks a significant departure from the traditional model of regulation that is more orientated towards a system of notice and consent embodied in fair information principles.[124]

The misuse model should be explored further as an alternative paradigm to handling the processing of personal data. The focus on actual breaches of privacy rather than on the risk of potential breaches of privacy might prove more realistic in light of technologies such as data mining.[125] By concentrating on the serious rather than technical infringements of privacy, it could also serve to limit the unrealistic burdens on legitimate commercial and other data processing.[126] It could further lead to a system more stepped in meaningful, substantive data protection laws rather than hollow, procedural rules.[127]

It is true that moving away from the fair information principles and towards a misuse model of data protection might not afford the individual the same theoretical level of control of his/her personal data because he/she will not be informed and consent to every use of his/her data. However, as noted at the outset of this article, the notion of privacy as control has been questioned by many commenters as resting on unsound foundations and thus, perhaps it is time to move away from it. Indeed, Schwartz' contention that the notion of privacy as control is deeply flawed because it incorrectly assumes that individuals have the ability to exercise control over their data in the first place is wholly accurate in the context of data mining.[128] It is also worth mentioning that just because an individual does not have control over his/her information does not necessarily mean that a privacy intrusion will be perpetrated.[129]

Another suggestion for handling the privacy threats posed by data mining would be to create a revised set of privacy principles that are more realistic in today's dynamic technological environment. For example, instead of placing the emphasis

---

[122] Kirchberger 2011.

[123] For more on how the law works, see, Kirchberger 2011.

[124] *See* Seipel 2001 (where Seipel explains that "(i)n short, the misuse model would mean freedom to process whereas the processing model would mean that processing requires some kind of permission.").

[125] Steele 2002.

[126] Steele 2002.

[127] Cate 2007.

[128] Schwartz 1999.

[129] Moor 1990 ("Although control of information is clearly an aspect of privacy, these definitions emphasizing control are inadequate for there are many situations in which people have no control over the exchange of personal information about themselves but in which there is no loss of privacy.").

on purpose limitation, the emphasis could be placed on accountability. This would mean that a data subject need not necessarily be told for what specific purpose his/her data will be processed but it would require that that the data controller verify and legitimize the reasons for its data processing after the processing. In this respect, privacy impact assessments could play a big role in explaining the specific privacy risks associated with particular data mining applications and to help public and private actors determine what controls are needed to mitigate those risks.[130] Likewise, the use of "accountable algorithms" could further accountability by allowing the individual to make sure that the algorithm, which may have caused some adverse decision to be made about him/her, was executed fairly and lawfully in a particular case.[131]

A related approach to achieving meaningful privacy legislation in light of data mining would be to create technological solutions. That is, instead of discharging the purpose limitation through the adoption of formal organizational policies and processes, it could be embedded into systems at the very outset of the technology's design phase.[132] The European Commission and the European Data Protection Supervisor have emphasized the need to design and develop information communication technologies in a way that respects privacy and data protection and encourage that this should be done from the very early design stage of ICT, right through to their deployment, use and ultimate disposal.[133] Essentially, the idea is that by building privacy in from the outset, it becomes possible to foster confidence and trust in the technology as being privacy-protective, and ideally avoiding costly future retrofits.[134]

Privacy by design could have a powerful impact in promoting the notion of purpose limitation. This is especially true because privacy by design could afford the principle with a more substantive character. For example, privacy by design could be applied to automatically limit the unfair use of the "new" data that arises from data mining for certain specified purposes or certain specified categories of data depending upon the specific context of the data mining.

14.7 Conclusion

The insight that is derived from the novel information flows that occur as a result of data mining may lead to many unforeseen human benefits ranging from the identification of chronic diseases[135] to the enhancement of e-government[136] to the better

---

[130] Vijayan 2007; see also, Wright and de Hert 2012.

[131] Felton 2012.

[132] Cavoukian and Jonas 2012; *see also*, Mulligan and King 2012.

[133] Opinion of the European Data Protection Supervisor on Promoting Trust in the Information Society by Fostering Data Protection and Privacy, Brussels (18 March 2010).

[134] Cavoukian 2012.

[135] *See generally*, Ponniah 2010.

[136] Hanumanthappa et al. 2012.

understanding of the nature of hate crime[137] to the facilitation of student learning[138] and even to the improvement of an individual's own sense of self.[139] As such, purpose limitation should not stand as obstacle to these developments. Rather, it must either be redefined in a way to allow for the positive benefits of data mining while simultaneously protecting privacy or abandoned altogether in favor a new approach that more aptly reflects the technological reality.

The proposed EU Data Protection Regulation attempts to address privacy challenges brought on by technological developments such data mining by introducing a number of new privacy rights such as the right to be forgotten and to data portability.[140] It also sets forth several new responsibilities on data controllers such as the requirement to build in "privacy by design" and to implement privacy impact assessments.[141] It still, however, relies extensively upon the purpose limitation principle, in a form unchanged from the current Directive, as a core principle for safeguarding privacy and in this respect reflects a naivety and disconnectedness with the technological reality.

References

Allen, Anita L. 1988. *Uneasy access: Privacy for women in a free society*. Totowa: Rowman and Littlefield.

Amoore, Louise. 2011. Data derivatives: On the emergence of a security risk calculus for our times. *Theory, Culture and Society* (*SAGE 2011*) 28(6):24.

Arthur, Charles. 2011. What's a zettabyte? By 2015, the internet will know, says Cisco. Technology Blog at *The Guardian UK Newspaper*.

Bennett, C. J. 2011. In defence of privacy: The concept and the regime. *Surveillance and Society* 8(4):485.

Berti-Equille, Laure. 2007. Measuring and modelling data quality for quality-awareness. *Data Mining, Quality Measures in Data Mining* 43:101–126.

Bertini, Enrico, and Denis Lalanne. 2009. Surveying the complementary role of automatic data analysis and visualization in knowledge discovery. Proceedings of the ACM SIGKDD workshop on visual analytics and knowledge discovery: Integrating automated analysis with interactive exploration (VAKD'09) 12–20. New York: ACM.

Bienkowski, Marie, Mingyu Feng, and Barbara Means. 2012. *Enhancing teaching and learning through educational data mining and learning analytics: An issue brief*. Washington, D. C.: US Department of Education.

Biersdorfer, J. D. 2006. Weeding out Windows fonts. *The New York Times*, February 16.

Calders, Toon, and Bart Custers. 2013. What is data mining and how does it work. In *Discrimination and privacy in the information society: Data mining and profiling in large databases*, eds. Bart Custers, Tal Zarsky, Bart Schermer and Toon Calders, p. 28. Berlin: Springer.

Cate, Fred H. 2006. The failure of fair information practice principles. In *Consumer protection in the age of the information economy*, ed. Jane K. Winn. Surry: Ashgate.

---

[137] Ozgul et al. 2012.

[138] Bienkowski et al. 2012.

[139] Eisenberg 2012; Rowan 2011

[140] *For more, see* Rubinstein 2013.

[141] Rubinstein 2013.

Cate, Fred H. 2007. The autonomy trap. The Privacy Symposium Cambridge, MA. http://www.fredhcate.com/Publications/The%20Autonomy%20Trap.revised.pdf. Accessed 24 Aug 2007.

Cavoukian, Ann. 1998. Data mining: Staking a claim on your privacy, Information and Privacy Commissioner/Ontario. http://www.ipc.on.ca/images/resources/datamine.pdf. Accessed 15 Sept 2013.

Cavoukian, Ann. 2012. Operationalizing privacy by design: A guide to implementing strong information and privacy practices. http://www.ipc.on.ca/images/Resources/operationalizing-pbd-guide.pdf. Accessed 4 Dec 2012.

Cavoukian, Ann, and Jeff Jonas. 2012. Privacy by design in the age of big data. 2012. http://privacybydesign.ca/content/uploads/2012/06/pbd-big_data.pdf. Accessed 8 June 2012.

Cohen, Julie E. 2000. Examined lives: Informational privacy and the subject as an object. *Stanford Law Review* 52:1373.

Colonna, Liane. 2012. The new EU proposal to regulate data protection in the law enforcement sector: Raises the bar but not high enough. IRI-memo, Nr. 2/2012.

Council of Europe Convention for the Protection of Individuals with regard to Automatic Processing of Personal Data (ETS No. 108) (Jan. 28, 1981).

Council Framework Decision. 2008. 2008/977/JHA of 27 November 2008 on the protection of personal data processed in the framework of police and judicial cooperation in criminal matters. *Official Journal L* 350:0060–0071.

Custers, Bart. 2013. Data dilemmas in the information society: Introduction and overview. In *Discrimination and privacy in the information society: Data mining and profiling in large databases*, eds. Bart Custers, Tal Zarsky, Bart Schermer and Toon Calders. Berlin: Springer.

David Rowan, "Personal data mining to improve your cognitive toolkit," David Rowans Blog http://www.wired.co.uk/news/archive/2011-01/18/edge-question. Accessed 18 January 2011

De Hert, Paul, and Rocco Bellanova. 2008. *Data protection from a transatlantic perspective: The EU and US move towards an International Data protection agreement?* Brussels: European Parliament's Committee on Civil Liberties, Justice and Home Affairs.

Directive 95/46/EC of the European Parliament and of the Council of 24 October 1995 on the protection of individuals with regard to the processing of personal data and on the free movement of such data *Official Journal L 281, 23/11/1995 P. 0031–0050.*

Duhigg, Charles. 2012. How companies learn your secrets. *The New York Times*, February 16.

Duhigg, Charles. 2013. Mine personal lives to get out vote. *The New York Times*, October 13.

Eisenberg, Anne. 2012. What 23 years of E-Mail may say about you, The New York Times (April 7, 2012)

Fayyad, U., and R. Uthurusamy. 2002. Evolving data into mining solutions for insights. *Communications of the ACM* 45(8):28–31.

Felton, ed. 2012. Accountable algorithms. In the blog, *Freedom to Tinker: Research and expert commentary on digital technologies in public life* available at https://freedom-to-tinker.com/blog/felten/accountable-algorithms/. Accessed 12 September 2012.

Fried, Charles. 1968. Privacy. *Yale Law Journal* 77:475.

Gandy, O.H. 2009. *Coming to terms with chance: Engaging rational discrimination and cumulative disadvantage*. Aldershot: Ashgate.

Government Surveillance. 2012. Little peepers everywhere. *The Economist* Issue 950.

Gunasekara, Gehan. 2009. The 'final' privacy frontier? Regulating trans-border data flows. *International Journal of Law and Information Technology* 17:147.

Han, Jiawei, and Micheline Kamber. 2001. *Data mining: Concepts and techniques*. San Diego: Academic Press.

Hanumanthappa, M., B. R. Prakash, and Manish Kumar. 2012. *Applications of data mining in e-governance: A case study of Bhoomi project in data engineering and management lecture notes in computer science. vol. 6411*, 208. Springer.

Imprecise Causality In Mined Rules. 2003. Proceedings: Rough sets, fuzzy sets, data mining, and granular computing: 9th International Conference, RSFDGrC 2003, Chongqing, China (Lecture Notes in Computer Science, Springer-Verlag Heidelberg, v 2639/ 2003), 581.

Inness, Julie C. 1996. *Privacy, intimacy, and isolation*. Oxford: Oxford University Press.
Jackson, Joyce. 2002. Data mining: A conceptual overview. *Communications of the Association for Information Systems* 8:267.
Jonas, Jeff. 2009. Data finds data. http://jeffjonas.typepad.com/jeff_jonas/2009/07/data-findsdata.html. Accessed 15 Sept 2013.
Keim, Daniel A. 2002. Information visualization and visual data mining. *IEEE Transactions on Visualization and Computer Graphics* 100.
Kirchberger, Christine. 2011. *Cyber law in Sweden*. The Netherlands: Kluwer Law International.
Korff, Douwe, and Ian Brown. 2010. *Comparative study on different approaches to new privacy challenges, in particular in the light of technological developments*. European Commission.
Kuner, Christopher, Fred H. Cate, Christopher Millard, and Dan Jerker B. Svantesson. 2012. The challenge of 'big data' for data protection. *International Data Privacy Law* 2(2):47.
Lloyd-Williams, Michael. 1997. Discovering the hidden secrets in your data—the data mining approach to information. *Information Research: An International Electronic Journal*.
Lyon, D. 1994. *The electronic eye: The rise of surveillance society*. Minneapolis: University of Minnesota Press.
Lyon, David. 2008. Surveillance society. Talk for Festival del Diritto, Piacenza, Italia.
Miller, Arthur. 1971. *Assault on Privacy*. Michigan: University of Michigan Press.
Moor, James H. 1990. The ethics of privacy protection. *Library Trends* 39(Fall):69.
Mulligan, Deirdre K., and Jennifer King. 2012. Bridging the gap between privacy and design. *University of Pennsylvania Journal of Constitutional Law* 14(4):989.
Nissenbaum, Helen. 2010. *Privacy in context: technology, policy, and the integrity of social life*. Stanford: Stanford University (Stanford Law Books).
Ozgul, F., M. Gok, A. Celik, and Y. Ozal. 2012. Mining hate crimes to figure out reasons behind. *Advances in Social Networks Analysis and Mining (ASONAM), 2012 IEEE/ACM International Conference* 887.
Peppet, Scott R. 2012. Privacy and the personal prospectus: Should we introduce privacy agents or regulate privacy intermediaries? *Iowa Law Review Bulletin* 7:77.
Ponniah, P. 2010. *Data mining basics in data warehousing fundamentals for it professionals, 2nd ed.* Hoboken: Wiley.
Ramasastry, Anita. 2006. Lost in translation? Data mining, national security and the 'adverse inference' problem. *Santa Clara Computer and High Tech Law Journal* 22(4):757.
Rosenzweig, Paul. 2010. Privacy and counter-terrorism: The pervasiveness of data, case western reserve. *Journal of International Law* 42(3):625.
Regan, Priscilla M. 1995. *Legislating privacy: technology, social values, and public policy*, p. 9. North Carolina: University of North Carolina Press.
Rubinstein, Ira S. 2013. Big data: The end of privacy or a new beginning? *International Data Privacy Law* 12–56.
Schermer, Bart W. 2011. The limits of privacy in automated profiling and data mining. *Computers Law and Security Review* 27:45.
Schoeman, Ferdinand. 1992. *Privacy and social freedom*. Cambridge: Cambridge University Press.
Schwartz, Paul M. 2000. Internet privacy and the state. *Conneticut Law Review* 32:815.
Schwartz, Paul M. 1999. Privacy and democracy in cyberspace. *Vanderbilt Law Review* ℜ52:1609.
Seifert, Jeffrey W. 2006. Data mining and homeland security: An overview, US Congressional Research Service Report.
Seipel, Peter. 2001. *Privacy and freedom of information in Sweden in nordic data protection law. 1st ed., ed. P. Blume, 124*. Copenhagen: DJØF Publishing.
Sivanandam, S. N. and S. Sumathi. 2006. *Introduction to data mining and its applications*. Springer.
Solove, Daniel J. 2002. Conceptualizing privacy. *California Law Review* 90:1087.
Solove, Daniel J. 2008. *Understanding privacy*. Boston: Harvard University Press.
Solove, Daniel J. 2011. *Nothing to hide: The false tradeoff between privacy and security*. New Haven: Yale University Press.
Steele, Jonathan. 2002. Data protection: An opening door? *Liverpool Law Review* 24(1–2):19.

Symeonidis, Andreas L., and Pericles A. Mitkas. 2005 Data mining and knowledge discovery: A brief overview. In *Agent intelligence through data mining multiagent systems, artificial societies, and simulated organizations*. Springer.

Taipale, K. A. 2003. Data mining and domestic security: Connecting the dots to make sense of data. *Columbia Science and Technology Law Review* 5:1.

Tavani, Herman T. 1999. Informational privacy, data mining, and the internet. *Ethics and information technology*. vol. 1, Issue 2, 137. Kluwer Academic Publishers.

Van den Hoven, J. 2007. *information technology, privacy and the protection of personal data. In Information technology, privacy and the protection of personal data*. Cambridge: University Press.

Vijayan, Jaikumar. 2007. DHS must assess privacy risk before using data mining tool, GAO say. *Computer World*.

Westin, Alan. 1967. *Privacy and freedom*. New York: Atheneum.

White, Martha C. 2012. Could that Facebook 'like' hurt your credit score? *Time Magazine*.

Whitehorn, Mark. 2006. The parable of the beer and diapers: Never let the facts get in the way of a good story, *The Register*.

Wiley, Steven. 2008. Hypothesis-Free? No such thing: Even so-called 'discovery-driven research' needs a hypothesis to make any sense, *The Scientist Magazine*.

Wright, David and Paul de Hert, eds. 2012. Privacy impact assessment. *Series: Law, governance and technology series*. vol. 6. Springer.

Yoo, Illhoi, Patricia Alafaireet, Miroslav Marinov, Keila Pena-Hernandez, Rajitha Gopidi, Jia-Fu Chang, and Lei Hua. 2012. Data mining in healthcare and biomedicine: A survey of the literature. *Journal of Medical Systems* 36(4):2431–2448.

Zarsky, Tal Z. 2011. Governmental data mining and its alternatives. *Penn State Law Review* 116(2):285.

Chapter 15
The Cost of Using Facebook: Assigning Value to Privacy Protection on Social Network Sites Against Data Mining, Identity Theft, and Social Conflict

Wouter Martinus Petrus Steijn

15.1 Introduction

A popularity amongst millions of users worldwide has rapidly befallen social network sites (SNS) which focus on social relationships and interaction, such as Facebook. This popularity is despite the different privacy risks users are exposed to at the same time. Not only is the shared information on SNSs subject to data mining (Andrews 2012), it also exposes the user to potential identity theft as well (Noda 2009; Timmer 2009), and users have to manage different social contexts (e.g. friends, family, and colleagues) to avoid social conflict (Binder et al. 2009; Skeels and Grudin 2009). The use of SNSs is therefore often seen as evidence that users no longer care about privacy (Johnson 2010) and that users could claim their privacy important is considered paradoxical.

The paradox quickly unravels though, if one takes the social merits SNS provide for its users into account. SNS provide social merits in the forms of new possibilities for self presentation and social interactions with friends (Ellison et al. 2007; Steinfield et al. 2008). These social merits depend on where one's social network is (e.g., where one's friends are) online, leaving users with little choice what SNSs they pick. As a result, participation on SNSs is not necessarily informative of actual privacy concern.

This study will not only provide new insight in SNS users privacy concerns by describing the relative importance they attribute to different privacy threats, but will also contribute to the ongoing privacy discussion by addressing the privacy paradox and by emphasizing the need for further development of new privacy policies and regulation. An innovative method will be used to determine the relative importance attributed by SNS users to the potential privacy threats of data mining, identity theft, and social conflict. Furthermore, the degree to which younger and older individuals differ in how they attribute importance to the various threats will be investigated. To this date no research exists, to the author's knowledge, which has explicitly compared

W. M. P. Steijn (✉)
Tilburg Institute for Law, Technology, and Society (TILT), Tilburg University,
Warandelaan 2, 5037 Tilburg, The Netherlands
e-mail: w.m.p.steijn@uvt.nl

S. Gutwirth et al. (eds.), *Reloading Data Protection*,
DOI 10.1007/978-94-007-7540-4_15,

what privacy threats users find most important. Next, some additional background will be given concerning the proposed methodology, before the research hypotheses are formulated based on related work.

15.1.1 Background

A choice based conjoint (CBC) design was chosen to investigate the relative importance SNS users attribute to different privacy threats[1]. CBC is a popular research design used in marketing to determine how a new product would best fit consumers' wishes (Curry 1996; Orme 1996). The strength of a CBC design is that it can determine the relative value respondents attribute to the features of a product avoiding direct questioning, but instead relying on respondents' actual decisions. In addition, as respondents are presented with a complete product, as opposed to for example paired wise comparison where respondents' decisions are based on only two features at the time, the decision making process can be considered more realistic

While potential privacy threats SNSs users encounter can't be included as features directly, it is easy to imagine how respondents could be presented with choice tasks between SNSs which vary on privacy protection features affecting these different potential privacy threats. This assumes that the importance users attribute to a certain privacy protection feature will be indicative of where their main privacy concerns lie. This assumption can be justified with Petronio's Communication Privacy Management (CPM) theory (2002).

CPM theory addresses the dialectical relationship between the need for privacy and the desire to share personal information with others. It describes how we create metaphorical boundaries to be able to share information with some, while excluding others to this information. These boundaries are signaled and maintained through an implicit rule-based system. For example, we whisper certain information not only to literally avoid being overheard, but also to signal to the other that what we say is private and do not wish for it to be shared with others. In effect, a boundary is formed surrounding that information including only the other to whom the information is whispered and who has become a co-owner of that information (Petronio 2002).

Although CPM focuses on the face-to-face context, the described dialectical need has become especially apparent on contemporary SNS. SNS have become an important medium and offer social merits to its users in the forms of new possibilities for self presentation and social interactions with friends (Ellison et al. 2007; Steinfield et al. 2008), resulting in large amounts of information being shared on these sites. Instead of being limited to metaphorical boundaries, SNS provide the technological tools to enforce the boundaries concerning personal information (Litt 2012). For example, by changing the settings of their posts to *friends* on Facebook, users restrict access to only their contacts and exclude anyone else who might be attempting to see what they posted on their profile. Maintaining these boundaries online has

[1] http://www.sawtoothsoftware.com/products/cbc/cbc_method

become increasingly important and necessary because the permanence and searchability of online information (boyd 2008b, p. 27) would otherwise make online shared information accessible to anyone at anytime.

The current study specifically focuses on the relative importance SNS users attribute to boundaries against the following three potential privacy threats on SNSs: data mining, identity theft, and social conflict. While users are exposed to all three privacy threats at the same time on SNSs, each threat has a different origin and other factors influence users' protection against each threat. Next, these three types of privacy threats will be briefly described in relation to SNSs and examples will be given of what privacy protection features SNSs could provide that affect that privacy threat.

Data Mining. This category concerns the potential privacy threat imposed by data mining and profiling by the SNS provider and third parties. Since the business model of SNSs is generally based on the use of the available personal information for commercial purposes, personal data placed on these sites often becomes available for companies. The scale on which data mining occurs is reflected in the economic value of Facebook as a company (Pékarek and Leenes 2009). SNSs can contain several features which affect the privacy protection of the user against data mining. First, it makes a difference whether the site owns the information posted on the site and whether information can be removed by users or remains in the database. Second, SNS generally have a policy concerning the access of third parties to personal information disclosed by the user as well.

Identity theft. This category concerns the potential privacy threats imposed by strangers with criminal intent, of which identity theft is the most familiar. Several features of SNSs can provide privacy protection against identity theft by strangers. One way is the privacy settings which only allow ones contacts to access ones full profile. However, this only provides partial protection as users are inclined to accept friend requests from strangers (Noda 2009). Another way to create boundaries against identity theft is by refraining from posting personally identifiable information online, since even posting seemingly innocent information such a the date of birth can have risks (Timmer 2009). Consequently, if SNSs require users to fill in identifying data (such as a name) or contact data (such as an email address) to verify their profile, the privacy threat of identity theft is increased.

Social conflict. This category concerns the potential privacy threat of social conflict. This is mainly a consequence of the mixed social contexts on SNSs: socializing with friends now occurs within reach of your family and (future) employers. Information shared with one group, is not necessarily appropriate or desirable to be disclosed to others and could lead to tension or conflict (Binder et al. 2009; Skeels and Grudin 2009; Lampinen et al. 2009). Several features of SNSs can affect the privacy protection of users against social conflict. First of all, being able to sort contacts into different groups and to discriminate in what information is available to what groups can help create boundaries between social context. Second, the possibility to tag pictures on SNSs can affect the boundary someone tries to protect in a negative way. An example for thiswould be a tagged picture (drunk at a party) posted by a friend becoming visible to family as well. Third, SNSs could enable users to track visitors to their profile which could expose frequent visitors. Facebook, however, does not support this option (Mongold 2010).

This study compares the relative importance attributed to privacy protection features against data mining, identity theft, or social conflict by users of SNSs of all ages. The current section has introduced the proposed methodology and grounded it in theory and operationalized the privacy threats of interest. Next, related work will be discussed in order to introduce the research hypotheses.

15.1.2 Related Work

Privacy has been a subject of research for many years, but in recent years the focus primarily lies on privacy and the internet. Not only the media (Andrews 2012; Noda 2009; Timmer 2009), but academia as well have given a lot of attention to the potential privacy threats of data mining and identity theft on SNSs. (Acquisti and Gross 2006; Debatin et al. 2009; Govani and Pashley 2005; Gross and Acquisti 2005). The studies generally concluded that their student samples of Facebook users appear not to care about the potential privacy threats on such sites. These conclusions were mainly driven by the amount of information shared by the students, despite the risks.

However, participation on SNS does not necessarily mean that users don't have any privacy concerns. The popularity of social network sites is a result of the possibilities they create for social interaction with friends (Ellison et al. 2007; Steinfield et al. 2008). Non-participation may even simply not be considered due to the related social costs in missing out on the social interactions amongst friends occurring on these sites (Raynes-Goldie 2010). As a result, even privacy concerned and aware individuals may join a SNS.

Indeed, SNSs users have proven to be creative concerning their privacy protection against social conflict (boyd and Marwick 2011, p. 14; Lampinen et al. 2009; Stutzman and Hartzog 2009), but when it comes to their boundaries against data mining and identity theft, they primarily have to rely on what the sites provide. This does not automatically suggest that they are not concerned about these potential privacy threats; there is simply little they can do if they want to socialize on these sites. The findings of Paine and colleagues (2007) support this notion. When asked, their 20-year-old and older respondents reported spam, spyware, hackers, access to personal information, and identity theft as their major privacy concerns in relation to the internet.

Respondents are therefore expected to be aware of the potential privacy threats of data mining and identity theft and attribute more importance to privacy protection features related to these aspects as opposed to features protecting against social conflict. In other words, it is expected that the fact that they are participating on SNSs does not diminish their concern about data mining and identity theft. Furthermore, respondents are expected to be unwilling to change to a different SNS provider, because they are bound to their SNS through their social network being present on that site. For this purpose the following hypotheses were formulated:

Hypothesis 1: Respondents will generally attribute more importance to privacy protection features against data mining and identity theft than privacy protection features against social conflict.

Hypothesis 2: Respondents will generally be unwilling to change to a different social network site provider.

As was discussed earlier, the fact that SNSs are most popular amongst younger individuals, does not necessarily say much since SNSs are a useful social tool for youth to accomplish several important developmental tasks: forming new friendships and creating their identity and reputation (Boneva et al. 2006; boyd 2008a; Ellison et al. 2007; Lampe et al. 2006; Madden and Smith 2010; Marwick et al. 2010). However, younger individuals are also pretty consistently found to be less concerned about their privacy compared to older individuals (Cho et al. 2009; Nowak and Phelps 1992; Paine et al. 2007).

One explanation given for this is that young and old differ in what they consider privacy to entail. Some studies reported that younger individuals might be more concerned with protecting their privacy in relation to social conflict (boyd and Marwick 2011; Livingstone 2008; Marwick et al. 2010; Raynes-Goldie 2010), as opposed to data mining and identity theft, which may become more important concerns only after individuals grow older. This would also be in line with CPM theory which states that as individuals grow older their desired privacy boundaries will evolve as well (Petronio 2002). Therefore the following hypothesis was formulated:

Hypothesis 3: Younger respondents, compared to older individuals, will attribute more importance to privacy protection features against social conflict.

CPM theory also states that in the case of turbulence, or privacy violations, individuals will be motivated to adjust their privacy boundaries (Petronio 2002). Indeed, several studies reporting a reactive attitude of users concerning their online privacy settings. Debatin and colleagues found that respondents who actually experienced a privacy violation, as opposed to hearing about it happening from others, were more likely to take steps to protect their online privacy (2009). Similarly, Govani and Pashley concluded that raising the awareness of the privacy threats is not enough to nudge people into protecting their privacy (2007). It is therefore likely that a relationship exists between the importance individuals attribute to different privacy protection features and any negative consequences they may have experienced on SNSs. Thus, the following hypothesis was formulated:

Hypothesis 4: Respondents who have experienced a negative consequence from using SNS will attribute more importance to privacy protection related to that experience.

15.2 Method

This study employed choice-based conjoint (CBC) analysis in order to be able to compare the relative importance attributed to various privacy protection features. In a traditional CBC design, respondents are given several discrete choice tasks of

Table 15.1 Overview of all features and levels used in the choice-based conjoint study

Feature	Level	
Data ownership	No ownership of data by SNS	O1
	Ownership of data by SNS, until deleting profile	O2
	Ownership of data by SNS, also after deleting profile	O3
Access by third parties	Third parties cannot access and use personal data	A1
	Third parties can only access and use personal data with permission	A2
	Third parties can access and use personal data without permission	A3
Real information	No obligatory information necessary	I1
	Real email-address must be entered, but not obligatory shown on profile	I2
	Real email address must be entered, and must be shown on profile	I3
	Real telephone number must be entered, but not obligatory shown on profile	I4
	Real telephone number must be entered, and must be shown on profile	I5
Private profile	Private profile and sorting of contacts	S1
	Private profile but no sorting of contacts	S2
	No private profile and no sorting of contacts	S3
Visibility of visitors	Profile visitors are not visible	V1
	Profile visitors are visible	V1
Tagging	Photos cannot be tagged.	T1
	Photos can be tagged, only with permission	T2
	Photos can be tagged, without permission	T3

selecting a concrete offering out of a selection of products with several features which differ over several levels. This could for example concern pizza's, which vary in the features price (e.g., with the levels cheap versus expensive), size (e.g., large versus small), toppings (e.g., cheese versus salami) and brand (e.g. unknown versus familiar). The respondents would be presented with several different pizza's and asked which they would be most likely to buy. When the resulting trade-off decision is repeated several times, the relative value of each feature can be determined; will people buy a pizza based on the price or the brand?In addition, it ccan be determined which level of these features is most preferred; do they rather have cheese or salami as a topping.

For the current study, respondents were presented with several scenarios depicting hypothetical SNSs and were asked on which SNS they would prefer to create a profile. The SNSs varied based on 6 features affecting online privacy: *data ownership, access by third parties, real information, private profile, visibility of visitors,* and *tagging.* An overview of the features and the levels in which the features will vary during the discrete choice tasks are shown in Table 15.1.

Generally, all features included a level for the presence or absence of a privacy protective setting or policy. An additional level was added for *data ownership*, *access third parties*, *private profile*, and *tagging* in which the user had control over the

If you have to choose between the three social network sites below, which would you choose?

Data ownership	Ownership data by SNS, also after deleting	Ownership data by SNS, also after deleting	No Ownership data by SNS
Access by third parties	Third parties can access and use personal data, without permission	Third parties cannot access and use personal data	Third parties cannot access and use personal data
Real Information	Real e-mail address must be entered and shown on profile	No obligatory information required	Real telephone number must be entered and shown on profile.
Private profile	You can shield your profile and sort you contacts	You cannot shield your profile	You can shield your profile but cannot sort your contacts
Visibility of visitors	Visitors are visible	Visitors are not visible	Visitors are visible
Tagging	Photos cannot be tagged	Photos can be tagged, only with permission	Photos can be tagged without permission
	O	O	O

Fig. 15.1 This is an example of the screen respondents were presented

feature. The feature *real information* included levels which varied in the sensitivity of the information required to be provided (i.e. an email-address versus a telephone number) and whether the obligatory information should also be visible on the profile.

Each of these features affects the privacy protection against data mining, identity theft, or social conflict differently. The features *data ownership* and *access by third parties* primarily concern the protection against data mining. The features *tagging* and *visibility of visitors* concern the protection against social conflict. The feature *private profile*, however, affects both identity theft—is the profile private or not- and social conflict—can the user sort his contacts in different groups or not. Similarly, the feature *real information* affects both data mining—is contact information obligatory or not- and identity theft—should contact information be shown on the profile or not.

The SNSs were presented to respondents in the form of an online survey. The online survey was conducted by the research institute TNS-NIPO[2] by means of the CAWI-method (computer assisted web interviewing), which allows respondents to participate from their own computer at home. The survey consisted of three parts.

The first part contained instructions that explained the content of the survey. All features and their levels were explained in the instruction, to get all respondents to have a similar understanding of what the different levels entail.

The second part consisted of the actual discrete choice tasks. Respondents were presented with 15 discrete choice tasks each. Each task consisted of three different SNSs from which respondents had to pick the one they preferred. Figure 15.1 shows

[2] www.tns-nipo.com

Table 15.2 Age gender, and profile of respondents across age groups

	12–13	14–15	16–19	20–25	26–30	31–40	41–50	50+	Total
N	66	68	66	77	67	67	71	78	560
Age	12.6	14.5	17.2	22.5	28.2	35.5	44.9	62.5	30.4
Gender (male)	42.4 %	42.6 %	40.9 %	40.3 %	31.3 %	44.8 %	39.4 %	50.0 %	41.6 %
Facebook	6.1 %	16.2 %	24.2 %	42.9 %	41.8 %	31.3 %	29.6 %	34.6 %	28.8 %
Hyves	48.5 %	13.2 %	9.1 %	6.5 %	3.0 %	4.5 %	11.3 %	19.2 %	14.3 %
Both	45.5 %	70.6 %	66.7 %	50.6 %	55.2 %	64.2 %	59.2 %	46.2 %	57.0 %

an example of a discrete choice task as presented to the respondents. All possible combination of levels were equally represented throughout the experiment.

The third part contained a short questionnaire with several follow-up questions to further explore the motivation behind the choices participants made. First, respondents were asked to indicate which of the features had been most important for them in making their decisions. Next, respondents were asked if they were willing to switch to another social network site provider. If so, they were subsequently asked what their primary reason would be. Lastly, two yes/no questions asked respondents whether they were specifically concerned about something when using their profile and whether they have had a negative experience due to using their profile. When answered with a yes, respondents were further prompted to describe what exactly they were concerned with or have experienced. Subsequent responses were categorized as *Misuse information, Privacy* (e.g. greater visibility or other general statements about privacy), *Criminals* (e.g. hackers or burglars), *undesired contact, social conflict* (e.g. bullying or fights), or *other* (e.g. technical problems). Four raters categorized the responses independently and inter-rater reliability were acceptable for both *concerns* (Kappa's ranging from .727 to .807) and *negative experiences* (Kappa's ranging from .626 to .694). Disagreements were resolved through discussion.

15.2.1 Participants

Respondents were recruited from participants of an earlier study concerning privacy and user behavior on social network sites and obtained by means of a stratified sampling procedure. Five hundred and sixty respondents (327 female, 233 male, $M_{age} = 30.36$, $SD = 16.83$) completed the survey. Table 15.2 provides an overview of age and gender distribution over all age groups. All respondents are members of Facebook or Hyves. Respondents were rewarded for their participation with credits through which they can obtain coupons at TNS-NIPO. Informed consent was obtained from all respondents and parents provided consent for respondents younger than 18-years-old.

15.2.2 *Analysis Plan*

TNS-NIPO makes use of the simulation tool 'Valuemanager' for conjoint analysis, which provides two statistics of interest: (1) *importance percentages*, (2) *utility scores* (see also Orme 2010, chap. 9). An importance percentage is calculated for all six features. This percentage is an estimation of how many decisions were primarily based on that feature. The importance percentages of all features will add up to 100. The utility score provides the relative importance for each level within a feature. This utility score can't be compared between features, but within a certain feature one can determine which level was preferred most (provided the most utility) by respondents in their decisions. The utility scores of the levels within a feature add up to 0. As a result, a negative utility does not necessarily mean that that a specific level was disliked; other levels within that feature were simply preferred.

In order to analyze the importance percentages obtained through conjoint analysis and other percentages, one sample t-tests between percents were used. For the comparison between groups one-way ANOVA's and χ^2 analysis were used. Bonferroni post hoc analysis were used to examine significant one-way ANOVA results, whereas the adjusted standardized residuals were compared for significant χ^2's. Some analysis only involved a sub group of the total sample and therefore violated the assumption of χ^2 analysis that each cell should hold a minimum of 5 individuals. To avoid this, dummy variables were made of separate answer categories and only three age groups were used: 12–19-year-olds, 20–30-year-olds, and 31 and older.

15.3 Results

15.3.1 *Importance Percentages and Utilities*

Before testing the first hypothesis, the utility scores were inspected to gain some insight in the decision patterns of respondents. Table 15.3 shows that rather than each feature having a level that was clearly preferred over the others, each feature has a level that is clearly less preferred compared to the other levels. This means that their decisions were primarily based on avoiding certain levels, as opposed to picking SNSs which contained at least a certain level of privacy protection.

Only for *tagging* a clear preference for a certain level seems to exist as well; tagging should be possible, but only with permission (T2). Concerning the other features, the respondents clearly disfavored SNS where: SNS had ownership over the data, also after deleting the profile (O3), third parties can access and use personal data without permission (A3), a real telephone number must be provided, and must be shown on profile (I5), access to the profile cannot be limited to contacts only, and where the contacts cannot be sorted into groups (S3), profile visitors are visible (V1), and photos can be tagged without permission (T3).

The utilities show that the levels that provide control to the respondent (in the form of having to give permission) were most preferred. Furthermore, the utilities concerning the feature *real information*, suggest that respondents were primarily

Table 15.3 Utility scores obtained through conjoint analysis

Features	Level	Utility
Data Ownership	O1	34,5
	O2	24,0
	O3	−58,5
Access by third parties	A1	26,0
	A2	36,4
	A3	−62,3
Real information	I1	36,1
	I2	46,5
	I3	−22,7
	I4	21,5
	I5	−81,4
Private profile	S1	32,2
	S2	21,9
	S3	−54,0
Visibility of Visitors	V1	0,5
	V2	−0,5
Tagging	T1	0,4
	T2	19,7
	T3	−20,1

See Table 15.2 for the content of the levels

concerned with having to show contact information on their profile rather than having to share a telephone number with the SNS per se. Requiring a telephone number to create a profile was preferred over the requirement of a visible email address, but a required (not visible on the profile) email address was preferred over no required information at all. This suggests that respondents had little problem with providing contact information to the SNSs.

Next, the importance percentages obtained through conjoint analysis were investigated to test the first hypothesis that respondents would attribute more importance to privacy protection from data mining and identity theft. *Real information* was deemed most important (26.4 %). Followed by *data ownership* (20.8 %), *access by third parties* (19.6 %), *private profile* (16.6 %), *tagging* (11.6 %), and *visibility of visitors* (5.0 %). These percentages support the hypothesis that respondents attribute most importance to privacy protection data mining and identity theft. The features concerning privacy protection against data mining and identity theft, i.e. *real information, data ownership,* and *access by third parties*, determined the decision of respondents in 66.8 % of all discrete choice tasks which is significantly more often than the remaining features which primarily concerned social conflict, $t(559) = 8.44$, $p < .001$.

The responses to the question which of the features had been most important for respondents in making their decisions, shows a rather different picture from the one presented by the importance percentages obtained through the conjoint analysis. Only 10 % of respondents reported *real information* to be the most important feature for their decisions whereas 44.3 % of all respondents reported *private profile* to be the most important feature.

Table 15.4 Importance percentages obtained through conjoint analysis and self reported importance attributed to features for decision making

	Importance percentage	Self reported importance (%)
Real information	26.4 %	10.0 %
Data ownership	20.8 %	26.4 %
Access by third parties	19.6 %	11.1 %
Private profile	16.6 %	44.3 %
Tagging	11.6 %	1.6 %
Visibility of visistors	5.0 %	3.6 %
None/Don't know	–	1.3 %

Table 15.4 provides both the importance percentages obtained through conjoint analysis and the percentage of individuals reporting what feature was most important for their decisions. When comparing the self reported importance of the features for decision making with the through conjoint analysis obtained importance percentages of the features, a clear discrepancy can be seen. While respondents claim that the feature *private profile* was most important for their decisions, the importance percentages suggest that three other features have actually been more important instead in their actual decisions.

15.3.2 Willingness to Switch SNS

Next, the second hypothesis was explored: that respondents would generally be unwilling to change to a different SNS provider. In total 201 (35.9 %) respondents indicated they would be willing to switch. Responses to the question when they would be willing to switch could be grouped in several categories. Most respondents willing to switch mentioned they would change only if their friends would change as well (36 %) or if their privacy was better protected at the other site (33 %). Alternatively, respondents would be willing to switch if the other site might be easier, better, or more fun (20.3 %), or they would switch for another reason (10.7 %). Of the 359 (64.1 %) respondents not willing to change, the most often heard reason was that they were satisfied with their current SNS (54.2 %), followed by, that it would cost too much time and effort (10.2 %), they are using their current profile little as it is (9.6 %), it would result in even more information on the internet (3.1 %), and other reasons (8.2 %).

These results provide support for the second hypothesis as the majority of respondents indicated to be unwilling to switch to a different SNS provider. Furthermore, a third of the respondents willing to change will only do so if their current social network (i.e. their friends) switches as well.

15.3.3 Age Based Differences for Importance Percentages

The third hypothesis stated that in comparison to older individuals, younger individuals would attribute more importance to privacy protection against social conflict. Investigation of the importance percentages did not provide support for this hypothesis. Although one-way ANOVA showed a significant age effect for *tagging*, F (7,552) = 2.307, p = .025, post hoc analysis did not indicate a significant difference between any of the age groups. Furthermore, no age effect was found for *real information*, F (7,552) = 1.817, p = .082, *data ownership*, F (7,552) = 1.271, p = .262, *access by third parties*, F (7,552) = .583, p = .770, *private profile*, F (7,552) = 1.275, p = .260, and *visibility of visitors*, F (7,552) = .778, p = .606.

Similarly, investigation of the self-reported importance of the features with χ^2 analyses showed little to no support for the hypothesis. A significant age effect was found for *access by third parties*, χ^2 (7, 560) = 16.60, p = .020, and for *visibility of visitors*, χ^2 (7, 560) = 21.51, p = .003. Significantly more 41–50-year-olds and respondents older than 50 reported these features to be most important for their decisions. No age effect was found for *data ownership*, χ^2 (7, 560) = 11.14, p = .133, *real information*, χ^2 (7, 560) = 5.99, p = .540, *private profile*, χ^2 (7, 560) = 11.57, p = .115, and *tagging*, χ^2 (7, 560) = 6.44, p = .489.

To summarize, no concrete differences were found between the age groups concerning the importance percentages obtained through analysis and the self reported importance of the features. Which means that not only all respondents of all ages similalry reported which feature was most important for their decision making, but they also made similar decisions during the discrete choice tasks resulting in similar importance percentages. Subsequently, the discrepancy between the importance percentages obtained through analysis and the self reported importance of the features is similar as well for respondents of all ages.

15.3.4 Concerns and Experienced Negative Consequences

First, respondents' responses to what they were concerned about when using SNSs were investigated. Of all respondents, 228 (40.7 %) reported to be concerned with something when using their profile. Table 15.5 provides an overview of what respondents were concerned with. Overall, significantly fewer 12–19-year-olds (32 %) reported to be concerned when using their profile compared to 20–30-year-olds (52.8 %) and respondents 31-years-old and older (41.2 %), χ^2 (4, 560) = 18.64, p = .001.

Privacy was mentioned as a concern most often (41.2 %) followed by *misuse information* (32.5 %), *criminals* (12.3 %), *undesired contact* (7.0 %), *social conflict* (4.8 %) and *other* (2.2 %). Significantly more respondents 31 and older and fewer 12–19-year-olds were concerned about *misuse information*, χ^2 (2, 228) = 9.72, p = .008. No age differences were found in the number of respondents who were concerned with *privacy*, χ^2 (2, 228) = 4.94, p = .085, *criminals*, χ^2 (2, 228) = 4.29, p = .117.

Table 15.5 Reported concerns or experienced negative consequences from using social network sites

	12–19	20–30	30+	Total
Concerns				
N	64	76	88	228
Misuse information	21,9 %	27,6 %	44,3 %	32,5 %
Privacy	34,4 %	51,3 %	37,5 %	41,2 %
Criminals	10,9 %	18,4 %	8,0 %	12,3 %
Undesired contact	17,2 %	1,3 %	4,5 %	7,0 %
Social Conflict	12,5 %	1,3 %	2,3 %	4,8 %
Other	3,1 %	0,0 %	3,4 %	2,2 %
Experienced				
N	35	15	31	81
Misuse information	0,0 %	13,3 %	16,1 %	8,6 %
Privacy	17,1 %	6,7 %	22,6 %	17,3 %
Criminals	2,9 %	20,0 %	3,2 %	6,2 %
Undesired contact	5,7 %	26,7 %	19,4 %	14,8 %
Social Conflict	57,1 %	20,0 %	22,6 %	37,0 %
Other	17,1 %	13,3 %	16,1 %	16,0 %

Due to the low number of respondents *undesired contact, social conflict,*and *other* could not be reliably analyzed, although a trend is visible in Table 15.5 that 12–19-year-olds more often reported the former two.

Eighty-one respondents (14.5 %) reported to have actually experienced a negative consequence from their presence in a SNS. Table 15.5 shows what negative consequences were experienced by respondents. No age differences were found in number of respondents reporting negative experiences, χ^2 (2, 560) = 3.39, $p = .183$.

Social conflict was the most reported negative experience (41.2 %), followed by *privacy* (17.3 %), *other* (16.0 %), *undesired contact* (14.8 %), *misuse information* (8.6 %) and *criminals* (6.2 %). The low number of respondents does not allow for a reliable comparison between the age groups. However, a higher percentage of 12–19-year-olds reported *social conflict*, while fewer 12–19-year-olds reported *undesirable contact* and *misuse information. Privacy* and *criminals* were reported by more 20–30-year-olds.

These results suggest that differences do exist between the online experience of privacy of younger and older individuals in line with the expectations of hypothesis 3. Respondents 12–19-year-old appear to be more concerned about social conflict, and report to have experienced it more often.

Finally, to investigate hypothesis 4 which predicted a relationship between the attribution of importance to privacy protection features and experienced negative consequences, the relationship between negative experiences and the importance percentages obtained from analysis was explored. A significant relationship was found between the reported negative experiences and the feature *data ownership.* Respondents who reported to have experienced a negative experience had a significantly higher importance percentage (23.6 %) for the feature *data ownership* than respondents who reported not to have experienced a negative experience (20.3 %), F (1,559) = 6.330, $p = 0.012$.

Further investigation showed that only respondents who experienced misuse of their information attributed more importance to *data ownership*. The importance percentage for these respondents was 35.1 % as opposed to the average importance percentage of 23.6 %. However, due to the low number of respondents involved in this analysis, the results did not achieve statistical significance. Therefore, only marginal support was found for the hypothesis. No statistical significant relationship was found between reported concerns and attributed importance to the various privacy protection features.

15.4 Discussion

This article's main objective was to compare the relative importance SNS users attribute to privacy protection against data mining, identity theft, and social conflict. The presented results show that respondents of all ages attribute most importance to privacy protection against data mining and identity theft. Furthermore, respondents display decision patterns primarily aimed at avoiding obvious privacy violations as opposed to achieving the best possible privacy protection. The implications of these results will be further discussed next.

As was stated in the first hypothesis, respondents were found to attribute most importance to privacy protection features against data mining and identity theft. Thus, SNS users' privacy concerns appear to match the privacy threats given most attention by academia and media alike (Acquisti and Gross 2006; Andrews 2012; Debatin et al. 2009; Govani and Pashley 2005; Gross and Acquisti 2005; Noda 2009; Timmer 2009). This suggests that all respondents were at least to some degree aware of the possible dangers and thus the importance of protection against these potential privacy threats.

In the introduction, it was argued that individuals' online behavior should not be used as gradient for their privacy concerns, because SNSs are primarily used for the possibilities they create for social interaction (Ellison et al. 2007; Steinfield et al. 2008). Indeed, only a third of respondents reported to be willing to switch in line with the second hypothesis. Furthermore, a third of those willing to switch reported explicitly that they would only switch if their social network (of friends) would switch as well, further supporting that participation is generally based on the social merits these sites provide and the choice of SNSs thus largely depends on where the social network of the individual is present.

In other words, the social utility of SNSs appears to be the primary reason individuals make use of the sites and thus have to accept the potential privacy threats as a cost for participation. Given the massive popularity of SNSs- Facebook has over 1 billion users[3]- non-participating may even be associated with social costs by individuals as they miss out on the social interaction (Raynes-Goldie 2010). As a result even privacy concerned individuals are likely to participate on SNSs. Since the business model of SNSs depends on their users sharing information as openly

[3] Statistic from newsroom.fb.com

as possible (Andrews 2012; Pékarek and Leenes 2009), safeguarding the privacy of their users can not be considered their priority.

No support was found for the third hypothesis which stated that younger respondents would attribute more importance to protection features against social conflict. Neither the importance percentages obtained through conjoint analysis, nor the self reported importance attribution differed significantly between younger and older respondents. The hypothesis was based on CPM theory predicting differences between desired privacy boundaries as age progresses (Petronio 2002) and previous studies suggesting youth to be more concerned about their privacy in relation to social conflict with known others (boyd and Marwick 2011; Livingstone 2008; Marwick et al. 2010; Raynes-Goldie 2010). Only the fact that more 12–19-year-olds reported fears or negative consequences related to social conflict provides some support that these privacy concerns play a bigger role for youth.

A possible explanation for the lack of differences between the age groups concerning the importance attributed to the privacy protection features could be that users have numerous other tools to safeguard their privacy concerning social conflict. Even without the features used in this study, youth can safeguard their privacy concerning social conflict by using multiple sites, or by using more private channels for more intimate interactions (boyd and Marwick 2011, p. 14; Lampinen et al. 2009; Stutzman and Hartzog 2009). Conversely, users are fully dependent on the settings and policies provided by the SNS platform concerning their privacy protection against data mining and identity theft, especially if deception is not possible or desirable. As a result, even if younger individuals are more concerned to avoid social conflict they may still have prioritized their privacy protection against data mining and identity theft in this study, because they have no control over these forms of privacy other than the features the SNS provides.[4]

The lack of differences between age groups does suggest that even young respondents are aware of the importance of privacy protection features on SNSs against data mining and identity theft. It is noteworthy that respondents of all ages (i.e. even 12-year-olds and adults) attributed similar importance to all privacy protection features. Especially since younger individuals are often considered to care less about their privacy than adults. These results suggest that SNS users of all ages still attribute importance to their privacy protection from data mining and identity theft, even though their use of SNSs makes them vulnerable to these threats.

A distinctive pattern was found in the utility scores. Instead of demonstrating a clear preference, respondents instead demonstrated a clear dislike (relative to the other levels) for a certain level of each feature. In other words, respondents' decisions during the discrete choice tasks were not necessarily based on obtaining a certain ideal SNS concerning privacy protection, but mainly on avoiding unacceptable privacy violations. When looking at *real information*, for example, respondents didn't primarily pick the SNSs in which they didn't have to fill in any information (in fact having to provide an email address, not shown on the profile was most preferred),

[4] New developments like the Google dashboard may give users more control in time in this respect. Google Dashboard promises users more transparency and control concerning the information linked to their google accounts. https://accounts.google.com

but mainly avoided those SNSs which required them to show the information on their profile. At that point the privacy situation apparently became unacceptable for respondents.

The previously described decision pattern seems related to the fact that individuals often take active steps to protect their privacy after an incident. A negative incident is often the first clear sign that the privacy protection was lacking. Therefore, individuals may be under the impression that their privacy protection is good enough until it is too late. Either because they lack of accurate knowledge on how well protected they are (Hoofnagle et al. 2010) or don't expect to be singled out amongst all the other SNS users (e.g. "Safety in numbers", see Grimmelman 2009, p. 1161). As a result, individuals can be expected to only take action once their lacking privacy protection has become visible through a negative experience, as opposed to continuously trying to obtain the best privacy protection.

Here, however, only marginal support was found for the fourth hypothesis that respondents importance attribution to privacy protection features would related to experienced negative consequences. Respondents who reported their data having been misused, did attribute more importance to the feature *data ownership,* but not statistical significance was obtained. This could be the result of the low number of respondents reporting to have experienced a negative consequence, or could simply mean that this relationship does not exist. Future studies may want to explore this more elaborately.

The focus on avoiding unacceptable privacy violations might also be related to the discrepancy found between what conjoint analysis produced as main features for decision making, i.e. *real information*, *data ownership*, and *access third parties,* and what respondents reported to be the main feature, i.e. *private profile*. Although the former three features were the most prominent for decision making according to conjoint analysis, the decision for these features may have been made by rules of thumb: if the level to be avoided was present, that SNS would not be chosen. As mentioned before, for the feature *private profile,* numerous alternatives exist for SNS users to protect their privacy, while for the other three features concerning data mining and identity theft they are primarily dependent on what the SNS provides.

In this study, respondents apparently first and foremost tried to avoid the undesired levels for *real information*, *data ownership*, and *access third parties*. This decision may have been made relatively easy in the respondents' mind. The subsequent decision if two or more SNSs would still be left, would have to be based on the remaining features. This aspect of decision making may have been more prominent for respondents and as a result indicate the feature *private profile* as most important for decision making.

15.4.1 Limitations

This study made use of a CBC design to assess the privacy attitudes of SNS users. The used levels in this experiment are rather long compared to usual discrete choice tasks making the decision making more difficult. When discrete choice models become too complex, respondents could resort to simplified decision strategies. Instead of assessing the entire scenario, they will mainly focus on one or two features that are important to them. In this case it could have occurred for *real information,* and especially for the level which required a telephone number to be provided which would be visible on the profile. This may have been particularly unacceptable, causing some respondents to base their decision primarily on this.

A second limitation is the fact that the presented SNSs consisted of only privacy related features. No features were included concerning the services a SNS provides (e.g., gaming or interaction possibilities), possible costs (e.g., monthly fee), or other concerns (e.g., safety). The aim of the current study was to distinguish the attributed importance to privacy protection from various risks. Additional non-privacy related features would have made the SNSs too big for respondents to make repeated concentrated rational decisions. Therefore, no conclusions can be drawn on how important privacy protection is in relation to the provided services, costs, or other concerns. Future studies might want to do a similar set up with non-privacy related features in order to investigate the relative importance of privacy protection in relation to these other factors.

15.5 Conclusion

The results showed that respondents of all ages attribute more importance to privacy protection features against data mining and identity theft than social conflict. Here it was argued, that SNSs are used for their social merits and individuals generally have little choice on what SNSs to use. As a results, users are dependent on the SNS platform to provide the necessary privacy protection against the various privacy threats of data mining and identity theft. Lack of these features does not necessarily mean that the users no longer care about these privacy threats, as the results here demonstrated.

A recent initiative, The Brussels Privacy Declaration[5], calls attention to the need for regulation of privacy rights online. The results presented here further support the urgency for the development of regulation of online privacy. Even though SNS users are aware of the importance of privacy protection against data mining and identity theft, they generally do not optimize their privacy protection. Instead, they appear to settle for what they perceive as good enough, which is avoiding the obvious and worst privacy violations. In addition, users are generally dependent on the service provider concerning what privacy protection is available. As such the need for regulation is great; SNS users cannot be expected to protect their own privacy optimally, certainly if the only way to perfectly maintain the online boundaries is by not participating.

---

[5] Brusselsdecleration.net

15.6 Funding

The research reported in this chapter was partially funded by the Netherlands Organization for Scientific Research (NWO).

References

Acquisti, A., and R. Gross. 2006. Imagined communities: Awareness, information sharing, and privacy on the facebook. Paper presented at the 6th Workshop on Privacy Enhancing Technologies, Cambridge, UK.

Andrews, L. 2012. Facebook is using you. The New York Times. http://www.nytimes.com. Accessed 5 Jul. 2012.

Binder, J., A. Howes, and A. Sutcliffe. 2009. The problem of conflicting social spheres: Effects of network structure on experienced tension in social network sites. Paper presented at the CHI, Boston, M.A.

Boneva, B.S., A. Quinn, R.E. Kraut, S. Kiesler, and I. Shklovski. 2006. Teenage communication in the instant messaging era. In *Computers, phones, and the internet: Domesticating information technology,* eds. R. Kraut, M. Brynin, and S. Kiesler 201–218. Oxford: Oxford University Press.

boyd, d.m. 2008a. Facebook's privacy trainwreck: Exposure, invasion and social convergence. *Convergence* 14 (1): 13–20.

boyd, d.m. 2008b. Taken out of context: American teen sociality in networked publics. Doctoral Diss., Berkeley, University of California. http://www.danah.org/papers/TakenOutOfContext.pdf.

boyd, d.m., and A. E. Marwick. 2011. Social privacy in networked publics: Teens' attitudes, practices, and strategies. A decade in internettime: Symposium on the dynamics of the internet and society. http://ssrn.com/abstract=1925128.

Cho, H., M. Rivera-Sánchez, and S.S. Lim. 2009. A multinational study on online privacy: Global concerns and local responses. *New Media Society* 11 (3): 395–416.

Curry, J. 1996. Understanding conjoint analysis in 15 minutes. *Sawtooth Software Research Paper Series.* Sequim, WA: Sawtooth Software.

Debatin, B., J. P. Lovejoy, A. Horn, and B. N. Hughes. 2009. Facebook and online privacy: Attitudes, behaviors, and unintended consequences. *Journal of Computer Media Communication* 15 (1): 83–108.

Ellison, N. B., C. Steinfield, and C. Lampe. 2007. The benefits of facebook "friends": Social capital and college students' use of online social network sites. *Journal of Computer Media Communication* 12 (4): 1143–1168.

Govani, T., and H. Pashley. 2005. Student awareness of the privacy implications when using facebook. http://citeseerx.ist.psu.edu/viewdoc/download?doi=2010.2001.2001.2095.6108&rep=rep2001&type=pdf. Accessed Sep. 2007.

Grimmelmann, J. 2009. Saving Facebook. *Iowa Law Review* 94:1137–1206.

Gross, R., and A. Acquisti. 2005. Information revelation and privacy in online social networks. Paper presented at the proceedings of the ACM Workshop on Privacy in the Electronic Society, Alexandria, Virginia, USA.

Johnson, B. 2010. Privacy no longer a social norm, says Facebook founder. *The Guardian.* http://www.guardian.co.uk/technology/2010/jan/11/facebook-privacy. Accessed 14 Nov. 2011.

Hoofnagle, C., J. King, S. Li, and J. Turow. 2010. How different are young adults from older adults when it comes to information privacy attitudes & policies? http://ssrn.com/abstract=1589864.

Lampe, C., N.B. Ellison, and C. Steinfield. 2006. A Face(book) in the crowd: social searching vs. social browsing. *Proceedings of CSCW-2006,* 161–170. New York: ACM Press.

Lampinen, A., S. Tamminen, and A. Oulasvirta. 2009. "All my people right here, right now": Management of group co-presence on a social networking site. Paper presented at the International Conference on Supporting Group Work (GROUP'09), Sanibel Island, Florida, USA.

Litt, E. 2012. Privy to privacy on social network sites: Another digital divide. Paper presented at the Amsterdam Privacy Conference, 2012, Amsterdam, The Netherlands.

Livingstone, S. 2008. Taking risky opportunities in youthful content creation: Teenagers' use of social networking sites for intimacy, privacy and self-expression. *New Media & Society* 10 (3): 393–411.

Madden, M., and A. Smith. 2010. *Reputation management and social media*. Pew Internet and American Life Project report. http://pewinternet.org/~/media//Files/Reports/2010/PIP_Reputation_Management.pdf.

Marwick, A. E., D. M. Diaz, and J. Palfrey. 2010. Youth, privacy and reputation. Berkman Center Research Publication No. 2010–2015; Harvard Public Law Working Paper No. 2010–2029. http://ssrn.com/abstract=1588163.

Mongold, B. 2010. Facebook Privacy—Can you track who visits your profile? *Five Free Apps*. http://www.fivefreeapps.com/2010/01/facebook-privacy-can-you-track-who-visits-your-profile.html. Accessed 16 Jul. 2012.

Noda, T. S. 2009. Facebook still a hotbed of identity theft, study claims. PCWorld. http://www.pcworld.com. Accessed 5 Jul. 2012.

Nowak, G. J., and J. E. Phelps. 1992. Understanding privacy concerns: an assessment of consumers' information-related knowledge and beliefs. *Journal of Direct Marketing* 6 (4): 28–39.

Orme, B. 1996. Which conjoint method should I use? Sawtooth Software Research Paper Series. Sequim, WA: Sawtooth Software.

Orme, B. 2010. Getting started with conjoint analysis: Strategies for product design and pricing research. Madison: Research Publishers LLC.

Paine, C., U-D. Reips, S. Stieger, A. Joinson, and T. Buchanan. 2007. Internet users' perception of 'privacy concerns' and 'privacy actions'. *Human-Computer Studies* 65:526–536.

Pekárek, M., and R. Leenes. 2009. *Privacy and social network sites: Follow the money*. Paper presented at the W3C workshop on the future of social networking, Barcelona, Spain.

Petronio, S. 2002. *Boundaries of privacy: Dialectics of disclosure*. Albany: State University of New York Press.

Raynes-Goldie, K. 2010. Aliases, creeping, and wall cleaning: Understanding privacy in the age of Facebook. *First Monday* 15 (1).

Skeels, M. M., and J. Grudin. 2009. When social networks cross boundaries: A case study of workplace use of Facebook and LinkedIn. Paper presented at the International Conference on Supporting Group Work (GROUP'09), Sanibel Island, Florida, USA.

Steinfield, C., N. B. Ellison, and C. Lampe. 2008. Social capital, self-esteem, and use of online social network sites: A longitudinal analysis. *Journal of Applied Developmental Psychology* 29 (6): 434–445.

Stutzman, F. D., and W. Hartzog. 2009. Boundary regulation in social media. http://ssrn.com/abstract=1566904.

Timmer, J. 2009. New algorithm guesses SSNs using date and place of birth. Arstechnica. http://arstechnica.com. Accessed 5 Jul. 2012.

Chapter 16
Strong Accountability: Beyond Vague Promises

Denis Butin, Marcos Chicote and Daniel Le Métayer

16.1 Introduction

Individuals share more and more personal data and are out of touch with what happens to their data after their release. The principle of *accountability*, which was introduced three decades ago in the OECD's guidelines (OECD 1980), has been enjoying growing popularity over the last few years as a solution to mitigate this loss of control by increasing transparency of data processing. For example, a consortium has been set up in 2009, with precisely the definition and analysis of accountability as one of its primary goals (CIPL 2009a). At the European level, the Article 29 Working Group published an opinion dedicated to the matter two years ago (Article 29 Working Party 2010) and the principle is expected to be enshrined in the upcoming European data protection regulation (EC 2012).

The very popularity of the word yields suspicion. Its widespread use, combined with the lack of a unique definition, begs the question of whether accountability can be characterised precisely enough to achieve consensus and bring sufficient protection. Can one leave behind questions of terminology and elucidate accountability in a way congruent with most interpretations?

In addition, the concept of accountability has been mentioned in so many different settings that it is legitimate to wonder whether a precise and consensual definition, assuming it can be established, would be as broadly applicable as the larger interpretation of the concept seems to be. Is the notion of accountability so diluted that trying

This work has been partially funded by the European FI-WARE project/ FP7-2012-ICT-FI. See http://www.fi-ware.eu/.

D. Butin (✉) · M. Chicote · D. Le Métayer
Inria, Université de Lyon, CITI, F-69621 Villeurbanne, France
e-mail: denis.butin@inria.fr

M. Chicote
e-mail: mchicote@dc.uba.ar

D. Le Métayer
e-mail: daniel.le-metayer@inria.fr

S. Gutwirth et al. (eds.), *Reloading Data Protection*,
DOI 10.1007/978-94-007-7540-4_16, © Springer Science+Business Media Dordrecht 2014

to pinpoint it would remove all the generality that caused its initial appeal as well as its expected virtues?

Finally, assuming accountability can be characterised precisely and is still a concept with broad applications, does it bear the capacity to deliver innovative solutions to long-standing problems such as loss of control over personal data? Could accountability turn out to be little more than an umbrella buzzword for a variety of old solutions merely rehashed under the guise of new terminology?

Even if all those concerns cannot be resolved easily, there is no reason to give accountability a blank check. Apprehensions over the possibility of an accountability strategy backfiring have been spelled out and need to be taken into consideration.

In this article, we will first review the reasons put forward to support accountability, as well as the criticisms raised against it (Sect. 2). It will become apparent how current and upcoming regulations are unsatisfactory in their way to address accountability when compared with requirements seen as essential by many sources.

Discussing accountability critically requires distinguishing between its application levels. We will emphasise what has sometimes been termed *accountability of practice*, the requirement that data controllers should be able to provide a statement (an account) showing that their actual data handling practice complies with their obligations. We contend that the resulting opacity of actual practices and excessive focus on procedures is harmful enough to derail the overall accountability approach. To overcome these limitations, we put forward *strong accountability*, which relies on precise legal requirements supported by effective tools (Sect. 3). We then show that such tools can be provided considering the state of the art in terms of technology and suggest an approach for *accountability by design* (Sect. 4). Of course, technical feasibility is only a prerequisite, not a sufficient condition for effective adoption. As expressed by Colin Bennett (Bennett 2012), "there is little evidence that market pressures alone will push this kind of external conformity assessment". To address this issue, we also provide suggestions for an overall architecture for strong accountability, including legal and economic dimensions (Sect. 5). Finally, we put strong accountability in perspective and discuss its complementary with other privacy instruments such as Binding Corporate Rules, Privacy Impact Assessments, privacy by design and privacy seals (Sect. 6).

16.2 The Meanings of Accountability and the Question of its Value

While accountability is no new idea, its use in the field of privacy and data protection has increased considerably lately. To set up the stage, we sketch in this section some reference documents on accountability in normative documents (Sect. 2.1), in the legal doctrine (Sect. 2.2) and in the computer science literature (Sect. 2.3). Let us note that the goal of this section is not to present a comprehensive survey of accountability[1], but to provide some background information before discussing the pros and cons of accountability in Sect. 3.

16.2.1 Accountability in Regulation and Guidelines

In this subsection, we start with a quick review of some landmarks in terms of accountability before discussing their reception in the legal doctrine in Sect. 2.2 and the computer science view in Sect. 2.3.

16.2.1.1 The United States' FTC FIPPs

Accountability in the context of data protection is currently not enshrined in US law. This is not entirely surprising given the general orientation of US data protection law, which tends to favour self-regulation and only reluctantly impose binding commitments. As far as soft law is concerned, the US Federal Trade Commission's Fair Information Practice Principles (FIPPs) (US Federal Trade Commission 1973), a set of non-binding[2] guidelines that have been used as a basis for specific, sectoral laws such as the Right to Financial Privacy Act (Title 12 of the U.S. Code 1978), do not list accountability in their principles even though they refer to related concepts.[3]

16.2.1.2 The 1980 OECD Guidelines

The introduction of accountability as a *basic principle* in the 1980 Organisation for Economic Cooperation and Development *Guidelines on the Protection of Privacy and Transborder Flows of Personal Data* (OECD 1980) is often cited as its first notable appearance. The precise wording of the Guidelines is the following: "A data controller should be accountable for complying with measures which give effect to the principles stated above." While the aim of these Guidelines is the effective protection of individuals' privacy, the additional goal of economic benefits through simplified data export procedures is evidenced by the second part of their title. As far as enforcement is concerned, one should note that the OECD cannot legislate but only issue *soft law* in the incarnation of guidelines or recommendations.

The *Detailed Comments* part of the Guidelines provides some details about accountability, even though the word itself is never defined. It is written that "Accountability under Paragraph 14 refers to accountability supported by legal sanctions,

[1] We refer the reader to Charles Raab (Raab 2012), Colin Bennett (Bennett 2012) and Daniel Guagnin et al. (2012) for a more complete review.

[2] Note however that the FIPPs have been used as a basis for the US Privacy Act of 1974.

[3] The fifth principle, *Enforcement/Redress*, states that "(...) the core principles of privacy protection can only be effective if there is a mechanism in place to enforce them."

as well as to accountability established by codes of conduct, for instance." An interesting precision is that accountability is still required from a data controller when it uses the services of a third party for data processing. However, the nature of the evidence and the entity receiving that account are not discussed. Some authors (Raab 2012) conclude that the sense in which accountability is used here is close to liability.

In 2011, the OECD published a report (OECD 2011) reviewing the principles of its original Guidelines, including accountability, in light of the new technological and regulatory landscapes. The rising role of the accountability principle is highlighted[4] and "reporting, audits, education, and performance appraisals" are mentioned as some of its components. However, the paragraph of the report[5] dedicated explicitly to accountability mainly addresses data export issues.

16.2.1.3 The Canadian PIPEDA

In terms of regulation, the 2000 Canadian federal[6] Personal Information Protection and Electronic Documents Act (PIPEDA) (Parliament of Canada 2000) also includes a principle of accountability. The stated intent of the act is to balance the protection of personal information with the support of electronic commerce. It is partly based on the Canadian Standards Association's Model Code for the Protection of Personal Information (CSA 1996) and was also heavily influenced by the aforementioned OECD Guidelines.

Of the ten *privacy principles* it includes, accountability is the first one. The principle states that "An organization is responsible for personal information under its control and shall designate an individual or individuals who are accountable for the organization's compliance with the following principles." The concrete measure of designating employees responsible for ensuring company-wide accountability is therefore central to PIPEDA's interpretation of the notion. The fact that this privacy principle refers to all other principles enumerated in the act gives it an overarching, prominent tone.

Another important aspect is the responsibility of organisations for data transferred to third parties: the same section of the act states that an organisation's responsibility includes "(...) information that has been transferred to a third party for processing."

PIPEDA also addresses the issue of practical compliance to some extent, even though its specifications in this respect remain broad: "Organizations shall implement policies and practices to give effect to the principles." The need to provide for means of redress is also mentioned. While not all facets of accountability are made explicit in the Canadian act, it globally remains comparatively precise in its integration of the

[4] Notably the fact that PIPEDA "used the OECD Guidelines as a starting point" while "moving the Accountability Principle to the beginning."

[5] Role of *accountability*, p. 52.

[6] In addition, provincial private sector privacy laws exist in Alberta, British Columbia and Quebec. The principle of accountability also appears in those provincial regulations, although in an implicit form.

principle. For instance, it was innovative in shifting the focus of accountability "from the legal regime to the actual protections afforded by the receiving organisation." (Bennett 2012).

16.2.1.4 The 2004 APEC Privacy Framework and the Data Privacy Pathfinder Program

The Asia–Pacific Economic Cooperation (APEC) forum, an international organisation of 21 countries[7], defined a Privacy Framework (APEC 2004) that includes accountability as one of its 9 *information privacy principles*. Its first mention states that "A personal information controller should be accountable for complying with measures that give effect to the Principles stated above". The document does not specify who should receive the evidence making accountability possible, and is not more explicit than the OECD Guidelines in terms of an actual definition of the concept.

Recently, the APEC started reconsidering the question of accountability in more detail in the scope of its Data Privacy Pathfinder (APEC 2009); in APEC terminology, a Pathfinder is a *cooperative project among participating APEC Economies*. This effort is mostly about facilitating regulated data exports. One of its prominent features is the *Cross-Border Privacy Rules* system, under which organisations, on a voluntary basis, can follow a set of rules with the goal of increasing the trust of consumers and partner organisations in their commitment to privacy. Applications are assessed by APEC-recognised *accountability agents*, "which may include trustmarks, seals, and other private bodies." (OECD 2011).

16.2.1.5 The Accountability Project

Launched by the Centre for Information Policy Leadership in 2009 and commonly termed simply *Accountability Project*, the *Accountability-Based Privacy Governance Project* is an ongoing collaboration between industry actors, non-governmental organisations and government representatives aimed at defining and disseminating components of a standardised accountability strategy. White papers are being released, and the fifth phase of the project, in 2013, discusses the specific challenges of distributed environments such as mobile applications and cloud computing.

Accountability is made more precise in the publications of the project. Notably, it adds the dimension of what could be called the *accountee* or entity receiving the evidence. Unlike the OECD Guidelines and the APEC Privacy Framework, the Accountability Project addresses this point. For instance, the white paper resulting from the second phase of the project (*The Paris Project*) mentions "Organizations may be accountable to three entities: data subjects/individuals, regulators, and business partners." (CIPL 2010).

---

[7] Generally speaking, as pointed out by Colin Bennett (2012), a number of countries engaging in APEC have no national data protection regulation, which makes the existence of this framework all the more important.

The necessity of the link between regulation and concrete measures is articulated in the Paris Project document: "Accountability promotes implementation of practical mechanisms whereby legal requirements and guidance are translated into effective protection for data." Charles Raab (Raab 2012) notes the frequent use of the notion of *demonstration* in the publications of the Accountability Project, reinforcing the idea that accountability implies the readiness by data controllers to show and explain their actions, possibly upon request—a kind of information transparency. The Project, like other think-tanks and regulations surveyed so far, however falls short of going into the details of acceptable practical mechanisms for demonstrable data protection. In addition, even though the role of third-party accountability agents is recognised, data controllers seem to keep the central role, which may cast doubts about the impartiality of the whole process.

16.2.1.6 European Law and the Upcoming Regulation

In European data protection law, there is no explicit principle of accountability of data controllers until now, even though one may argue that the accountability obligation is implicitly present. In its *Opinion on the principle of accountability* (Article 29 Working Party 2010), the Article 29 Working Party has advocated the introduction of an accountability principle defined as "showing how responsibility is exercised and making this verifiable." The verifiability aspect of this definition is important: it implies an audit, which opens the possibility of finding that a data controller did not comply with its obligations. The draft of the new regulation released in 2012 by the European Commission (EC 2012) indeed includes an article about accountability[8], even though the word itself is not used in the article, and the provisions are rather vague. Article 20 states that "The controller shall adopt policies and implement appropriate measures to ensure and be able to demonstrate that the processing of personal data is performed in compliance with this Regulation." and "The controller shall implement mechanisms to ensure the verification of the effectiveness of the measures referred to [in earlier paragraphs]". The draft regulation envisions delegated acts to further specify appropriate measures.

16.2.2 Legal Doctrine

The legal doctrine discusses both the fundamental values underlying the accountability principle and its effectiveness for data protection. As far as principles are concerned, Paul De Hert (De Hert 2012) establishes a link between accountability for privacy and human rights law by pinpointing the duty for EU member states to require effective data protection measures from organisations. More generally, he associates the concept of accountability with external scrutiny—implying the need

[8] Article 20.

for a recipient of the account—and with account giving—the keeping of a record, and its transmission to an authority.

Other authors see accountability as a focusing lens with the potential to address a range of issues with an integrated approach: it "can form the focus for dealing with issues of scale in regulation, privacy risk assessment, self-regulation (. . .) and foster an environment for the development of new technologies for managing privacy" (Guagnin et al. 2012).

Charles Raab subscribes to this idea that accountability is a multifaceted notion, stating that not all of its aspects have been exploited yet: "There are unused dimensions in the concept of accountability that need to be examined and developed" (Raab 2012). He furthermore quotes the interpretation of accountability as *stewardship*: the entrusting of "resources and/or responsibilities" from one party to another. The importance of transparent data sources, the *accounts*, which empower *audiences* to come to their own conclusions regarding the interpretation of data, which should not be left solely to data controllers, is also emphasised. Indeed, many actions tend to be invisible, or at least do not leave a trace in event histories. This combination of facts and descriptions justifies seeing accounts as *stories*, possibly carrying elements of propaganda or bias. He also contends that not only the final account but also the process of manufacturing it should be visible by the audience if full transparency is the ultimate goal.

Taking an operational approach, Colin Bennett clarifies the different levels of accountability by distinguishing three layers: *accountability of policy*, *of procedures*, and *of practice*. He emphasises that excessive focus is often placed on the first one, resulting in only superficial guarantees; and furthermore states that few organisations provide accountability at the practice level, and that this level requires external audit to be credible. The ultimate onus is on data protection authorities to specify a coordinated list of acceptable verification mechanisms. In addition, the incomplete description of actors and evidence in existing regulations and guidelines is pinpointed: for instance, the OECD guidelines do not mention who is expected to receive the evidence.

Joseph Alhadeff, Brendan Van Alsenoy and Jos Dumortier support the need for an accountability of practice as defined by Colin Bennett and an explicit account as stated by Charles Raab by declaring that *common accountability measures* as defined by the Article 29 Working Party "are mainly articulated in the language of principles and concepts" and not linked to practical mechanisms (Alhadeff et al. 2011).[9]

Many points made in this section are discussed in greater detail in the recent volume (Guagnin et al. 2012) resulting from the European Privacy Awareness through Security Organisation Branding (PATS) project.[10] We can see from the above examples that accountability in legal doctrine is often considered as requiring a more

---

[9] They however emphasise that too strict regulations would be a burden and an unacceptable cost for budding companies.

[10] See the project website: http://pats-project.eu/.

concrete, practice-oriented aspect where the nature of the evidence is made explicit. As we will see now, research in computer science concentrates heavily on this practical facet.

16.2.3 Accountability in Computer Science

Accountability in computer science is generally associated with very specific properties. An example of formal property attached to accountability is *non-repudiation*: for example, in an analysis of a certified email protocol, Giampaolo Bella and Lawrence Paulson (2006) see accountability as a proof that a participant cannot deny that he has taken part in the protocol and performed certain actions. The proof of non-repudiation relies on the presence of specific messages in the network history of security protocols. A complementary concept in this work is *fairness*: it is not possible that one agent obtains what they seek while the other does not.

Jan Cederquist et al. (2005) introduce another concept of *agent accountability*: in a data usage control system, an audit authority in possession of evidence should be able to check the formal proofs that entities have to provide to justify themselves. The focus here is on establishing a kind of evidence that is unforgeable, thereby guaranteeing the detection of inappropriate data usage.

Jagadeesan et al. (2009) define accountability as a set of mechanisms based on "after-the-fact verification" by auditors for distributed systems. Mathematics-based methods are used to rigorously check properties of "accountability-based systems" where the interaction between entities, including auditors, is modelled and trade-offs between "potentially conflicting design parameters" are explored. As in (Schneider 2009), blame assignment based on evidence plays a central role in this framework. Integrity (the consistency of data) and authentication (the proof of an actor's identity) are integral to the communication model. Together with non-repudiation (Bella and Paulson 2006), these rather technical concepts are often seen as pillars of the concept of accountability in computer science literature.

On the practical side, (Haeberlen 2009) outlines the challenges and building blocks for accountable cloud computing. Accountability is seen as a desirable property both for customers of cloud services, who need to know whether something went wrong, and for cloud service providers, who can handle complaints and resolve disputes more easily. The building blocks of accountability are defined as *completeness*, *accuracy* and *verifiability*[11]. Technical solutions to enable these characteristics on cloud computing platforms have been devised by the authors.

[11] Those characteristics are defined as follows: completeness means that all agreement violations lead to reports and supporting evidence; accuracy signifies that no violation reports are created if nothing went wrong; and verifiability means that evidence is checkable independently.

16.2.4 Conclusion: Overgenerality Versus Overprecision

The above discussion of the perceptions of accountability in normative texts on the one hand, and in computer science on the other, show that there is quite a shift of emphasis between the two views: normative texts mostly focus on what Colin Bennett calls *accountability of policy* and *accountability of procedures* (internal rules, existence of a data protection officer, corporate training, organizational issues, etc.) while computer scientists place more emphasis on very specific technical requirements for *accountability of practice*. To fill this gap and ensure that technical means can effectively contribute to the implementation of accountability in a broader perspective, more interdisciplinarity is needed: *we need to get together,* as pointed out in (Guagnin et al. 2012). In the following section, we discuss in more general terms the potential benefits and limitations of accountability for privacy protection before suggesting ways to move forward considering technical, legal and economic aspects in Sects. 4 and 5.

16.3 Accountability for Privacy Protection: Promises and Pitfalls

In the previous section, we have reviewed the definitions of accountability in a somewhat neutral way, considering the differences in terms of scope, level of precision and interpretation in the definitions proposed by different communities and authors. The key issue that we want to address now is the potential impact of accountability rules on privacy protection. In Sect. 3.1 we will analyse the reasons to support the view that accountability should play a key role in future privacy protection regulations before discussing the potential pitfalls of accountability for privacy in Sect. 3.2. In Sect. 3.3, we will build on these arguments to argue that (1) accountability principles should indeed become a pillar for privacy protection but, (2) for accountability to be able to play this role, its must meet an absolute requirement of precision at all levels[12]; in default thereof, accountability might turn into a deceptive packaging and a way to further weaken privacy protection.

16.3.1 Accountability as a Key Privacy Enabler

One commonality among the definitions reviewed in Sect. 2, which is at the core of the accountability concept, is its introduction of a set of obligations bearing on controllers: in other words, accountability is complementary to the a priori controls

[12] Definitions of the roles of all stakeholders, their respective commitments, the accounts, the audit procedures, sanctions, etc.

provided by most *privacy enhancing technologies* which make it possible for subjects to limit their release of personal data (e.g. through selective disclosure or the restriction of the disclosure to anonymised or sanitised data). The first and foremost motivation for accountability in the context of privacy is the issue that, after the disclosure of their personal data, subjects are powerless—they have no choice but to trust controllers to handle their data appropriately. But subjects do generally not have any reason to trust data controllers blindly—one could even argue that subjects often have good reasons to distrust them because many companies have strong economic interests in the exploitation of personal data. The potential benefits of accountability appear exactly in such situations where an actor has a sufficient amount of trust in another actor to rely on him for a given action (e.g. to collect his personal data and use it for a given purpose), but is still not completely sure that his confidence is not misplaced. Accountability provides further means to check what happens on the side of the controller when the data has been released and therefore to move from *blind trust* to *proven trust* (De Hert 2012). Actually, considering the ever-growing collection and flow of personal data in our digital societies, a priori controls will be less and less effective for many reasons, and accountability will become more and more necessary to counterbalance this loss of ex ante control.[13]

The reasons why a priori controls lose effectiveness are varied: first, more and more data is collected without the subject knowing it (through various logs, web cookies, surveillance systems, mobile phone applications leaking personal data to application providers or third parties, etc.). Even when the subject is aware of the data collection and asked to provide his consent, this consent has become a fictitious protection because he generally does not take the time to read the privacy notice provided by the controller[14], does not understand its implications[15], or gives his consent for lack of a real alternative (because he needs to get access to information or to a service). Even in situations where the consent of the subject could be considered free and well informed, the privacy notice on which it is based is by no means a proof of actual behaviour of the controller. A privacy notice is a declaration of a controller at a point in time, but the relation between what is announced and the actual mechanics of personal data processing is invisible. Strong discrepancies can be observed between privacy policies and actual practices, which can be due to different causes: the data controller may provide misleading policies from the start,

[13] As stated in the Article 29 Data Protection Working Party Opinion 3/2010 on the principle of accountability (Article 29 Working Party 2010): "Firstly, we are witnessing a so-called 'data deluge' effect, where the amount of personal data that exists, is processed and is further transferred continues to grow. Both technological developments, i.e. the growth of information and communication systems, and the increasing capability for individuals to use and interact with technologies favour this phenomenon. As more data is available and travels across the globe, the risks of data breaches also increase."

[14] In the survey Privacy Notices Research by the Privacy Leadership Initiative, only 3 % of respondents declared to "carefully read" privacy notices "most of the time".

[15] The sheer length of this type of document and their convoluted language often prevents users from finding straightforward answers to simple questions such as a promise not to share personal data with third parties or, in case of share, the precise list of third parties which can receive the data.

the system may evolve without maintaining its original privacy protection, certain controls may rely on actions of the personnel of the controller or on subcontractors, the staff of the controller or his subcontractors may not be well aware or informed about privacy commitments, etc. In addition, the controller himself is not immune to privacy breaches from malicious (or curious) insiders or external attackers. As a result, data subjects have no clear knowledge of how much privacy they give up, do not know what actually happens to their data, and have no way of noticing whether the data controller breaches his obligations. As distributed systems such as mobile or cloud computing become ubiquitous, data subjects lose touch even more with what happens to their personal data.

Even though accountability should by no means be seen as an alternative to substantive data protection requirements (Bennett 2012) or an encouragement to weaken principles such as data minimality, it can help mitigating this loss of control, firstly by making actual behaviour visible and verifiable. Indeed, another common thread in the definitions of Sect. 2 is that accountability relies on the creation of accounts and their audits. Regardless of when and by whom these audits are conducted, their goal is to provide more transparency in data processing and therefore to increase the level of trust that the subject can place on the data controller. Another major benefit of accountability is that it can act as an incentive for data controllers to take privacy commitments more seriously and put appropriate measures in place, especially if audits are conducted in a truly independent way and possibly followed by sanctions in case of breach. As pointed out by Paul De Hert (De Hert 2012), "the qualitative dimension of accountability schemes may not be underrated".

16.3.2 Objections Against Accountability

Accountability is not a principle that receives unanimous support, though. The criticisms of accountability can be based on three types of arguments:

1. Objections from the legal point of view: some lawyers argue that accountability does not bring anything new to the existing notions and legal instruments; others claim that accountability could even accentuate the imbalance of powers between data controllers and data subjects by providing deceptive protections.
2. Reservations based on technical arguments: the very implementation of accountability measures might introduce further risks of personal data breaches.
3. Warnings based on economic arguments: accountability rules would impose unacceptable burdens on the industry.

Let us consider each of these categories of criticisms in turn.

The manifold nature of accountability, combined with currently vague definitions in legal instruments, may lead some data controllers to promote accountability in the hope of avoiding more constraining and comprehensive regulations. An example of such trends is described in a recent report (Ernst and Young 2012): "To avoid greater regulation, organizations in the retail and consumer products industries and

GS1, a supply chain standards organization, are working with privacy commissioners to voluntarily set guidelines that address the privacy implications of using radio frequency identification (RFID) technology in their operations". In the worst case, accountability could be implemented by light organisational measures, for instance by just having in place a data protection officer, an awareness plan and some executive oversight. When audits are conducted by the companies themselves or business associations, the subject may also be concerned about their neutrality: after all, why should he be more confident in self-audits than in self-declarations of privacy policies? For these reasons, accountability has been criticised for offering companies a cheap "data protection favourable" reputation even if their actual practices and accountability rules actually offer limited guarantees, amounting to "privacy greenwashing" (Guagnin 2012). In the same vein, the 2009 white paper (CIPL 2009b) from the Accountability Project also draws criticisms from certain lawyers (Bennett 2012), as it mentions that an accountability strategy allows companies to reach data protection goals in a way "that best serves their business models."

More generally, accountability is often associated with self-regulation, which is a controversial approach. The main benefits of self-regulation are its flexibility and its wider acceptance in the industry: because the rules can be tailored to a given business sector and controlled by the concerned actors, these actors are more likely to follow them. More generally, considering the difficulty to regulate the Internet in the international context, self-regulation is often presented as an adequate solution to face the "disintegration of traditional sovereignty paradigms" (Poullet 2001). However, the validity of self-regulation as a norm has to be assessed against traditional criteria such as the legitimacy of its authors, the conformity of its content with respect to other legal rules and its effectiveness, including the possibility of sanctions (Ibid.). In the context of accountability, one could argue that the second criterion should generally be satisfied (it is to be hoped that the accountability rules defined by e.g. an industrial sector would comply with applicable laws), but the first one is not really satisfied unless a data protection authority officially endorses the rules (or the rules are defined in collaboration with the authority, which could be seen as a form of co-regulation), nor does generally the last one. It should be clear that the lack of real consequences for data controller breaching the code or the lack of effective control would seriously weaken the assurance provided by self-regulated accountability schemes.

Another critical view of accountability relies on the idea that it is just a superfluous notion because it is already implicitly covered by existing instruments. For example, Colin Bennett (Bennett 2012) argues that there is an "unfortunate tendency" to believe that "new constructs for privacy" are needed. According to him, the essential principles of privacy do not need reformulating to allow for accountability: its key aspects can be integrated in existing frameworks. To support this view, one may argue that legal wording such as "Article 22 takes account of the debate on a principle of accountability and describes in detail the obligation of responsibility of the controller to comply with this Regulation and to demonstrate this compliance, including by way of adoption of internal policies and mechanisms for ensuring such compliance." in the draft General Data Protection Regulation released by the European

Commission in January 2012[16] does not add very much to existing obligations. Even more striking are the following comments in the Working Party 29 Opinion on the principle of accountability (Article 29 Working Party 2010): "One may also suggest that accountability refers to the implementation of data protection principles", and "The Article 29 Working Party wishes to highlight that most of the requirements set out in this new provision actually already exist, albeit less explicitly, under existing laws."[17]

From the technical point of view, tensions can also arise between accountability and privacy: the accounts which form the basis of the accountability procedure can themselves involve personal data; enforcing the implementation and storage of these accounts therefore introduces an additional risk for these data. This can be the case, for example, when the accounts take the form of execution logs. Obviously these logs should be subject to strong security measures but, as experience has shown too often, there is no absolute security protection. Data minimisation should therefore be encouraged: only information essential for compliance checking should be recorded in logs. Efficiency is one reason but the main one is to avoid further spreading of personal data. Another concern on the technical side is the authenticity of the accounts. Because they are, by definition, built and stored by (or under the control of) data controllers, how can the auditor and the subject be convinced that they provide a faithful representation of the actual data processing? The accounts could have been forged by the controller to cover up privacy breaches or they could have been tampered with by external actors. Again, technical means can be implemented to enhance the trustworthiness of the accounts (Bellare and Yee 1997; Schneier and Kelsey 1999), but they cannot provide an absolute guarantee, which might become a problem if the accounts are to be used as evidence in legal proceedings.

Needless to say, binding accountability rules are not necessarily welcome in the industry because they would introduce additional obligations and potential costs. As discussed above, this fear can actually turn into a support for a weak form of accountability (focusing on light organisational measures adopted on a voluntary basis). This economic argument should be taken seriously though, as it would be illusory to believe that strong accountability measures could be imposed in any country if they had to result into unacceptable burdens on the industry, especially at a time when personal data has become the "oil of the new economy". We investigate promising paths to address these issues in the following sections.

16.3.3 Beyond Vague Promises: Need for Precise Commitments

We believe that the criticisms discussed in the previous subsection deserve great attention. First, the fact that accountability could turn into deceptive promises providing erroneous expectations to data subjects is of great concern. Indeed, if this grim

[16] Section 3.4.4.1.

[17] Even though this suggestion is not exactly the definition adopted by the Working Party 29 in the rest of the document.

prediction became a reality it would undermine the very value that accountability is supposed to restore, namely trust. To analyse the reasons why an accountability system could be misleading and provide to the subjects a false sense of protection, let us consider the characterisation of accountability proposed by the Article 29 Data Protection Working Party (Article 29 Working Party 2010): "its emphasis is on showing how responsibility is exercised and making this verifiable". To achieve this objective, it is necessary to know precisely: (i) what the responsibilities are, (ii) what pieces of evidence will make the verification possible and (iii) who will be in charge of the verification and in what conditions. Each of the objections in Sect. 3.2 can be related to a failure in one of these steps:

(i) If the commitments of the data controller are not well defined (and properly understood by the data subject) the guarantees provided by the accountability mechanisms are illusory. These commitments should obviously include all applicable legal obligations, but also any industry standards and declarations made by the data controller in his privacy statements.
(ii) If the pieces of evidence are not sufficient to establish that the commitments have been fulfilled, the verification process will not be reliable. This may be the case in particular if the evidence is incomplete or if no guarantee is provided about its integrity and authenticity.
(iii) If the actor in charge of the verification is not trusted by the subject, the whole accountability process will suffer from the same distrust. This would obviously be the case if the audits were conducted by the data controllers themselves or by representatives of their business sector.

The solutions to avoid these failures in the accountability process necessarily blend legal, technical and economic ingredients: the commitments of the controller involve legal obligations; the definition and analysis of the accounts have to rely on technical means; and the roles of all the stakeholders in the process must be integrated within a viable ecosystem. But the keyword and true imperative for all these aspects of accountability is *precision*: any doubt or uncertainty in the process would cause mistrust and subvert the whole approach.

Precision can also be an answer to the second criticism discussed above, i.e. the fact that accountability is a superfluous notion because it is already covered by existing instruments. Indeed, one may agree that if accountability remains a vague obligation as stated in Article 22 of the Draft General Data Protection Regulation (EC 2012)[18], it does not add very much to existing measures. Except for the designation

[18] § 1. The controller shall adopt policies and implement appropriate measures to ensure and be able to demonstrate that the processing of personal data is performed in compliance with this Regulation. § 2. The measures provided for in paragraph 1 shall in particular include: (a) keeping the documentation pursuant to Article 28; (b) implementing the data security requirements laid down in Article 30; (c) performing a data protection impact assessment pursuant to Article 33; (d) complying with the requirements for prior authorisation or prior consultation of the supervisory authority pursuant to Article 34(1) and (2); (e) designating a data protection officer pursuant to Article 35(1). § 3. The controller shall implement mechanisms to ensure the verification of the

of a privacy officer in certain circumstances[19] and the reference to a Privacy Impact Assessment (PIA) which is required only when "processing operations present specific risks to the rights and freedoms of data subjects by virtue of their nature, their scope or their purposes", one may wonder whether this article does not merely make explicit obligations that data controllers already have to meet in case of control by a supervisory authority. In addition, the verification by "independent internal or external auditors" is required only "if proportionate", which can hardly inspire confidence to data subjects. This article adds very little because it lacks precision: the only mandatory items in the records[20] do not include any information that would allow an auditor to check that the controller has processed the personal data in a way consistent with his obligations and declarations. In other words, the Draft General Data Protection Regulation introduces no more than a form of *accountability of procedures*, in Colin Bennett's classification. As a matter of fact, it is significant that Article 22 heavily relies on references to other articles of the draft, which reinforces this impression of redundancy.

As far as technical issues are concerned, solutions have been proposed in the computer science community to enhance the integrity and authenticity of execution logs. For example "forward integrity" (Bellare and Yee 1997) ensures that an attacker taking the control of a computer in which the logs are stored cannot tamper with existing logs (even though he would obviously be able to delete them or to fake future logs). Similarly, techniques have been proposed to authenticate the log entries and to set up a selective access to them, e.g. for external auditors. Again, these techniques can provide strong guarantees if the requirements and assumptions (types of attackers, level of trust between the stakeholders) are precisely defined.

In the remainder of this contribution, we make the point that accountability, to yield real added value for data subjects in terms of trust, should:

---

effectiveness of the measures referred to in paragraphs 1 and 2. If proportionate, this verification shall be carried out by independent internal or external auditors.

[19] Article 35 § 1: The controller and the processor shall designate a data protection officer in any case where: (a) the processing is carried out by a public authority or body; or (b) the processing is carried out by an enterprise employing 250 persons or more; or (c) the core activities of the controller or the processor consist of processing operations which, by virtue of their nature, their scope and/or their purposes, require regular and systematic monitoring of data subjects.

[20] Article 28. § 2 : The documentation shall contain at least the following information: (a) the name and contact details of the controller, or any joint controller or processor, and of the representative, if any; (b) the name and contact details of the data protection officer, if any; (c) the purposes of the processing, including the legitimate interests pursued by the controller where the processing is based on point (f) of Article 6(1); (d) a description of categories of data subjects and of the categories of personal data relating to them; (e) the recipients or categories of recipients of the personal data, including the controllers to whom personal data are disclosed for the legitimate interest pursued by them; (f) where applicable, transfers of data to a third country or an international organisation, including the identification of that third country or international organisation and, in case of transfers referred to in point (h) of Article 44(1), the documentation of appropriate safeguards; (g) a general indication of the time limits for erasure of the different categories of data; (h) the description of the mechanisms referred to in Article 22(3).

- be defined precisely, in all aspects, including the contents of the accounts and the rules to decide if an account is compliant;
- include *accountability of practice* (in Colin Bennett's terminology), i.e., apply not only to declared policies or procedures but also to the actual data processing;
- be supported by independent audits to avoid any risk of accommodating attitudes of the auditors and mistrust from the subjects.

In the following section, we show that this kind of *strong accountability* can be supported by appropriate tools and in Sect. 5, we make some suggestions on the overall accountability architecture, including organisational, legal and economic aspects.

16.4 Technical Solutions for Accountability of Practice

The first condition for the advent of *strong accountability* is that it can be supported by effective tools. In this section, we outline the key components of an accountability system (Sect. 4.1) and illustrate them with a practical application, which allows us to draw some recommendations for *accountability by design* (Sect. 4.2). We also discuss the limitations of these solutions (Sect. 4.3).

16.4.1 Key Accountability Components

The first step of the accountability process should be a clear definition of the privacy policy that the controller has to comply with. *Privacy policy languages* are a technical solution for specifying privacy policies in a machine-readable format. By using a well-defined (formal) syntax, these languages are amenable to automated processing. A number of such languages have been around for some time, such as P3P (W3C 2006), EPAL (IBM 2003) or SIMPLE (Le Métayer 2009). Other languages, such as XACML (OASIS 2013) and UCON (Park and Sandhu 2002; Lazouski et. al. 2010) can also be used to define privacy policies, even though they are more general purpose. A distinction is usually made between *data access* and *data usage* languages; the former makes it possible to set fine-grained permissions for the initial access to data, while the latter can also be used to specify what can happen to the data after it has been accessed. Common examples are the use of data for a specific purpose, its deletion, its anonymisation or its forwarding to a third party. Some languages, such as XACML, are restricted to data access control[21]; others, such as UCON and PPL[22], combine both aspects. In practice, privacy policy languages make it possible to translate the wishes of a data subject, the promises of a data controller and their

[21] XACML deals with access control.

[22] PPL (PrimeLife Policy Language), based on XACML for its access control aspect, also includes many usage control features. It was developed by SAP (Trabelsi et al. 1978) as part of PrimeLife,

common agreement about the use of the personal data into a format that can be processed automatically. Therefore, privacy policy languages are the first building block of accountability of practice: by formalising agreements about the authorised uses of the data, they help structure the evidence (accounts), which is at the core of the principle of accountability. From the privacy policy, it is possible to derive the information that must be present in the accounts to establish their compliance.

Accounts for accountability of practice can typically take the form of *log files*. A log is essentially a detailed history of the events of the system, often in the form of a chronological list. Such files can be generated automatically and in real time by the execution environment of the system. Assuming the mechanism generating them is tamper-proof (Schneier and Kelsey 1999; Waters et al. 2004), logs make up the core of the evidence against which accountability is to be assessed for data handling systems.

The next requirement of an accountability architecture is the possibility of conducting audits. The formal nature of privacy policy languages makes it possible to design tools to conduct automated and rigorous checks of the logs. Such a *log analyser* compares the actual sequence of data handling operations (events) represented in the log with the predefined agreement between the data controller and the data subject as included in a joint policy. After having processed the log, the tool outputs a conclusion about the compliance of the actual events with the initial agreement. If the implementation of the tool itself is transparent[23], this process provides real guarantees and confidence about the analysis of the accounts. If the log is deemed non-compliant, such a tool can automatically pinpoint which event (or absence of event) caused the breach.

16.4.2 Illustration With PPL

To illustrate the framework suggested in the previous subsection, we focus now on an example of a privacy policy language that includes both data access and data usage features: the PPL language.[24]

16.4.2.1 Specifics of the PPL Language

The PPL engine includes a negotiation feature, which allows the data subject and the data controller to express their preferred policies separately before comparing

---

a 36 month long European project with the goal of investigating "...how to protect privacy in emerging Internet applications such as collaborative scenarios and virtual communities".

[23] For example if its source code is available or can be checked by an independent third party.

[24] The a posteriori compliance checking approach is not tied to any particular privacy policy language, but we present the specific example of PPL to give a clearer idea of how the strategy can look like concretely.

them automatically to decide whether they are compatible. If it is the case, a joint agreement (strict enough to accommodate both parties) called *sticky policy* (Karjoth et al. 2002)[25] is generated and attached to the data; in case of incompatibility, a report detailing the mismatch is generated, allowing each party to reassess his privacy policy or to abort the interaction if no compromise can be reached.

In PPL, two central features are available to express privacy policies: authorizations and obligations. They are used both to express policy preferences (for subjects and for controllers), and to define the resulting joint agreements (sticky policies). The PPL *authorizations* feature a notion of *purpose* formalised by keywords such as "marketing" or "identity checking". Data controllers specify in their policies for which purposes they intend to use the personal data they would collect; data subjects, on their side, specify explicitly the purposes they would approve. For the controller and subject policies to be compatible and generate a joint agreement, all purposes listed by the controller must be part of the subject's list. Authorizations also state whether *downstream usage*, i.e. the processing of the data by a third party, is allowed.

The core mechanism of PPL is the *obligations* concept. An obligation consists of a *trigger* and an *action*. Triggers are specific events or circumstances (e.g. data being used for a specific purpose, or forwarded to a third party). Actions are the events that are meant to take place once the trigger has fired, i.e. when the specific event or circumstance has taken place. For instance, a policy may mention that the phone number of a data subject should be deleted after it has been used for the purpose of identity checking. In this case, the deletion of data is the action event, and the trigger is the use of the number for identity checking. To prevent data controllers from claiming that they will fulfil an obligation in an indefinite future, triggers include a maximum delay. A number of trigger events are predefined in PPL[26] and new ones can be added. For action events, which specify what should happen if the associated trigger fires, the same flexibility applies. Default actions include the sending of a notification to the data subject, the deletion of personal data, and its anonymisation.[27]

16.4.2.2 Compliance Checking and Log Design Guidelines

PPL logs include both trigger and action events. Trigger events can be seen as promises, arising from the sticky policy, to be fulfilled in subsequent events by the data controller. If the log is compliant, the trigger event will be followed, at some point, by the action event imposed by the corresponding obligation. Temporal parameters are taken into account to check whether the action event was performed before the agreed deadline. Because of this constraint, all PPL log entries are timestamped.

---

[25] Sticky policies have also been used in the field of digital rights management; however, they play a very different role in our context because here they are checked a posteriori (rather than on the fly) and the process is audited by third parties.

[26] Such as the use of personal data for a specific purpose, its forwarding to a third party, its access by the subject, etc.

[27] Anonymisation is technically realized through cryptography.

Furthermore, trigger events must carry identification tags so they can be referenced from action events. Without this tag, ambiguities may arise and propagate to the global compliance checking.

It must be emphasised that the structure of the logs must be considered carefully to ensure that a privacy policy is accountable. First and foremost, all relevant data handling operations must be represented precisely enough to prevent any ambiguity; the decision of what to include in the logs thus requires careful consideration. In case of insufficient expressiveness, one log entry may refer to several data handling events, yielding different consequences on overall compliance. This precision requirement is complicated by the potential need for the data controller to minimise the amount of data stored in the logs for reason of efficiency or intellectual property protection.

The frequent subcontracting of data handling to third parties raises other issues: not only have the outsourced data handling operations also to be logged but sufficiently detailed information must be kept in the logs to settle disputes in case of malfunctions or breaches of obligations on the third-party side. Log architecture design and precise definitions of accountability are intertwined, and evolving circumstances can alter the distribution of responsibilities—these changes ought to be reflected in logging systems. Both the contents of the logs and their format are directly influenced by the way responsibilities are distributed among the main data controller and (possibly multiple) third parties.

Another source of complication may be the need to support *break-glass situations*[28], which refer to circumstances under which exceptional access to data should be granted to an entity that does not possess the required privileges (NEMA/COCIR/JIRA 2004).[29] This type of situation should be part of the scenarios supported by compliance checking mechanisms; hence the structure of logs must support them. Complementary human assistance may be required to prevent abuse of such mechanisms. Nevertheless contextual data ought to be included in the logs in conjunction with data handling events so as to accurately express the combination of modalities characterising break-glass situations.

The guidelines sketched in this subsection result from the experience gained while developing an accountability system for the PPL language. More detailed illustrations of these issues are described in (Butin et al. 2013).

16.4.3 Challenges and Limitations of Technical Solutions

Since the technical framework outlined here is based on the analysis of logs, these logs must be truthful. More precisely, they ought to display the following properties:

---

[28] Referring to the breaking of glass to trigger an alarm.

[29] Common examples include the exceptional access to medical records in life-threatening situations, credit card fraud scenarios and military information classification systems (Feigenbaum et al. 2012).

1. It should not be possible for a DC to create fake logs: in other words, logs should reflect the actual execution of the system, especially in terms of personal data processing (*unforgeability*).
2. Once logs are generated, it should not be possible to alter them without detection (*integrity*).
3. It should be impossible to access logs without proper credentials (*confidentiality*).

Confidentiality can be achieved by encoding logs with cryptographic tools but care must be taken to allow for selective access: one cannot simply encrypt all logs at once, since different entities (e.g. auditors, subjects) should be granted access to different parts of the logs. The second property, integrity, can be supported by techniques such as the ones proposed by Bellare (Bellare and Yee 1997).

Unforgeability is the most challenging objective because it depends on the whole architecture of the system. Ideally, the architecture should be designed with accountability requirements in mind, so that verifying unforgeability can be made easier. This kind of architecture, for instance featuring a single decision point for all access requests to personal data, should make it easier to check informally whether logs reflect the actual events. The highest level of assurance would be attained through the application of mathematical modelling (*formal methods*). In this approach, all components playing a role in personal data processing and log generation must be accounted for. However, formal methods tend to be costly and could be applied only to the most critical parts of the system.

Great care should also be taken to minimize the ambiguities of log contents. Consider the example of ontologies in PPL: one of the available data handling events corresponds to the use of personal data for a specific purpose. A list of purposes can be agreed on, but simply defining a list seems insufficient: the ontology could be misused by stretching the meaning of words, claiming that the different available purposes where never clearly defined. This could be addressed by attaching informal statements of intent by the data controller to corresponding data handling events. Requiring data controllers to word their intentions in more detail should increase the pressure on them not to misbehave.

A different limitation is that some obligations defined by policy languages may not be checkable automatically, requiring human intervention. Integration of this aspect within an interactive verification tool is feasible but not straightforward; this kind of tool would produce hybrid compliance arguments involving both mechanical and manual steps.

Generally speaking, most of the necessary tools for the implementation of accountability already exist, but they must be used and combined carefully to yield a credible framework. Many challenges of this approach are therefore as much organisational as technical. On the other hand, no bullet-proof solution exists and the very purpose of accountability is to make it more difficult and more risky for data controllers to misbehave, not to enforce correct behaviours. In the next section, we take a closer look at non-technical challenges and solutions for an integrated accountability approach.

16.5 Accountability Architecture: Legal and Economic Aspects

In the previous section, we have shown that *strong accountability* is possible from a technical point of view and we have suggested practical means to support it. Obviously, it is not because strong accountability principles are technically feasible that they will actually be implemented. The next questions to address are therefore: should they be adopted on a voluntary basis (and why would this happen?) or should they be enforced by the law (and how)? What should the roles of the stakeholders (data controller, data subject, data protection authorities, third parties) be? What would be the costs and benefits for the industry?

First, following Colin Bennett (Bennett 2012)[30], it is unlikely that large-scale accountability can be adopted on a voluntary basis. Regulation should therefore impose binding accountability requirements. But such regulation should take into account two essential requirements:

- As argued in Sect. 3.3, just recalling general or vague accountability principles is not enough, and it could even provide a false sense of protection. Legal uncertainty would undermine the very principle of accountability.
- As stated by the Article 29 Working Party (Article 29 Working Party 2010), accountability should not impose "cumbersome new legal requirements upon data controllers, particularly given the current, challenging EU economic situation."

To solve this tension between the need for precise legal obligations on the one hand and economic acceptability on the other hand, we should stress that precision does not necessarily mean lack of flexibility. Indeed, it should be clear that a one-size-fits-all approach would not make sense in this area and different factors, such as the type of personal data at stake and the size and activities of the company, have to be taken into account to determine the required level of accountability and the associated measures. Also, because laws (and European regulations) should remain at a sufficient level of abstraction to be of general application and to avoid quick obsolescence, they should not go into the details of the accountability process but rather provide high level requirements imposing the necessary level of precision[31]. For example, following the recommendations of Sect. 4, they should state that any information or event which could have an impact on the data protection requirements must be recorded in the accounts, without defining what these events are and how they should be recorded. They should define the requirements for audits (periodicity, level of detail) depending on the situation. Such a flexible, multi-tier approach does not contradict the precision requirement: it should always be possible for the data subject to know, for a given controller, his privacy policy, the precise accountability measures implemented, the auditors, as well as the way to interact with them to be informed of the results of their audits.

---

[30] "Privacy audits have been around for a long time, but there is little evidence that market pressure alone will push this kind of external conformity assessment around the international economy".

[31] "Technology neutrality has long been held up as a guiding principle for the proper regulation of technology, particularly the information and communications technologies" (Reed 2007).

This combination of legal requirements, flexibility and transparency is instrumental to restore the trust of the data subjects. It is also the key to economic viability of strong accountability: each data controller could decide to opt for the minimal requirements imposed by the law (both in terms of privacy policy and accountability measures) or to provide higher guarantees and use them as a business differentiator to get a competitive advantage.

As far as the extra costs incurred by the mandatory accountability requirements are concerned, they can be separated in three parts:

(i) Organisational costs: for staff training, privacy officer activities, documentation keeping, etc.
(ii) Technical costs: to build, store and secure the accounts.
(iii) Audit costs.

Category (i) should not represent significant additional costs, as it mostly corresponds to tasks already carried out by data controllers. Otherwise, they represent true sources of improvement of the quality of data handling procedures and overall internal organisation of the company.[32]

Category (ii) can be reduced to marginal costs if accountability obligations are considered in the design of the system itself, following an *accountability by design* approach as suggested in Sect. 4.

As far as Category (ii) is concerned, the frequency of the audits and the associated costs should be proportionate to the level of sensitivity of the data and the size and type of activities of the controller. Technical tools such as the log analyser sketched in Sect. 4 can also help reducing audit costs.

In any case, as stated in Sect. 3.3, audits should be conducted by independent third parties: this is an essential condition for accountability to play its trust enhancing role. As mentioned by Colin Bennett (Bennett 2012), "the 'trust me, my account is the truth' approach will not be sufficient for many organizations". Furthermore, one may argue, following Paul De Hert (De Hert 2012), that external review is at the core of the concept of accountability: "It was brought into twentieth century public administration literature to denote the external scrutiny process, as opposed to the inner responsibility processes of the individual as per his or her conscience or moral value". Both high-level aspects of accountability such as company policies and practice-oriented aspects (through data handling log compliance checking) should be subject to audit.

But how should this independence be established and what kind of actor could play this role? We believe that in this matter inspiration could be taken from certification schemes, in particular information technology security schemes such as the Common Criteria for Information Technology Security Evaluation (Common Criteria 2013) in which national authorities can deliver accreditations to independent evaluators who are themselves in charge of conducting the evaluations. Similarly, data protection authorities, which do not have the resources to conduct large scale, country-wide

[32] To this respect, it would be advisable to introduce accountability as a new requirement of Information Security Management Systems (ISMS).

audits could deliver accreditations to data protection auditors. A first step in this direction has been made in France with the introduction of the CNIL audit procedure seals in 2011 (CNIL 2011). The number of auditors approved by the CNIL is not very large yet but this business would obviously grow if strong accountability with independent audits became mandatory. Lobbying could prove to be a challenge in this area, and solutions such as anonymous auditing ought to be explored.

As far as efficiency is concerned, such an ecosystem of auditors could also help data protection authorities facing growing needs for controls, considering that their own resources cannot be extended ad infinitum. Of course, data protection authorities should keep the power to supervise on a regular basis the activities of the auditors themselves, to ensure that they keep a high evaluation standard, but auditors are necessarily much less numerous than data controllers. This monitoring of the whole process by data protection authorities would be essential, especially if the choice of the auditor is made by the data controller itself, which could otherwise lead to a quality dumping race among auditors.

Another benefit of accountability for data protection authorities is pointed out by the Article 29 Working Party: "putting the accountability principle into effect will provide useful information to data protection authorities to monitor compliance levels. Indeed, because data controllers will have to be able to demonstrate to the authorities whether and how they have implemented the measures, very relevant compliance related information would be available to authorities. They will then be able to use this information in the context of their enforcement actions."

Last but not least, for accountability to fully play its deterrence role, data protection authorities should have powers of sanction, not only to punish data controllers who have breached substantive data protection principles but also those who do not meet their accountability obligations. Penalties should be especially severe if the accounts provided by the data controller are proved to be inaccurate or forged, the same way organisations manipulating their financial accounts are severely sanctioned.

16.6 Accountability and Perspectives

In this paper, we have argued that *strong accountability* should be a cornerstone of future data protection regulations. By "strong accountability" we mean a principle of accountability which

- applies not only to policies and procedures, but also to practices, thus providing means to oversee the effective processing of the personal data, not only the promises of the data controller and its organisational measures to meet them;
- is supported by precise binding commitments enshrined in law;
- involves audits by independent entities.

As discussed in Sect. 5, we believe that this quest for precision is critical to ensure the effectiveness of accountability, and therefore of substantial data protection principles, and it should not be contradictory with the need for flexibility that is required by the industry. Generally speaking, a system where data controllers are audited by officially recognised third parties that are themselves accredited by data protection authorities would provide a consistent and efficient integrated accountability approach featuring a chain of trust all the way between supervisory authorities and data subjects.

Strong accountability should benefit all stakeholders: data subjects, data controllers, and even data protection authorities whose workload should be considerably streamlined. Indeed, if standardised accountability mechanisms become widespread, it would be far more efficient for data protection authorities to evaluate data controllers against well-defined criteria. Here, a form of standardisation would benefit both data protection authorities, which would enjoy a reduced workload, and data controllers, who would know in advance and more precisely to which metrics they must conform.

A further question could be the relationship between strong accountability and other instruments for privacy protection which have received a lot of attention during the last decade such as *Binding Corporate Rules* (BCRs), *Privacy impact assessments* (PIAs), *privacy by design* (Cavoukian 2012) and *privacy seals*.

The European Commission defines BCRs as "internal rules (such as a code of conduct) adopted by multinational group of companies which define its global policy with regard to the international transfers of personal data within the same corporate group to entities located in countries which do not provide an adequate level of protection."[33] The 1995 Directive's adequacy model (EU 1995), whereby permissions to export data depend on the country of destination, is ill-fitted to current data transfer practices. However, its derogation[34] permits transfers to countries deemed inadequate if "the controller adduces adequate safeguards." A working document by the Article 29 Working Party (Article 29 Working Party 2003) states that Binding Corporate Rules (BCR) can be considered as an acceptable safeguard to this respect. But BCRs have shown some limitations, in particular in terms of enforceability. As stated in (Alhadeff et al. 2011), "the integration of accountability mechanisms could be used to extend the existing adequacy regime. Our experience with Directive 95/46/EC has shown that the applicability of legislation offering 'adequate' safeguards does not by itself ensure that appropriate guarantees are implemented in practice." Indeed, it may be argued that the additional protection provided by accountability is even more necessary in case of international transfers of personal data.

PIAs (Wright et al. 2011; Wright and De Hert 2012) constitute a fundamental approach to evaluating risks: potential issues should be foreseen and analysed in a collaborative and interactive way before the design and deployment of a new system. As stated by Gary Marx (Marx 2012), "It anticipates problems, seeking to prevent, rather than to put out fires." PIAs have thus to be conducted at the earliest stages, before a system is deployed. They should result into recommendations and requirements about the system and organisational measures. These recommendations should be taken as input to a privacy by design process resulting in an implementation

[33] http://ec.europa.eu/justice/data-protection/document/international-transfers/binding-corporate-rules/index_en.htm.

[34] Article 26 (2).

of the system. This implementation can be evaluated by independent experts to get a privacy seal, which provides some guarantees about the fact that the system meets well-defined privacy requirements (including legal obligations) in terms of privacy. Strong accountability, in contrast with PIAs and privacy by design, concerns the practices, hence the effective exploitation of the product or system. In other words, it is an a posteriori rather than an a priori control. PIA do offer benefits, but as an ex ante analysis: they offer no guarantee regarding the actual processing of data.

However, as shown in Sect. 4, accountability does not emerge spontaneously. A system has to be designed with accountability requirements in mind, and these requirements should arise from the PIA. Indeed, the feasibility of accurate and comprehensive a posteriori verifications depends directly on the architecture of the technical platform under consideration. The privacy by design approach should thus include an *accountability by design* component, to ensure that accountability will indeed be feasible. This accountability component could also be evaluated as part of a privacy seal mechanism.[35] More generally, we should envisage in the long term a continuum between privacy seals and the regular audits required by strong accountability: the privacy seal would be the original certificate, providing well defined guarantees about the design of the system and the organisation in place, while accountability certificates would complement the original seal with guarantees about the effective use of the system. In this architecture, strong accountability could take the form of continuous maintenance of the original privacy seal. This maintenance could also have an impact on risk assessment (for example through the identification of new risks) leading to a new iteration of PIA and a virtuous improvement process.

References

Alhadeff, Joseph, Brendan Van Alsenoy, and Jos Dumortier. 2011. The accountability principle in data protection regulation: Origin, development and future directions. Paper presented at privacy and accountability, Berlin, Germany, April 5–6, 2011.

Article 29 Data Protection Working Party. 2003. Working document on transfers of personal data to third countries: Applying article 26 (2) of the EU Data Protection Directive to Binding Corporate Rules for International Data Transfers. http://ec.europa.eu/justice/policies/privacy/docs/wpdocs/2003/wp74_en.pdf. Accessed 28 Feb 2013.

Article 29 Data Protection Working Party. 2010. Opinion 3/2010 on the principle of accountability. http://ec.europa.eu/justice/policies/privacy/docs/wpdocs/2010/wp173_en.pdf. Accessed 28 Feb 2013.

Asia-Pacific Economic Cooperation, Electronic Commerce Steering Group (ECSG). 2004. APEC Privacy Framework. http://www.apec.org/Groups/Committee-on-Trade-and-Investment/~/media/Files/Groups/ECSG/05_ecsg_privacyframewk.ashx. Accessed 28 Feb 2013.

Asia-Pacific Economic Cooperation, Electronic Commerce Steering Group (ECSG). 2009. APEC data privacy pathfinder projects implementation work plan—revised. http://aimp.apec.org/Documents/2009/ECSG/SEM1/09_ecsg_sem1_027.doc. Accessed 28 Feb 2013.

Bella, Giampaolo, and Lawrence C. Paulson. 2006. Accountability protocols: Formalized and verified. *ACM Transactions on Information and System Security* 9:138–161.

[35] See for example EuroPriSe, the European privacy seal: https://www.european-privacy-seal.eu.

Bellare, Mihir, and Bennet Yee. 1997. Forward integrity for secure audit logs. Technical Report CS98-580, Department of Computer Science and Engineering, University of California at San Diego.

Bennett, Colin. 2012. The accountability approach to privacy and data protection: Assumptions and caveats. In *Managing privacy through accountability*, ed. Daniel Guagnin et al., 33–48. Basingstoke: Palgrave Macmillan.

Butin, Denis, Marcos Chicote, and Daniel Le Métayer. 2013. Log design for accountability. Proceedings of the 4th international workshop on data usage management. Washington, D.C.: IEEE Computer Society.

Canadian Standards Association. 1996. Model code for the protection of personal information (Q830-96). Mississauga: CSA.

Cavoukian, Ann. 2012. Privacy by design [Leading edge]. *IEEE Technology and Society Magazine* 31:18–19.

Cederquist, JG, Ricardo Corin, M. A. C. Dekker, Sandro Etalle, and J. I. den Hartog. 2005. An audit logic for accountability. Proceedings of the 6th international workshop on policies for distributed systems and networks. Washington, D.C.: IEEE Computer Society.

Centre for Information Policy Leadership. 2009a. Global discussion on the commonly-accepted elements of privacy accountability. http://www.huntonfiles.com/files/webupload/CIPL_Galway_Conference_Summary.pdf. Accessed 28 Feb 2013.

Centre for Information Policy Leadership. 2009b. Data protection accountability: The essential elements. http://www.huntonfiles.com/files/webupload/CIPL_Galway_Accountability_Paper.pdf. Accessed 28 Feb 2013.

Centre for Information Policy Leadership. 2010. Demonstrating and measuring accountability: A discussion document. http://www.huntonfiles.com/files/webupload/CIPL_Accountability_Phase_II_Paris_Project.PDF. Accessed 28 Feb 2013.

Commission Nationale Informatique et Libertés (CNIL), Label CNIL procédures d'audit de traitements. 2011. http://www.cnil.fr/la-cnil/labels-cnil/procedures-daudit/. Accessed 28 Feb 2013.

Common Criteria for Information Technology Security Evaluation. 2013. http://www.commoncriteriaportal.org/cc/. Accessed 28 Feb 2013.

De Hert, Paul. 2012. Accountability and system responsibility: New concepts in data protection law and human rights law. In *Managing privacy through accountability*, ed. Daniel Guagnin et al., 193–232. Basingstoke: Palgrave Macmillan.

Ernst & Young. 2012. Privacy trends 2012. The case for growing accountability.

European Commission. 2012. *Proposal for a regulation of the European parliament and of the council on the protection of individuals with regard to the processing of personal data and on the free movement of such data (general data protection regulation)*. Brussels: European Commission.

European Parliament and the Council of the European Union. 1995. *Directive 95/46/EC of the European parliament and of the council of 24 October 1995 on the protection of individuals with regard to the processing of personal data and on the free movement of such data*. Brussels: European Parliament.

Feigenbaum, Joan, James Hendler, Aaron Jaggard, Daniel Weitzner, and Rebecca Wright. 2011. Accountability and deterrence in online life. Paper presented at *ACM* Web Science Conference 2011, Koblenz, Germany, June 14–17, 2011.

Guagnin, Daniel et al., ed. 2012. *Managing privacy through accountability*. Basingstoke: Palgrave Macmillan.

Haeberlen, Andreas. 2009. A case for the accountable cloud. Proceedings of the 3rd ACM SIGOPS international workshop on large-scale distributed systems and middleware. New York: ACM.

IBM. 2003. The enterprise privacy authorization language (EPAL). http://www.zurich.ibm.com/security/enterprise-privacy/epal/. Accessed 28 Feb 2013.

Jagadeesan, Radha, Alan Jeffrey, Corin Pitcher, and James Riely. 2009. Towards a theory of accountability and audit. Proceedings of the 14th European conference on Research in computer security. Berlin: Springer.

Joint NEMA/COCIR/JIRA Security and Privacy Committee (SPC). 2004. Break-glass: An approach to granting emergency access to healthcare systems.

Karjoth, Günter, Matthias Schunter, and Michael Waidner. 2002. Platform for enterprise privacy practices: Privacy-enabled management of customer data. Proceedings of the 2nd workshop on privacy enhancing technologies. Berlin: Springer.

Lazouski, Aliaksandr, Fabio Martinelli, and Paolo Mori. 2010. Usage control in computer security: A survey. *Computer Science Review* 4:81–99.

Le Métayer, Daniel. 2009. A formal privacy management framework. Proceedings of formal aspects in security and trust. Berlin: Springer.

Marx, Gary. 2012. Privacy is not quite like the weather. In *privacy impact assessment*, ed. David Wright and Paul De Hert. Berlin: Springer.

Organisation for Economic Cooperation and Development. 1980. Guidelines on the protection of privacy and transborder flows of personal data.

Organisation for Economic Cooperation and Development. 2011. Thirty years after the OECD privacy guidelines. http://www.oecd.org/sti/ieconomy/49710223.pdf. Accessed 28 Feb 2013.

Organization for the Advancement of Structured Information Standards (OASIS). 2013. eXtensible Access Control Markup Language (XACML) version 3.0 OASIS standard. http://docs.oasis-open.org/xacml/3.0/xacml-3.0-core-spec-os-en.pdf. Accessed 28 Feb 2013.

Park, Jaehong, and Ravi S. Sandhu. 2002. Towards usage control models: Beyond traditional access control. Proceedings of ACM symposium on access control models and technologies. New York: ACM.

Parliament of Canada. 2000. Personal information protection and electronic documents act.

Poullet, Yves. 2001. How to regulate Internet: new paradigms for Internet governance Self-regulation: value and limits. In *Variations sur le droit de la société de l'information*, ed. Claire Monville, Cahiers du Centre de Recherches Informatique et Droit. 79–114. Bruxelles: Bruylant.

Raab, Charles. 2012. The meaning of 'accountability' in the information privacy context." In *Managing privacy through accountability*, ed. Daniel Guagnin et al., 15–32. Basingstoke: Palgrave Macmillan.

Reed, Chris. 2007. Taking sides on technology neutrality. *SCRIPTed* 263:263–284.

Schneider, Fred. 2009. Accountability for perfection. *IEEE Security and Privacy Magazine* 7:3–4.

Schneier, Bruce, and John Kelsey. 1999. Secure audit logs to support computer forensics. *ACM Transactions on Information and System Security* 2:159–176.

Title 12 of the United States Code. 1978. Right to Financial Privacy Act.

Trabelsi, Slim, Gregory Neven, and Dave Raggett. 2011. PrimeLife Deliverable D5.3.4: Report on design and implementation.

US Federal Trade Commission. 1973. Fair Information Practice Principles.

W3C. 2006. The platform for privacy preferences 1.1 (P3P1.1) specification. http://www.w3.org/TR/P3P11/. Accessed 28 Feb 2013.

Waters, Brent, Dirk Balfanz, Glenn Durfee, and Diana Smetters. 2004. Building an encrypted and searchable audit log. Proceedings of the network and distributed system security symposium. Reston: The Internet Society.

Wright, David, and Paul De Hert, ed. 2012. *Privacy impact assessment*. Berlin: Springer.

Wright, David, Raphaël Gellert, Serge Gutwirth, and Michael Friedewald. 2011. Minimizing technology risks with PIAs, precaution, and participation. *IEEE Technology and Society Magazine* 30:47–54.

MIX
Papier | Fördert
gute Waldnutzung
FSC® C083411

Kolibri 360 GmbH
Ferdinand-Jühlke-Straße 7
99095 Erfurt, Deutschland
produktsicherheit@kolibri360.de